Microfluidics and Microfabrication

Suman Chakraborty
Editor

Microfluidics and Microfabrication

Springer

Editor
Suman Chakraborty
Department of Mechanical Engineering
Indian Institute of Technology
Kharagpur-721302
India
suman@mech.iitkgp.ernet.in;
sumanchakraborty_iitkgp@yahoo.com

ISBN 978-1-4419-1542-9 e-ISBN 978-1-4419-1543-6
DOI 10.1007/978-1-4419-1543-6
Springer New York Dordrecht Heidelberg London

Library of Congress Control Number: 2009941404

Printed on acid-free paper

Springer is part of Springer Science+Business Media (www.springer.com)

Preface

Conversion of the naturally available materials and minerals into useful products has been one of the major contributing factors towards the progress of human civilization. A substantial proportion of all engineering and scientific activities are, in fact, devoted to this endeavor and from the dawn of civilization many processes and technologies have been developed with this objective. Characteristically, most such traditional manufacturing processes are based upon a 'top down' approach in which a piece of material is worked upon to produce the required shape, size, feature and finish. As and when engineers and scientists face a challenging task because of the unfavorable material properties, extreme level of precision and finish required or a high level of miniaturization, new and novel approaches and processes are developed. Of late miniaturization has assumed an extremely important place in engineering activities because of the revolution in micro electronics and emergence of Micro Electro Mechanical Systems (MEMS). With the turn of the century another major development has started taking place – integration of life science with engineering. As the complexity of the new emerging systems and devices is reaching unbelievably high levels, engineers have started following the principles of life science in artificially created systems and devices. This has become possible because of the tremendous progress made in material science, manufacturing techniques and life sciences. A major deviation in fabrication techniques has been from the age old 'top down' approach to the recently conceived 'bottom up' approach that is essential for achieving extreme degree of complexity and miniaturization. Still in most situations such processes are controlled from outside agencies like computes etc., but it is already being noticed that much better results can be achieved if material elements self assemble into the desired products (like living objects grow). Though such processes are still at their infancy it is hoped that during the current century the 'bottom up' processes based on 'self assembly' of material at molecular, nanometer or micron levels will reach matured levels. Miniaturization is important even for developing systems and devices at macroscopic levels because the advantages of the scaling laws can be effectively utilized. Thus, many new-generation machines and devices will be based on massive parallelism of micro or nano units. As a result new approaches and new subjects like 'microfluidics' have started playing a centre stage role in engineering. All these new developments will bring increasing degree of interdisciplinarity leading to a grand synthesis of life science, engineering and physical science. Devices and machines may look more and more like living organisms and 'synthetic biology' is emerging as an important subject. In not-too-distant future one may find a micro-pump to run using photosynthesis as the energy source or a micro robot may function inside human body extracting energy from the consumed food through metabolism.

The ability to create structures and patterns on microscopic and sub-microscopic length scales has triggered a wide range of scientific investigations, leading to the development of novel miniaturized devices and systems for transporting and manipulating fluidic samples in a rapid, efficient and controllable manner. Microscale transport processes and microfluidics are becoming increasingly important in several emerging applications due to their inherent advantages such as high transfer coefficients (on account of large surface area to volume ratios), efficient process management, miniaturization of devices for specific applications, and addressability of cellular length scales. The applications are many in diverse high-technology areas including biotechnology and biomedical engineering, inkjet printing, and thermal management of electronic devices/systems. However, it needs to be appreciated that microfluidics does not become functional by itself; it requires a strong interfacing with advanced miniaturized fabrication protocols to make the devices and fluidic systems functional and effective. Unprecedented advancements in the science and technology of microfabrication and nanofabrication achieved over the past few years have indeed paved new pathways for more efficient integration of microfluidics with fabrication over small scales, which is indeed the broad vision behind designing this Edited volume.

This book contains nine Chapters, encompassing several fundamental as well as advanced issues of Microfluidics and Microfabrication in comprehensive details. These Chapters are essentially based on the invited keynote lectures presented by various renowned speakers in the Indo-US Workshop on 'Microfluidics and Fabrionics', which was organized by the Departments of Mechanical and Chemical Engineering, IIT Kharagpur, India, during January 9–11, 2009. Without compromising rigorousness, the present book is designed for maximum readability by a broad audience starting from senior undergraduate students to advanced researchers and industrial personnel working in this emerging field. Although several authors are involved in designing the independent and self-content Chapters of this book, efforts have been devoted to maintain a consistent depth, breadth, and writing style, despite disparate acceptable styles and norms universally considered on different topical areas (for instance, one may note contrasting styles of articles commonly published in the areas of Life Sciences and Microfabrication). The integrated Chapters of this book should act as effective references to scientists and engineers with pre-exposure to this field, as well as to budding researchers whose initiation of activities in this area has just been kindled.

The Editor expresses his sincere gratitude to all the contributing authors for their dedicated endeavor in making the Chapters happen, despite stringent deadlines from the publisher's end. The enthusiasm and indefatigability of Mr. Steven Elliot from the Springer also played a key role in making the integrated volume happen within a scheduled time-frame. The Editor also expresses his earnest gratitude to the

'Indo-US Centre for Research Excellence in Fabrionics' for financing the lecture series that acted as a prelude to this Edited volume. Special acknowledgements are due to Dr. Arabinda Mitra, Executive Director, Indo-US Science and Technology Forum, for his support in creating the Centre. Noticing the trends of far reaching consequences of microfluidics and microfabrication in engineering a large number of eminent scientists and engineers from India and USA attended a workshop in March 2004 at IIT Kanpur, which was also funded by the Indo-US Science and Technology Forum. A major point that was taken up for discussion in a subsequent similar Indo-US Workshop in October 2007 at IIT Kanpur was to evolve a suitable terminology to designate/identify this new era engineering that is very distinct from the traditional term 'manufacturing science'. After a considerable amount of discussion and consultation the participating faculty from India and USA converged to the term 'fabrionics'. It is expected to indicate the whole branch of activities in which the required pattern, shape, features and characteristics can be achieved through material manipulation at molecular, nanometer and micrometer levels. Such manipulation can be externally motivated or can be based upon the principle of 'self assembly'. Subsequently the Indo-US Science and Technology Forum, New Delhi provided funds for creating a 'Centre for Research Excellence in Fabrionics' which presently involves IIT Kanpur, IIT Kharagpur, Bengal Engineering & Science University-Shibpur, Central Mechanical Engineering Research Institute-Durgapur, University of Illinois at Urbana-Champaign and Chicago, Northwestern University Evanston, University of California Irvine and University of Missouri-Columbia. The Editor whole-heartedly acknowledges the contribution of this group towards conceptualizing the Chapters of this book.

Last but not the least, special thanks are due to Prof. Amitabha Ghosh (Senior Scientist of the Indian National Science Academy, New Delhi & Honorary Distinguished Professor, Bengal Engineering and Science University, Shibpur West Bengal, India), who has literally played the most significant role in conceptualizing the lecture series on Microfluidics and Fabrionics based on which this Edited volume has been designed, with a pioneering vision that can match the ranks of only a topmost level scientific and technological researcher set by his own exceptional high standards. His overall guidance and motivation was indeed very contagious, and has percolated through all the Chapters of this Book that have acted as backbones behind this ambitious endeavor. He has also been kind enough to write the inaugural paragraph of this preface, which the Editor believes would greatly enhance the readability and effectiveness of all the Chapters of this book to the interested readers, by binding those together with a common central theme.

Kharagpur, India Suman Chakraborty

Contents

Contributors

G.K. Ananthasuresh Mechanical Engineering, Indian Institute of Science, Bangalore, India, suresh@mecheng.iisc.ernet.in

Debapriya Chakraborty Department of Mechanical Engineering, Indian Institute of Technology, Kharagpur, Kharagpur-721302, India, debapriya.chakraborty@gmail.com

Suman Chakraborty Department of Mechanical Engineering, Indian Institute of Technology, Kharagpur, Kharagpur-721302, India, suman@mech.iitkgp.ernet.in

Tamal Das Department of Biotechnology, Indian Institute of Technology Kharagpur, Kharagpur-721302, India, tamal.iitkgp@gmail.com

Sunando Dasgupta Department of Chemical Engineering, Indian Institute of Technology, Kharagpur, Kharagpur-721302, India, sunando@che.iitkgp.ernet.in

R. Martinez-Duarte Department of Mechanical and Aerospace Engineering, University of California, Irvine, CA, USA, drmartnz@gmail.com

Animangsu Ghatak Department of Chemical Engineering, Indian Institute of Technology, Kanpur, Kanpur-208016, India, aghatak@iitk.ac.in

Sandip Ghosal Department of Mechanical Engineering, Northwestern University, Evanston, IL, USA, s-ghosal@northwestern.edu

S.R. Kajale Department of Mechanical Engineering, College of Engineering, Pune, India, kajale@gmail.com

Q. Kang Los Alamos National Laboratory, Los Alamos, NM, USA, qkang@lanl.gov

M.J. Madou Department of Mechanical and Aerospace Engineering, University of California, Irvine, CA, USA, mmadou@uci.edu

Nandan Maheswari Mechanical Engineering, Indian Institute of Science, Bangalore, India, nandaniitm@gmail.com

Abhijit Majumder Department of Chemical Engineering, Indian Institute of Technology, Kanpur, Kanpur-208016, India, majumder@iitk.ac.in

P. K. Mishra Department of Mechanical Engineering, College of Engineering, Pune, India, pk.pkmishra@gmail.com

P. P. Mukherjee Los Alamos National Laboratory, Los Alamos, NM, USA partham@lanl.gov; mukherjee.pp@gmail.com

S.S. Pardeshi Department of Mechanical Engineering, College of Engineering, Pune, India, ssp@mech.coep.org.in

S.B. Patil Department of Mechanical Engineering, College of Engineering, Pune, India, sbp@mech.coep.org.in

A. Narayana Reddy Mechanical Engineering, Indian Institute of Science, Bangalore, India, anreddy@mecheng.iisc.ernet.in

Deepak Sahu Mechanical Engineering, Indian Institute of Science, Bangalore, India, deepak.sahu@ge.com

Ashutosh Sharma Department of Chemical Engineering, Indian Institute of Technology, Kanpur, Kanpur-208016, India, ashutos@iitk.ac.in

Jayant K. Singh Department of Chemical Engineering, Indian Institute of Technology, Kanpur, Kanpur-208016, India, jayantks@iitk.ac.in

G. Turon Teixidor Department of Mechanical and Aerospace Engineering, University of California, Irvine, CA, USA, gturon@gmail.com

Chapter 1
Microfluidic Transport and Micro-scale Flow Physics: An Overview

Debapriya Chakraborty and Suman Chakraborty

Abstract In this article, we delineate some of the distinctive and demarcating fundamental aspects of microscale fluid flows as compared to their macroscale counterparts, and illustrate the utilization of these principles towards exploiting new functionalities in devices and systems of emerging importance. In particular, we emphasize on some of the important flow actuation mechanisms (pressure-driven, surface tension-driven, centrifugal, electrokinetic, magnetohydrodynamic, optical, and acoustic) in microfluidics and their implications, and outline the fundamental physical and mathematical principles that govern their implementations in practice.

Keywords Interfacial slip · slip length · Kundsen number · Reynolds number · capillary number · Bond number · continuum hypothesis · hydrophobic interaction · nanobubble · friction factor · continuity equation · momentum equation · Stokes hypothesis · Navier–Stokes equation · laminar flow · fully developed flow · surface tension driven flow · contact angle · Young–Laplace equation · capillary filling · Marangoni effect · electrocapillary · continuous electrowetting · electrowetting on dielectric (EWOD) · Young Lippman equation · optofluidics · rotational microfluidics (lab on a CD) · electrokinetics · electrical double layer (EDL) · electroosmosis · streaming current · streaming potential · Poisson equation · Poisson Boltzmann equation · steric effect · AC electroosmosis · electrophoresis · dielectrophoresis · electrothermal effect · electro-magneto-hydrodynamics (EMHD) · acoustic streaming · droplet based microfluidics

1.1 Introduction

Microfluidics deals with the science and technology of fluid flows over micron or sub-micron length scales, pertaining to the actuation, precise control and

---

S. Chakraborty (✉)
Department of Mechanical Engineering, Indian Institute of Technology Kharagpur, Kharagpur-721302, India
e-mail: suman@mech.iitkgp.ernet.in

S. Chakraborty (ed.), *Microfluidics and Microfabrication*,
DOI 10.1007/978-1-4419-1543-6_1, © Springer Science+Business Media, LLC 2010

manipulation of small volumes of liquids or gases (commonly, in femto-litre to micro-litre precision) through miniaturized conduits of varied geometrical shapes and practical functionalities. The emergence of microfluidics, primarily triggered by the phenomenal advancements in generating small-scale geometrical features through micro- and nano-fabrication technologies, has simultaneously rekindled the interests in several classical areas of fluid dynamics, including creeping flows, with a fusion and agglomeration of concepts from molecular physics, surface chemistry and life sciences. Traditionally, silicon micromachining methods have been used to fabricate microfluidic channels from silicon and glass. Of late, other types of materials such as Polydimethyl Dimethylsiloxane (PDMS) and Polymethyl methacrylate (PMMA) have been successfully employed for generating microfluidic structures, because of their advantages in terms of faster design times, low cost, the ability to seamlessly impregnate nanoscale features, and the possibilities of obtaining flexibly deformable shapes.

The applications of microfluidics are truly diverse in every respect, ranging from multifarious facets of biotechnology and biomedical engineering (commonly classified under the broad theme of biomicrofluidics or more broadly, BioMEMS) and biological weapon detection on one side to inkjet printing and thermal management of electronic devices on the other. Developments in many of these applications have often been facilitated by the technological advancements in optical and other detection technologies, thereby enabling the researchers to probe the device functionalities with unprecedented accuracies, precisions and sensitivities [1]. With the successes of the existing micro-analytical methods, it has become an obvious challenge to develop new, more compact and more versatile sensing, actuating and diagnostic protocols, and to look for expanded applications of microfluidics in areas beyond traditional chemistry and life sciences. Microfluidics, in that respect, has truly emerged as an interdisciplinary science and holds the potential in influencing subject areas from interfacial physics, surface chemistry, chemical synthesis and biological analysis to optics and information technology. Nevertheless, the heart and soul of microfluidics appears to be revolving around the transport phenomena and flow physics in micro and nano scale systems. In this Chapter, we briefly elucidate the fundamental aspects of fluid flows over microscopic scales, flow actuation mechanisms in microscale and their implications, and finally outline some representative applications as examples.

1.2 Microfluidics Versus Traditional Fluidics

To begin with, it is imperative to emphasize on highlighting some of the key demarcating features of microscale fluid mechanics, as compared to its macroscale counterparts. These distinctive issues are critical, and in many respects determine the functionalities of microfluidic devices to a significant extent, as indicated in our subsequent discussions.

(i) As a system reduces in size, its surface area to volume ratio increases. Typically, for a micro-device, this ratio turns out to be of the order of 10^6 m, resulting in the dominance of surface effects over volumetric effects. Thus, over microscopic length scales, inertia forces may often turn out to be negligible in comparison to viscous forces, electrostatics/ electrodynamics forces, or surface tension forces. Effects of many of the surface forces, which are not otherwise felt very prominently over macroscopic scales, may thus play decisive roles towards regulating the functionalities of microfluidic devices.

In order to adjudge the relative importance of different competing forces/interactions in microfluidics, various dimensionless numbers are commonly invoked. These numbers are typically indicatives of the relative strengths of the various forces acting on a particular system. For example, Reynolds number, *Re*, often relates the ratio of the inertial forces to viscous forces. In most cases of liquid microflows, *Re* is typically less than 10, so that viscous forces play significant roles towards dictating the fluid flow characteristics. Further, because of characteristic low values of *Re*, momentum transport is essentially diffusion dominated, which renders efficient mixing in microfluidics an ever-threatening potential challenge, particularly considering the exploitation of turbulence as a prohibitively difficult proposition over reduced length scales. Apart from *Re*, several other dimensionless numbers are also commonly invoked in the context of micro-scale transport, as summarized in Table 1.1.

Table 1.1 Some dimensionless numbers commonly invoked in microscale fluid mechanics

Re	Reynolds number	Inertial/viscous
Pe	Peclet number	Advection/diffusion
Pr	Prandtl number	Momentum diffusivity/thermal diffusivity
Ca	Capillary number	Viscous/surface tension
Bo	Bond number	Gravity/surface tension
Kn	Knudsen number	Molecular mean free path/characteristic system length scale

(ii) Definition of fluid properties over micron or sub-micron scales is not often free from ambiguities. Such ambiguities stem from the fact that traditional fluid mechanics is normally concerned with the behaviour of matter over dimensions that are significantly larger as compared to the molecular length scales (may be characterized loosely by average intermolecular distances, or more rigorously, by the molecular mean free path). The behaviour of fluids, under such conditions, may be idealized as the same as if the fluid were a continuous medium (continuum description). The local fluid property at a point may then be defined as the average property of all the molecules occupying a 'sensitive' elemental volume chosen in the neighbourhood of the point under concern [2]. The sensitive volume should be small enough for the measurement to be local enough; so that further reduction in its size does not change the value of the property. Considering a large elemental volume

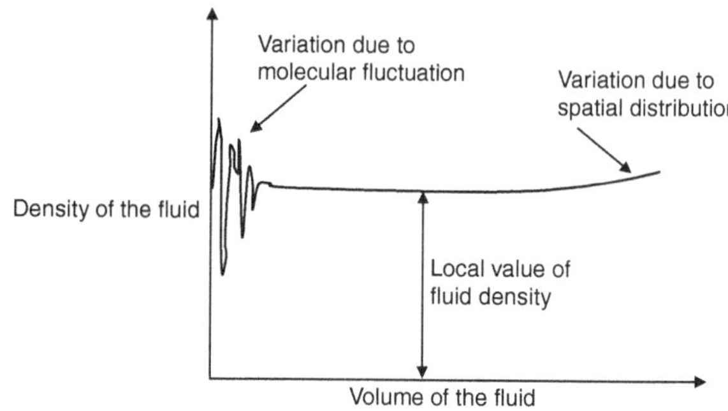

Fig. 1.1 Variation of the density of the fluid on the volume considered

would invariably include the variations associated with the spatial distribution of the property, as demonstrated in Fig. 1.1, preventing the analyzer in capturing the trends in variations of properties over the system scale.

On the other hand, if the volume is considered to be too small to contain only a few number of molecules, then statistical fluctuations with regard to the relative occupancies of the molecules in the elemental volume may give rise to locally oscillating natures in the predicted fluid properties (see Fig.1.1). If the system length scale is itself of comparable extent as that of the characteristic length scales of these local oscillations, continuum considerations may not be applicable altogether. Issues of microfluidics concerning the flow behaviour over such regimes are generally addressed by molecular modelling considerations, and will not be discussed in this introductory Chapter. However, at this stage we would simply reiterate the fundamental consideration that fluid is said to be in a continuum when the measured fluid property is constant for sensitive elemental volumes that are small as compared to the system scale but large as compared to the local scale, and would essentially consider the suitability of this consideration for our foregoing discussions.

(iii) Microfluidic transport is often characterized by sharply demarcating variations in the fluid properties over the interfacial regions as compared to those in the bulk. As the confinement size becomes narrower, interactions between the wall and fluid atoms (typically van der Waals, electrostatic, structuration and solvation forces) tend to play more prominent and decisive roles towards dictating the interfacial phenomena, primarily by giving rise to strong fluctuations in the near-wall number densities. Such strong local density fluctuations may be observed for several reasons, for example due to layering of fluid atoms parallel to the atomic layers adjacent to the solid boundary. On the contrary, there may also be an appreciable depletion of a denser fluid phase close to the walls in preference to a less dense one, under the influences of hydrophobic interactions in narrow confinements. Accordingly, the effective transport properties as well as diffusion coefficients of the fluid may be remarkably different in the near-wall region, as compared to the far stream. In several instances, anomalous transport close to the wall may render the validity of the classical notion or paradigm of no-slip boundary condition (i.e., zero relative velocity between the fluid and the solid at their points of contact) somewhat questionable. The criticality of this issue is immense; without an appropriate boundary condition no accurate estimates of the flow field may become possible, despite the employment of correct governing equations. The underlying issues, however, are by no means trivial and demand more involved discussions beyond a mere summarization. We discuss further on this matter in the subsequent section.

1.3 Interfacial Boundary Condition: Slip Versus No-Slip

1.3.1 General Considerations

Solid–fluid interface in microscale exhibits interesting and complex physical phenomena. The traditionally applied no-slip boundary condition at the fluid–solid interface is an idealized paradigm, which assumes moderately strong attractive forces between the fluid and wall, thereby trivially disallowing the fluid atoms in attaining momentum and energy states different from those of the solid boundary atoms in proximal contact. However, effects of surface tension, liquid evaporation, porosity, osmotic transport, van der Waals forces, electrostatic forces, etc. may potentially result in true or apparent deviations from this classical picture. Even from a pure mathematical viewpoint, slip at the interface appears to be a more acceptable general notion than that of no-slip, since no-slip is a special case of slip with the magnitude of slip equated to zero! This fact was recognized by Navier more than a century back (in fact, way back in 1823), who first introduced the general notion of boundary slip, by defining a slip length (L_S) as the distance behind the interface at which the fluid velocity extrapolates to zero (see Fig. 1.2). However, it needs to be recognized at this stage that this description of slip is more mathematical than physical. Nevertheless, one may refer to the essential flow physics towards developing an insight on the slip length characteristics as a function of the interfacial transport, which we would elucidate further through a couple of simple illustrative examples.

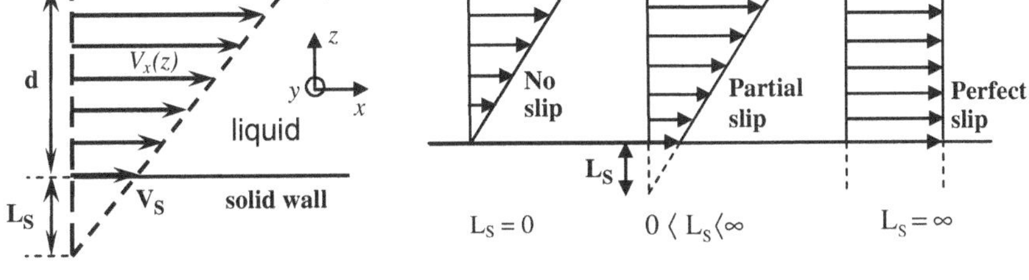

Fig. 1.2 Concept of slip parameterized by the slip length, L_S

1.3.1.1 Gas Flows

Recent developments in the field of MEMS and NEMS have raised several interesting questions with regard to the suitability of the traditionally accepted no-slip boundary condition for gas flows occurring over miniaturized length scales. Perhaps the most important consideration in this regard is the applicability of the classical continuum hypothesis, which is based on the following two important postulates:

(a) The number of molecules is sufficiently large in any chosen elemental volume for fluid property predictions at a given position, so that the statistical uncertainties with regard to their respective positions and velocities do not perceptibly influence the fluid/flow property predictions, as well as the predictions in the local gradients of properties through well-known rules of differential calculus, and
(b) The system does not deviate significantly from local thermodynamic equilibrium.

Although the condition (a) is more commonly satisfied except for highly rarefied gas flows, the condition (b) may be more ominously violated, in a strict sense. This may be attributed to the fact that even in a slightly rarefied medium, gas molecules transfer their momentum and energy to the walls only partially, through a sequence of 'infrequent' collisions. These molecules are temporarily adsorbed to the surface and subsequently ejected from the same. If the gas adjacent to the solid boundary undergoes 'sufficiently' large numbers of collisions with the wall over a given period of time, the exchange of momentum and energy between the gas molecules and the wall is considered to be virtually complete, resulting in the tangential component of the gas velocity equal to the corresponding component of the boundary velocity. However, in a 'non-compact' system with 'insufficient' numbers of gas molecules, the collision frequency may not be 'adequate' to equilibrate the properties of the incipient gas with the wall, resulting in consequent deviations from local thermodynamic equilibrium. This may result in a 'slip' between fluid and the solid boundary. This phenomenon may be pronounced by the presence of strong local gradients of temperature (commonly termed as 'thermophoresis') or density (known as 'diffusophoresis'). As a result, the molecules tending to 'slip' on the walls experience a net driving force. The extent of this deviation is dependant on the Knudsen number ($Kn = \lambda/L$), which refers to the comparability of the mean free path (λ) to the characteristic system length scale (L). This ratio is the most important decisive factor for the applicability of a particular flow modelling strategy as against the extent of rarefaction of the fluid medium, pertinent to gas microflows. Four distinct gas flow regimes are characterised from this parameter – when the value of Kn is less than 0.001, the gas flow may be modelled using Navier–Stokes equation along with no slip boundary condition at the walls; as Kn exceeds 0.001 and below 0.1, the gas molecules may start slipping at the walls; beyond the values of 0.1–10 there is an apparent transition of even the bulk flow from continuum to free molecular regime ($Kn > 10$), which becomes increasingly more significant as Kn is progressively increased to higher orders.

Deviations from the no-slip boundary condition for gases, as manifested through enhanced mass flow rates obtained under experimental conditions in comparison to the traditionally-based predictions, have been addressed by several authors theoretically, by employing the continuum equations in conjunction with the application of a slip boundary condition. Maxwell [3] originally introduced a first-order slip model, which assumes that the slip velocity is equal to the velocity at a mean free path away from the wall. The extent of slip was modelled to be proportional to the local mean free path, for partially specular and partially diffuse walls.

The extent of interaction of the wall with the gas molecules, expressed in terms of the fraction of the transfer of the momentum from the incident molecules to the gas molecules reflected from the wall, was defined by Maxwell as

$$\sigma = \frac{\tau_i - \tau_r}{\tau_i - \tau_w} \tag{1.1}$$

where σ is the tangential momentum accommodation coefficient, τ_i, τ_r and τ_w are the tangential component of the incident, reflected and wall velocity, respectively (see Fig. 1.3). The value of σ lies between 0 and 1. Physically, σ=0 represents the case in which the tangential momentum of all the incident molecules equals to the same of the reflected molecules (specular reflection). As a result, the gas molecules do not exchange any net tangential momentum with the wall on collision. On the other hand, σ=1 represents the case in which the incipient gas molecules assume the same tangential momentum as that of the wall (diffuse reflection). In reality, one may imagine that only a fraction of the molecules is specularly reflected and the remaining is diffusely reflected, so that the reflected molecules have a tangential momentum that is intermediate of the two limiting cases possible. To express this understanding in simple mathematical terms, let us consider three layers, one coincident with the wall (= u_{w}) with corresponding tangential molecular velocity on reflection (= u_r) and the other two layers located at one and two mean free paths away from the wall, (with respective velocities of the corresponding gas layers as u_{g} and u_λ; see Fig. 1.3). Following Maxwell, it may be conjectured that half of the velocity u_{g} is contributed by the incipient gas molecules from its upper layer and the remaining half from the lower layer (the specularly + diffusely reflected molecules from the surface). Accordingly, one may write:

$$u_{\mathrm{g}} = \frac{u_\lambda}{2} + \frac{[(1-\sigma)u_\lambda + \sigma u_{\mathrm{w}}]}{2} \tag{1.2}$$

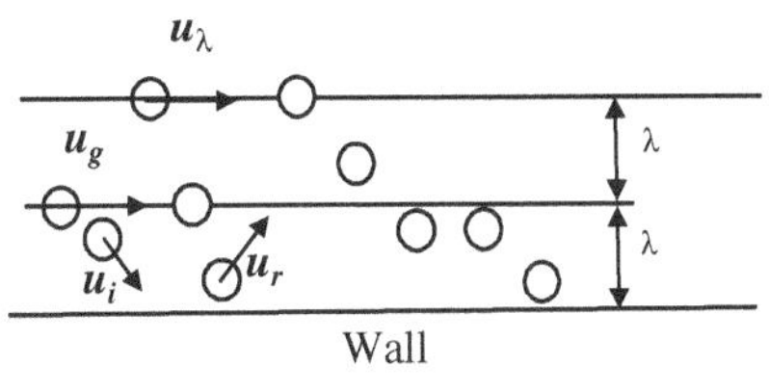

Fig. 1.3 Gas molecules near the wall

Further, in a linearlized framework, the normal gradient of the tangential velocity component at the wall may be expressed as

$$\sigma \left(u_{\lambda} - u_{\mathrm{w}}\right) = 2\lambda \left.\frac{\mathrm{d}u}{\mathrm{d}y}\right|_{\mathrm{w}} \tag{1.3}$$

For further generalization, the velocity scales and the length scales may be non-dimensionalised using the normalization parameters U_{ref} and L (a characteristic system length scale), respectively. Accordingly, from Eqs. (1.1), (1.2)_and (1.3), it follows:

$$\bar{u}_{\mathrm{g}} = \bar{u}_{\mathrm{w}} + \left(\frac{2-\sigma}{\sigma}\right) Kn \left.\frac{d\bar{u}}{d\bar{y}}\right|_{\mathrm{w}} \tag{1.4}$$

where $Kn = \lambda/L$. Equation. (1.4) essentially describes a first order slip model, representing that the slip velocity is proportional to the local Knudsen number. Maxwell-slip model was subsequently modified to incorporate the contribution of stream-wise temperature gradients associated with thermal creep phenomenon, by considering that the slip velocity is simultaneously caused by the velocity gradient normal to the wall and a temperature gradient tangential to the wall. Similar considerations were also invoked for modeling the temperature jump boundary conditions at the walls by Smoluchowski [4].

Following the seminal work of Maxwell, several slip models (many representing higher order corrections and non-linear effects; for example see [4]) have been proposed in the literature for modeling microscale gas flows. However, it has also been argued by many researchers that for cases in which the continuum description may be used to model the bulk flow, an apparent invalidity of the no-slip boundary condition may need to be re-looked upon, perhaps by extending the scopes of the continuum conservation equations towards capturing certain special effects. In this context, it needs to be mentioned that the discrepancy in the apparent inapplicability of the classical continuum-based mathematical models for such cases does not essentially originate from the fundamental conservation principles as such, but may be attributable to an apparent abstraction from the seemingly non-trivial dependences of the fluid flow characteristics on the pertinent molecular level interaction mechanisms with characteristic sharp gradients adjacent to the confining boundaries. The derivative terms in the continuum conservation equations often tend to get invalid in such circumstances, since those may turn out to be incapable in capturing strong local gradients of fluid properties adjacent to the solid substrate. In this respect, it has been suggested [5–8] that the standard forms of the Navier-Stokes-Fourier equation may need to be modified, while considering the effects of fluid flows with strong local gradients of density and temperature. These modifications have been motivated from a thoughtful insights from experimentally observed thermophoretic motion, the hierarchical reordering of the Burnett terms in the Chapman-Enskog expansion of the viscous stress tensor, and the velocity/thermal creep coefficients introduced by Maxwell to account for non-zero slip velocities and temperature jumps at the fluid–solid interface (irrespective of the fact that the continuum hypothesis might otherwise be satisfied in the bulk). Brenner [5–7] first attempted to modify the Navier–Stokes equations, and extended them based on the concept of volume diffusion. In a strict sense, the concept of continuum cease to exist in reality because of the presence of discrete molecules and the contained empty spaces between themselves which are heavily linked with the extent of rarefaction or compression in the system. Presence of strong gradients of density leads to ambiguous definition of the velocity in continuum and hence one may refer to fluxes, instead of velocities. Description of the term ‘velocity’ is redefined to a normalized flux density of a system of particles with a fixed mass and identity. In an Eulerian control volume, it is interpreted as an advective flux density of mass across an element of the pertinent control surface. Due to the above mentioned strong local gradients of density and/or temperature, a phoretic or diffusive transport of mass across the control surface is superimposed on this advective flux. A classical example of phoretic transport is the phenomenon of thermophoresis, in which heat-conducting, force-free and torque-free tracer particles (typically, spherical) move from hotter to colder regions in the fluid (usually, a gas), against an externally imposed temperature gradient, without necessitating the aid of any externally imposed pressure gradient. Physically, this diffusive flux can be manifested by the Lagrangian velocity of a passive and neutrally-buoyant non-Brownian tracer particle introduced into the flow, relative to the advective fluid motion. The diffusive transport of fluid particles may result in a net diffusive transport of mass across the control surface, apparently inconsistent with the conservation of mass from a continuum perspective.

Based on the local gradient-driven diffusive transport considerations, researchers have proposed the employment of modified constitutive relationships in the continuum conservation equations, by considering a material frame that advects with the fluid flow. The term ‘velocity’ in this connection turns out to be nothing but the description of a normalized flux density that advectively transports mass, momentum or energy within the fluid envisaged as a continuum. Accordingly, the net ‘velocity’ is the summation of the advective and the diffusive components, to yield [5–8] $U_i^{\mathrm{T}} = U_i + u_i^{\mathrm{D}}$, where U_i^{T} is the total velocity, U_i is the advective component of velocity and u_i^{D} is the diffusion velocity. In terms of the kinetic theory, mass diffusion phenomenon can be explained as a net flow in the direction in which gradients of the thermodynamic properties are present because of the difference in the thermal velocity of molecules at different locations [9]. With constitutive relationships framed based on the total velocity (instead of the advective component of the velocity alone), the continuum conservation equations for fluid flow and heat transfer can be somewhat extended to take into account the considerations of phoretic transport mechanisms associated with strong local gradients of density and/or temperature within the flow domain. However, it also needs to be appreciated that such measures may not turn out to be adequate as we progressively move towards very high Knudsen number regimes (for example, free molecular flows), for which continuum hypothesis appears to become invalid altogether, because of the inadequacy of the gradient terms involving the continuum field variables in capturing the characteristic variations in the flow domain.

1.3.1.2 Slip Boundary Condition for Liquids

The physics of liquid–solid and liquid–liquid interactions in small-scale systems gives rise to many apparent anomalies, primarily attributable to seemingly non-trivial and complex dependences of the liquid flow characteristics on the molecular level interaction mechanisms and down-scaled topographical features of the confining boundaries, in a manner that is substantially more non-trivial than gas flows. For example, it may be rather intuitive to expect that the roughness elements of the solid boundaries in micro/nano-channels are likely to offer additional flow resistances, resulting in a possible reduction in the fluid flow rates. On the contrary, in reality, a 'super-fluidity' has been observed to occur in small scale fluidic confinements with 'designed' surface roughness conditions, apparently violating the classical principles (such as the celebrated 'no-slip' boundary condition). Till recent times, the underlying fluid dynamic mechanisms remained to be poorly understood, especially within the purview of experimentally tractable spatio-temporal scales. This deficit stemmed from the complexities in describing the underlying thermo-fluidic interactions at physical scales that are substantially larger than those addressed in the pertinent molecular-scale simulations.

For liquids, intermolecular attractions are substantially stronger than gases. As a result, liquid is intuitively expected to remain stationary relative to the solid boundary at their points of contacts. An exception to this may be observed at ultra-high shear rates, for which the liquid molecules may be forcibly dragged over the solid boundary under appropriate straining conditions [10]. In fact, it has also been shown by using Molecular Dynamics Simulations that high shear rates (typically realizable only in extremely narrow confinements of size roughly a few molecular diameters) may indeed move the liquid molecules adhering to a solid boundary by overcoming the van der Waals forces of attraction. The corresponding variation in slip length as a function of the shear rate has been shown to obey the following functional form [10]:

$$L_{\mathrm{s}} = L_{\mathrm{s}}^{0}\left(1 - \frac{\dot{\gamma}}{\dot{\gamma}_{\mathrm{c}}}\right)^{-\frac{1}{2}} \tag{1.5}$$

where L_{s} is the slip length as a function of the shear rate $\dot{\gamma}$, $\dot{\gamma}_{\mathrm{c}}$ is the critical shear rate above which the linear relationship between the slip length and the rate of deformation ceases to hold, and L_{s}^{0} is the constant slip length for shear rates below $\dot{\gamma}_{\mathrm{c}}$.

For more common circumstances in which the shear rates are substantially lower than the above, it has been hypothesized by many researchers that the no-slip boundary condition arises because of the microscopic asperities on the surface. The liquid molecules may get locally trapped in the surface asperities and thus may not be able to escape from an intimate contact with the solid boundary because of their otherwise compact intra-molecular packing. Following this argument, however, it may be speculated that a molecularly smooth boundary would allow the liquid to slip, because of the non-existence of the surface asperity barriers.

From an engineering perspective, the possibility of having a slip-based boundary condition, instead of a more intuitively acceptable no-slip boundary condition for liquids, may bear far-ranging consequences on the functionalities of microfluidic devices. This is because of the fact that the wall shear stress (or equivalently, the pressure drop over a given length for a fully developed flow), and hence the pumping effort, is explicitly linked with the nature of the boundary condition at the liquid–solid interface. In fact, numerous experimental investigations [11–14] have been executed by the researchers to pinpoint the anomalies of predictions in pressure drop-flow rate characterization in the context of liquid flows in microchannels (though, not necessarily systematic and repetitive) and deviations in the frictional behaviour of pressure-driven liquid flows through microchannels, in perspective of the classical theory. These studies have witnessed routinely a deviation in the form of increase in friction factor with increase in effective surface roughness. Several semi-empirical models have been proposed based on the concept of the so-called 'roughness viscosity' [15], 'porous medium layer' [16], etc., in an effort to fit the theoretical predictions with the observed experimental data. In parallel, a number of research studies have revealed reductions [17–19] in friction for pressure-driven liquid microflows, in comparison to the classical theory of Poiseullian flows, despite encountering rough microchannel-walls. Lack of fundamental understanding of the relative competing influence of friction-enhancing and friction-reducing effects have often lead to the apparently contradictory observations, which at occasions have tended to threaten the amazing acceptability of the paradigm of the celebrated 'no-slip' boundary condition for liquid flows.

Recent studies have demonstrated that the textbook assumption of 'no slip at the boundary' can fail greatly not only when the fluidic substrates are sufficiently smooth, but also when they are sufficiently rough. The reasons behind such apparently anomalous behavior lie in fundamental interfacial interactions (such as wettability) which can perhaps be tuned to some extent through the exploration of a novel 'physics-based' manufacturing paradigm. Majumdar et al. [20] first demonstrated that the rate of liquid flow through a membrane composed of an array of aligned carbon nanotubes might turn out to be 4–5 orders of magnitude faster than that can be predicted from classical fluid-flow analysis. They attributed this phenomenon to an apparently frictionless interfacial condition at the carbon-nanotube wall. Such observations were contrary to the common consensus that fluid flow through nanopores having chemical selectivity is rather slow [21]. However, from fundamental physical considerations, water is likely to be able to flow fast through hydrophobic single-walled carbon nanotubes; the primary reason being the fact that the process creates ordered hydrogen bonds between the water molecules [22]. Accordingly, ordered hydrogen bonds between water molecules and the weak attraction between the water and smooth carbon nanotube graphite sheets, as well as the rapid diffusion of hydrocarbons [23, 24] may be qualitatively attributed to the fundamental scientific origin of reduced frictional resistances encountered in such systems.

Detailed investigations [25] have attempted to resolve the apparent anomaly of 'reduced' fluid friction encountered in 'rough-walled' narrow confinements of characteristic dimensions even somewhat upscaled (ranging from a few nanometers to even upto a few microns) than those of the carbon nanotubes mentioned as above, by appealing to the thermo-fluidics over experimentally-resolvable spatio-temporal scales. By executing comprehensive theoretical and experimental studies, it has been established that combinations of surface roughness, hydrophobicity, and reduced confinement sizes may decrease the level of interfacial friction to a significant extent [26, 27]. Confining rough surfaces made of hydrophobic (water-disliking) materials may trigger the formation of nano-scale bubbles adhering to the walls in tiny channels (see Fig. 1.4a). The incipient vapour layer acts as smoothening blanket, by disallowing the liquid on the top to feel the effects of the rough surface asperities. In such cases, the liquid is not likely to feel the presence of the wall directly and may smoothly sail over the intervening vapor layer shield. Thus, instead of 'sticking' to a rough channel surface, the liquid may effectively 'slip' on the same (see Fig. 1.4b). The effects of microscopic roughness of the solid surface, however, may also obstruct the motion of the adhering fluid by promoting the stick-flow, at surface locations where nanobubbles are not formed. The resultant effect is therefore of slip-stick nature. Depending on the extent of surface coverage with nanobubbles, the slip effect can dominate considerably over the sticking influence, even for highly rough surfaces. Consequently, rough surfaces, as often considered to be undesirable from frictional considerations, but otherwise unavoidable artefacts of the fabrication processes, may play positive and decisive roles in imparting an apparently frictionless nature to the flow behaviour, by promoting the formation of surface-adhering nanobubbles. Fundamentally, nanobubbles may be nucleated when the driving force

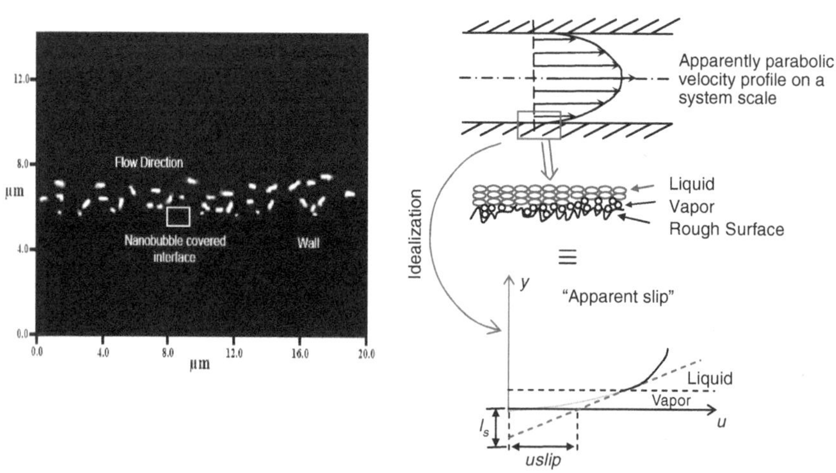

Fig. 1.4 (**a**) Nanobubbles on a microchannel substrate, (**b**) Apparent slip due to nanobubble formation

required to minimize the area of liquid–vapor interface is smaller than the forces that pin the contact line of the substrate. Thermodynamically, hydrophobic regions do not favour formation of hydrogen bonds with water molecules and hence give rise to excluded volume regions encompassing the locations characterized with sharply diminishing number density of water molecules. In such cases, the hydrophobic surface acts like a nucleation site for the formation of thin vapour layers. Such interfacial fluctuations may destabilize the liquid further away from the solid walls, leading to a pressure imbalance. In confined fluids, long-ranged interactions may also trigger separation-induced phase transitions. Such separation-induced cavitations physically originate from an increase in the local molecular field due to the replacement of polarizable fluids by solid walls.

It must be emphasized in this context that despite realizing apparent low-friction behaviour for substrates covered with nanobubbles, the no-slip boundary condition still remains to be a valid proposition at the walls (except for walls covered with rarefied gas layers). However, it is only an apparent inability in capturing and resolving the steep velocity gradients within the ultra-thin vapour (or gas) layers that prompts one to interpret experimental results by extrapolating the velocity profiles obtained in the liquid layer (which exists above the vapour blanket, see Fig. 1.4b) to the adjacent solid boundary, yielding an elusive deviation from the no-slip condition. Such conditions, in which the liquid layers sail over thin vapour layers covering the solid boundaries, may be termed as 'apparent slip' (see Fig. 1.4a), to mark an apparent deviation from the no-slip boundary condition at the wall. Influences of stochastic fluctuations on this amazing physical behaviour have also been addressed in the literature [26, 27], although have by no means been well-resolved. Interestingly and counter-intuitively, spontaneous roughening effects on liquid flows occurring over smooth bubble layers have also been recently observed [28], thereby raising further doubts on the universality of the above-mentioned physical conjecture.

In order to accommodate the stochastic nature of many input model parameters, as attributed to the uncertainties in the prevailing surface conditions, numerous random experiments have been performed with chosen ranges of these data, to come up with a generalized formula for friction factor in rectangular microchannels [26, 27], which may act as an important design basis for microfluidic systems of practical interest. Considering the stick-slip influences of surface roughness elements on fluid friction behaviour, a generalized model for friction factor (β, which, effectively, is a non-dimensional pressure drop) in pressure driven liquid microflows has been developed in the literature in the form of modified Poiseulli formula [26, 27], as:

$$\beta = \frac{8\left[\frac{\alpha}{\gamma} + \frac{3(1-\alpha)}{1-\frac{48}{\sqrt{\pi}}\left(\frac{\varepsilon}{D_h}\right)^2\frac{D_h}{l}}\right]}{\text{Re}} f_{HK} \tag{1.6}$$

where $\gamma = 1/3(1-\kappa)^3 + \frac{\Omega\kappa}{3}(\kappa^2 - 3\kappa + 3)$ with κ is the mean thickness of the nanobubble dispersed layer which is nondimensionalised with respect to the half of the channel height, D_h is the hydraulic diameter, ϵ is surface roughness, α is the extent of the stick-slip motion, l is the correlational length and Re is the Reynolds number. The parameter f_{HK} is the Hartnett–Kostic polynomial correction factor that accommodates a noninfinite extent of the rectangular microchannel into account as:

$$f_{\mathrm{HK}} = 1 - 1.3553\chi + 1.9467\chi^2 - 1.7012\chi^3 + 0.9564\chi^4 - 0.2537\chi^5 \quad (1.7)$$

where χ is the aspect ratio defined as the ration between the height to the width of the channel.

Utilizing some of the recent advances in this regard and addressing some of the concerned 'open' questions in the near future, researchers may potentially develop engineered rough surfaces with triggered hydrophobic interactions and consequent inception of friction-reducing vapour phases [29], so as to achieve a super-fluidic transport over small scales. Narrow confinements capable of mimicking the selective and rapid fluidic transport attainable in biological cellular channels but designed on the basis of such newly-discovered surface roughness-hydrophobicity coupling would open up a wide range of new applications, such as transdermal rapid drug delivery systems, selective chemical sensing and mixing in nano-scales and several other lab-on-a-chip based applications.

1.4 Liquid Micro-flow Actuation in Continuous Systems: Fundamental Principles

In this section, we briefly discuss about the fundamental continuum considerations and bring those in the perspectives of various flow actuation mechanisms commonly employed in microfluidics applications.

1.4.1 Conservation Equations

Traditionally, the basic laws of classical mechanics and thermodynamics were developed from a 'closed system' (control mass) perspective, considering the fundamental element of analysis as an entity of fixed mass and identity. However, for fluids, identification of tractable control masses (Lagrangian approach) may be rather cumbersome, since fluid is inherently a continuously deformable medium. Therefore, it may be more appropriate to describe the basic conservation laws for fluids by invoking the concept of a 'control volume' (Eulerian approach), which is nothing but an identified region in the flow-field over which attention may be focused, in terms of the transport of conserved quantities. The Reynolds Transport Theorem [30] is a general mathematical description of the transformation from a system (abbreviated as 'sys') to a control volume approach (abbreviated as 'cv'), so that one may apply the classical laws for fluids with respect to an Eulerian reference.

Generically, the Reynolds Transport Theorem may be described as

$$\left.\frac{DA}{Dt}\right|_{\mathrm{sys}} = \frac{\partial}{\partial t}\int_{\mathrm{cv}} \rho a d\forall + \int_{\mathrm{cs}} \rho a \bar{v}_r.\hat{n} dS \tag{1.8}$$

where A is an extensive property, a is A per unit mass, $\bar{v}_r$ is the velocity of the fluid relative to the control surface, $d\forall$ is a differentially small element in the chosen control volume, dS is an elemental area on the control surface with a unit outward normal vector denoted by $\hat{n}$, and ρ is the density of the fluid.

As an example for illustration, we may consider the conservation of mass, for which $A = m$ (total mass of the system), and $a = 1$. Since total mass of the system is conserved by definition, $\left.\frac{Dm}{Dt}\right|_{\mathrm{sys}} = 0$. Further, considering a non-deformable and stationary control volume, applying the Gauss divergence theorem for conversion of surface integral to volume integral, and noting that $d\forall$ is arbitrary, a differential form of the conservation equation for the total mass may be readily obtained, which is commonly known as the continuity equation in continuum fluid mechanics:

$$\frac{\partial \rho}{\partial t} + \nabla.(\rho \vec{u}) = 0 \tag{1.9}$$

where $\vec{u}$ is the velocity of flow. As an alternative, the same equation may also be written in Cartesian index notation as, $\frac{\partial \rho}{\partial t} + \frac{\partial}{\partial x_j}(\rho u_j) = 0$.

In a manner similar to the conservation of mass, one may also apply the Reynolds Transport Theorem for the conservation of linear momentum. The external forces acting on the system can be described using the summation of surface and body forces. Assuming a stationary and non-deformable control volume, and considering $A = m\bar{V}$ (so that $a=\bar{V}$), the linear momentum conservation principle (essentially, Newton's second law extended to control volumes) yields (in Cartesian index notations):

$$\frac{\partial(\rho u_i)}{\partial t} + \frac{\partial(\rho u_i u_j)}{\partial x_j} = \frac{\partial \tau_{ij}}{\partial x_j} + b_i \tag{1.10}$$

where τ_{ij} are the components of the stress tensor, and b_i is the body force per unit volume in ith direction. For fluids, which deform continuously under the action of even an infinitesimally small stress, τ_{ij} is related to the strain rate yielding a relationship which is constitutive in nature. It is a function of the rate of deformation and is given by:

$$\tau_{ij} = -p\delta_{ij} + \tau_{ij}^{\mathrm{dev}} \tag{1.11}$$

where p is the thermodynamic pressure, δ_{ij} is the Kronecker-delta symbol and τ_{ij}^{dev} is the deviatoric component of the stress vector. For Newtonian fluids, the deviatoric stress varies linearly with the strain rate (rate of deformation). For an isotropic fluid, it may be shown that this deviatoric component of stress may be expressed as:

$$\tau_{ij}^{\text{dev}} = \lambda e_{kk} \frac{\partial u_k}{\partial x_k} \delta_{ij} + 2\mu e_{ij} \tag{1.12}$$

where λ and μ are the volume dilation coefficient and the viscosity coefficient respectively, and e_{ij} is the rate of deformation (which is the symmetric part of the strain rate $\partial u_i/\partial x_j$). For such fluids, the stress tensor yields the following form:

$$\tau_{ij} = -p\delta_{ij} + \lambda \frac{\partial u_k}{\partial x_k} \delta_{ij} + \mu \left(\frac{\partial u_i}{\partial x_j} + \frac{\partial u_j}{\partial x_i} \right) \tag{1.13}$$

It needs to be noted here that the thermodynamic pressure, p, is different from the mechanical pressure, p_m (which is negative of the arithmetic mean of the normal stress components). These two pressures are thus related by:

$$-p_m = -p + \kappa \frac{\partial u_k}{\partial x_k} \tag{1.14}$$

where $\kappa(= \lambda + \frac{2}{3}\mu)$ is known as the bulk viscosity. It may be hypothesised that p_m and p are equal when κ is equal to zero; this is known as the Stokes hypothesis. This hypothesis holds true when the characteristic time scales in the system are large compared to the molecular relaxation time or if the fluid in question is a dilute monoatomic gas. The equality of these two pressures is also valid when the flow is incompressible ($\partial u_k/\partial x_k = 0$). Combining Eqs. (1.9), (1.10), (1.11), (1.12), (1.13) and (1.14), the celebrated Navier–Stokes equation follows, and is given by:

$$\rho \left(\frac{\partial u_i}{\partial t} + u_j \frac{\partial u_i}{\partial x_j} \right) = -\frac{\partial P}{\partial x_i} + \frac{\partial}{\partial x_j} \left(\mu \frac{\partial u_i}{\partial x_j} \right) + b_i \tag{1.15}$$

1.4.2 Pressure-Driven Flow Actuation and Its Microfluidics Perspective

We consider a simple illustrative example of pressure-driven flow in a parallel plate channel. For such cases, Navier–Stokes equation, under the approximations of steady, two-dimensional flow with constant physical properties, yields:

$$\rho \left(u \frac{\partial u}{\partial x} + v \frac{\partial u}{\partial y} \right) = -\frac{\partial P}{\partial x} + \mu \left(\frac{\partial^2 u}{\partial x^2} + \frac{\partial^2 u}{\partial y^2} \right) \tag{1.16a}$$

$$\rho \left(u \frac{\partial v}{\partial x} + v \frac{\partial v}{\partial y} \right) = -\frac{\partial P}{\partial y} + \mu \left(\frac{\partial^2 v}{\partial x^2} + \frac{\partial^2 v}{\partial y^2} \right) \tag{1.16b}$$

Equation 1.16(a, b) are essentially the statements of conservation of momentum in x, y directions with u, v as the respective velocity components. The characteristic length scales in x and y-directions are L and H respectively (see Fig. 1.5). With

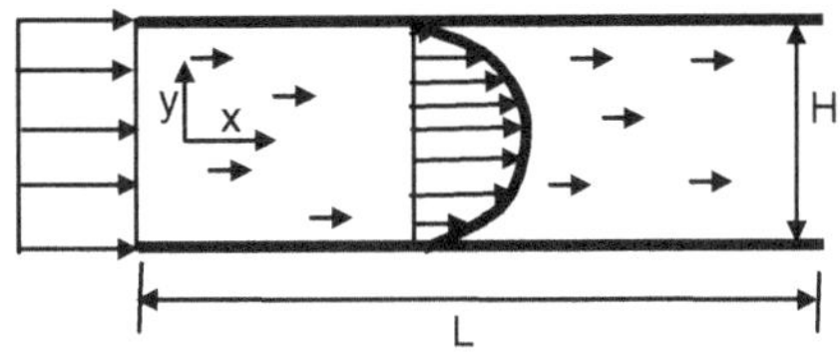

Fig. 1.5 Pressure driven flow in a parallel plate channel with height H and length L

$L>>H$, $\mu\frac{\partial^2 u}{\partial x^2} \ll \mu\frac{\partial^2 u}{\partial y^2}$ and $\mu\frac{\partial^2 v}{\partial x^2} \ll \mu\frac{\partial^2 v}{\partial y^2}$. Further, non-dimensionalising $\frac{u}{U} \to \bar{u}$, $\frac{v}{U} \to \bar{v}$ (U being the centreline velocity), $\frac{x}{L} \to \bar{x}$, $\frac{y}{L} \to \bar{y}$ and $\frac{P}{\mu U/L} \to \bar{P}$, Eq. (1.16a) yields:

$$Re\left(\bar{u}\frac{\partial \bar{u}}{\partial \bar{x}} + \bar{v}\frac{\partial \bar{u}}{\partial \bar{y}}\right) = -\frac{\partial \bar{P}}{\partial \bar{x}} + \frac{\partial^2 \bar{u}}{\partial \bar{y}^2} \tag{1.17}$$

where Re is the Reynolds number ($= \rho U_{av} H/\mu$). For low Re flows, the left hand side of Eq. (1.17) is negligible in comparison to the right hand side. Interestingly, the left hand side of Eq. (1.17) becomes identically equal to zero irrespective of the Re (if the flow is turbulent, the instantaneous velocities and pressures may be replaced with the corresponding Reynolds-averaged quantities towards satisfying this condition), if the flow is hydrodynamically fully developed (i.e, u is not a function of x). Under such circumstances, Eq. (1.17), in conjunction with the no-slip boundary condition at the walls ($\bar{y} = \pm 1$), may be integrated twice to yield:

$$\bar{u} = -\frac{1}{2}\frac{d\bar{P}}{d\bar{x}}\left(1 - \bar{y}^2\right) \tag{1.18}$$

The quantity dP/dx is essentially the axial pressure gradient acting on the flow, which is a constant for fully developed flows. Expressing in terms of the cross-sectionally averaged velocity (U_{av}), Eq. (1.18) may be re-written as

$$\frac{u}{U_{av}} = \frac{3}{2}\left(1 - \frac{y^2}{H^2}\right) \tag{1.18a}$$

Using Newton's law of viscosity, the wall shear stress (τ_w) may be evaluated as

$$\tau_w = \left|\mu\left(\frac{\partial u}{\partial y} + \frac{\partial v}{\partial x}\right)_{wall}\right| = \frac{3\mu U_{av}}{H} = \frac{3\mu Q}{2H^2} \tag{1.19}$$

where Q is the volumertic flow rate per unit channel width. A non-dimensional measure of τ_w, namely the friction factor (C_f) may be expressed as

$$C_f = \frac{\tau_w}{\frac{1}{2}\rho U_{av}^2} = \frac{6}{Re} \tag{1.20}$$

Similar expressions may be obtained for channels of various other cross sectional shapes as well [30, 31]. Irrespective of the quantitative differences associated with channels of different cross-sectional shapes, following general conclusions may be drawn from the above analysis, bearing particular significance to microfluidic transport:

(i) From Eq. (1.18), it is revealed that the axial velocity varies qudratically across the channel height, with the maximum velocity at the centerline. This implies a significant dispersion due to considerable variations between wall-adjacent and centerline velocities. For cases in which the end objective is to transport chemical or biological samples for wall-bounded reactions, the above may be a serious deterring factor, since the sample may have a tendency to be 'swept' away along the centerline instead of converging onto wall-bounded reactive sites. Clearly, more uniform velocity profiles would perform better in cases of such specific requirements.

(ii) From Eq. (1.19), it may be observed that the wall shear stress (and hence the pressure drop) increases perpetually as the cross-sectional dimensions of the channel are progressively reduced. This, in turn, implies the requirements of huge pumping power in driving the flow for reduced cross-sectional dimensions, and acts as one of the limiting constraints for pressure-driven flow actuation in microfluidic applications.

(iii) From Eq. (1.20), it is evident that the product of friction factor and Reynolds number is a constant for fully developed laminar flows, from a classical perspective. Essentially, this holds true until the no-slip boundary condition may be applied at the walls. However, micro-scale flows are commonly featured with surface roughness elements having characteristic length scales not of trivially negligible order as that of the system length scales. Consequently, interesting interfacial phenomena may occur, overruling the satisfaction of the no-slip boundary condition, either in a true sense or in an apparent sense (see Section 1.2 for details). Under such circumstances, surface roughness dependent frictional characteristics may be observed in microchannels, even for fully developed laminar transport.

1.4.3 Surface Tension Driven Flow

Surface tension driven flows refer to the actuation and control of fluid transport through a manipulation of the surface tension forces. The manipulation may be hydrodynamic, thermal, chemical, electrical or optical in nature. Since surface tension forces scale with the linear dimensions, these become progressively more dominant with reduction of system length scale from macro to micro and further to nano. Surface tension (γ) is the force per unit length acting along the interface of immiscible phases. In a liquid–gas system, for example, molecules in the bulk of the liquid are pulled equally in all directions by the neighboring liquid molecules, resulting in no net force. At the interface, however, the molecules experience a net attractive force by other molecules inside the liquid, since these are not attracted as intensely by the molecules in the gaseous phase located on the other side, because of a denser molecular packing in the liquid phase than in the gas phase and a consequent stronger intermolecular force of attraction offered by the liquid molecules. To maintain interfacial equilibrium, molecules at the interface rearrange themselves to diminish the surface area (in order to minimize the surface energy), and a meniscus is formed in the form of a surface resembling a stretched elastic membrane. The pressure difference on either sides of the meniscus leads to development of net normal force (pressure difference times the projected surface area). This normal force acting on the meniscus is balanced by the surface tension force in equilibrium, leading to a curved meniscus.

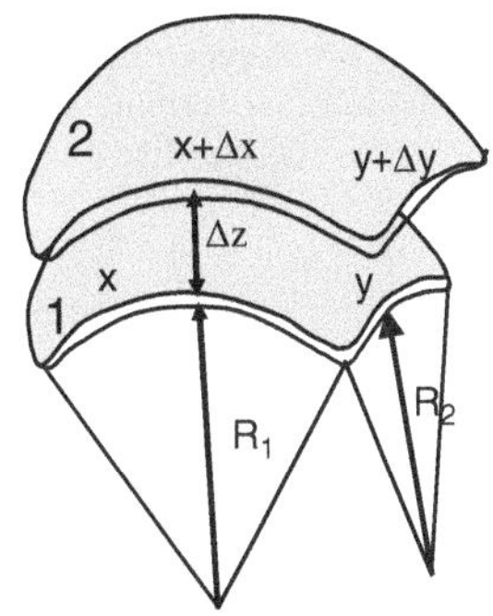

Fig. 1.6 Virtual displacement Δz of a surface from 1 to 2

For a simple quantification of the pressure difference across the two sides of the interface, we consider a surface *1* having area A (given by $= xy$) and the radii of curvatures as R_1 and R_2 (as shown in Fig. 1.6). Under a small virtual displacement of Δz normal to the surface, the surface assumes a new configuration (surface *2*) with area $A + \Delta A$, given by $(x + \Delta x)(y + \Delta y)$. From similarity considerations, it may be easily shown that $\frac{R_1}{R_1+\Delta z} = \frac{x}{x+\Delta x}$ and $\frac{R_2}{R_2+\Delta z} = \frac{y}{y+\Delta y}$, which imply

$$R_1 = \Delta z \frac{x}{\Delta x} \text{ and } R_2 = \Delta z \frac{y}{\Delta y} \quad (1.21)$$

The work done by pressure in displacing the interface from position 1 to 2 is equal to: $\Delta P\,(A\Delta z)$, which is equivalent to the work done in stretching the interface by overcoming surface tension: $\gamma \Delta A$. Balancing these two and using Eq. (1.21), one may obtain a relationship between the pressure difference (ΔP) and the curvature along two principle radii (R_1 and R_2), which is known as the Young–Laplace equation, and is given by:

$$\Delta P = \gamma \left(\frac{1}{R_1} + \frac{1}{R_2} \right) \quad (1.22)$$

Additionally, it is important to mention that the interface is the junction of two or more intersecting phases and the angle at the junction formed by these phases is known as the contact angle. The contact angle is specific for any given system and is determined by the interactions across the three interfaces. Mostly, the contact angle involving liquid–solid–gas interfaces is determined by a small liquid droplet over a solid surface, thereby creating a three phase contact line – solid–liquid, liquid–gas and solid–gas. The shape of the droplet (see Fig. 1.7) is determined by the Young's Law, and is given by:

$$\gamma_{\mathrm{lg}} \cos\theta = \gamma_{\mathrm{sl}} - \gamma_{\mathrm{sg}} \tag{1.23}$$

where γ_{lg}, γ_{sl} and γ_{sg} is the surface tension of liquid–gas, solid–liquid and solid–gas system, θ is the contact angle.

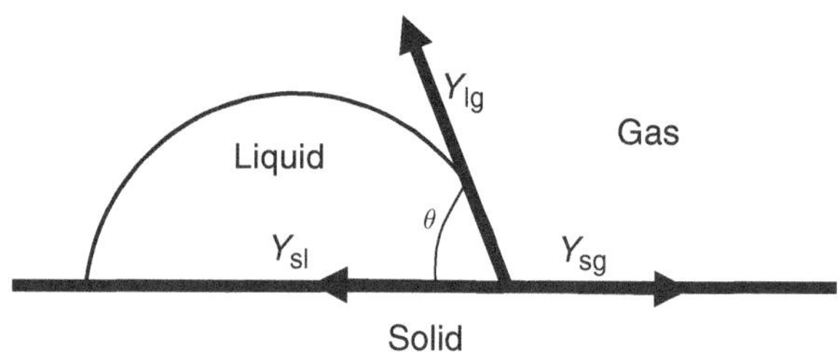

Fig. 1.7 A droplet resting on a solid substrate forming a three phase contact line

Equation (1.23) may follow simply from the balance of surface tension forces along the horizontal direction (Fig. 1.7). Interestingly, it may be observed that the vertical component $\gamma_{\mathrm{lg}} \sin\theta$ elusively appears to remain as an unbalanced force. This, however, is impossible for a system under equilibrium. This apparently unbalanced force, in reality, is balanced by high local stresses at the solid surface. This, in extreme cases, may even lead to elastic or plastic deformation, which may cause the surface to bulge upwards. It may also lead to molecular rearrangements that alter the local surface energies in order to relieve these local stresses. The contact angle is indicative of the nature of wetting characteristic of different substrates. Zero contact angle implies the surface is perfectly wetting with respect to the liquid in question; values of contact angle between 0° and 90° confer to highly wetting substrates; contact angles between 90° and 180° imply low wetting substrates, and a contact angle of 180° signifies perfectly non-wetting substrates. Equation. (1.22) and (1.23) are the necessary conditions for the equilibrium although not sufficient.

The equations for interfacial equilibrium may also be alternatively derived from the energy minimization considerations. The thermodynamic description in terms of the free energy (E) of a droplet with the constraint of the fixed volume ($\forall$) of the droplet is given by [32]:

$$E = \sum_{i \neq j} A_{ij}\gamma_{ij} - \lambda\forall \tag{1.24}$$

where A_{ij} is the interfacial area that demarcates the phases i and j, with the corresponding surface energy being designated as γ_{ij} and λ is the Lagrange multiplier. A droplet with radius of curvature R, making an angle θ with the solid substrate has a volume: $\forall = \pi R^3 \left(\frac{2}{3} - \frac{3}{4}\cos\theta + \frac{\cos 3\theta}{12}\right)$. The surface areas of the liquid–vapor and solid – liquid interfaces are given as: $A_{\mathrm{lv}} = 2\pi R^2 (1 - \cos\theta)$ and $A_{sl} = \pi R^2 \sin^2\theta$. Hence, the free energy as written in Eq. (1.24) may be written as:

$$\begin{aligned} E = {} & \pi R^2 \sin^2\theta\,(\gamma_{\mathrm{ls}} - \gamma_{\mathrm{sv}}) + \gamma_{\mathrm{lv}}\, 2\pi R^2 (1 - \cos\theta) \\ & - \lambda \left\{ \pi R^3 \left(\frac{2}{3} - \frac{3}{4}\cos\theta + \frac{\cos 3\theta}{12} \right) \right\} \end{aligned} \tag{1.25}$$

The free energy should be minimized in terms of the independent variables – R and θ, which is obtained by partially differentiating with respect to these variables and setting them to zero, i.e.,

$$\begin{aligned} \frac{\partial E}{\partial R} = {} & 2\pi R \sin^2\theta\,(\gamma_{\mathrm{ls}} - \gamma_{\mathrm{sv}}) + \gamma_{\mathrm{lv}}\, 4\pi R (1 - \cos\theta) \\ & - \lambda \left\{ 3\pi R^2 \left(\frac{2}{3} - \frac{3}{4}\cos\theta + \frac{\cos 3\theta}{12} \right) \right\} = 0 \end{aligned} \tag{1.26a}$$

$$\begin{aligned} \frac{\partial E}{\partial \theta} = {} & 2\pi R^2 \sin\theta\cos\theta\,(\gamma_{\mathrm{ls}} - \gamma_{\mathrm{sv}}) + \gamma_{\mathrm{lv}}\, 2\pi R^2 \sin\theta \\ & - \lambda \left\{ \pi R^3 \left(\frac{3}{4}\sin\theta - \frac{\sin 3\theta}{4} \right) \right\} = 0 \end{aligned} \tag{1.26b}$$

Upon eliminating λ from Eqs. (1.26a) and (1.26b), we may obtain the Young's Law, as given by Eq. (1.23). Eliminating θ from Eqs. (1.26a) and (1.26b), we obtain the Young–Laplace equation given by Eq. (1.22). Interestingly, the Lagrange multiplier physically turns out to be the interfacial pressure difference.

1.4.3.1 Capillary Filling

Surface tension driven flows can take place in open conduits like microgrooves or in closed systems like capillary tubes. As an illustrative example, we may consider the later case, in which a liquid meniscus advances through a micro-capillary, by displacing a gas or vapor phase (see Fig. 1.8). In a generic 'lumped-mass' form, one can write the corresponding equation of motion describing the capillary advancement (just a statement of the Newton's second law), as

$$\frac{d\,(MV_{\mathrm{c}})}{dt} = \sum F \tag{1.27}$$

where M is the total mass of the fluid system, V_{c} is the velocity of its centre of mass of the system, and F is the resultant force acting on the same. For illustration, we consider three different forces acting on the system – surface tension force, viscous

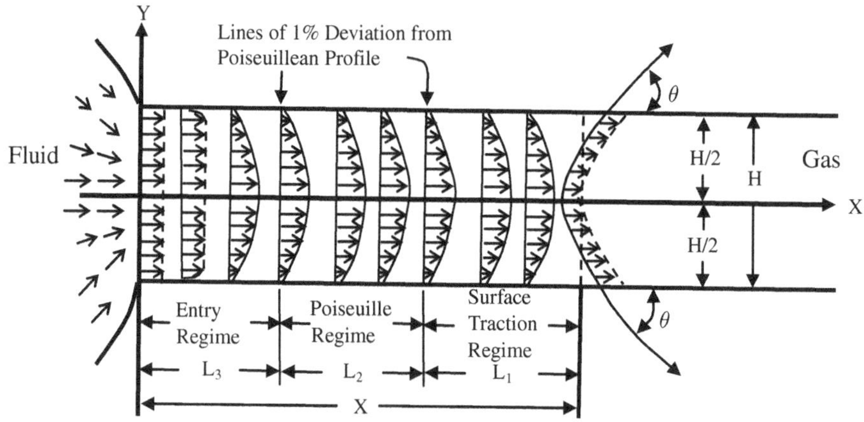

Fig. 1.8 Velocity profiles corresponding to the flow into a capillary tube

force and the gravity; these three sum up to the net force, F. For simplicity in illustration, let us consider viscous, incompressible and Newtonian flow taking place against the direction of gravity along a long cylindrical capillary of radius r. For this case, the mass of fluid enclosed in the capillary may be given by: $M = \rho\pi r^2 z$, corresponding to an instantaneous meniscus advancement upto a length of z from the inlet. The velocity of the centre of mass of the volume, V_c, is $\dot{z}$. The surface tension force is given by $F_{\text{surface tension}} = 2\pi r\gamma \cos\theta$, where γ is the liquid–vapor surface tension coefficient. The viscous force is given by $F_{\text{viscous}} = -8\pi\mu z\dot{z}$ (assuming fully developed flow) with μ as the viscosity coefficient. The force of gravity is given by $F_{\text{gravity}} = -\rho g z$, where z is axial displacement of the capillary front. This leads to the following equation of capillary motion:

$$\rho[z\ddot{z} + \dot{z}^2] = \frac{2}{r}\gamma\cos\theta - \frac{8}{r^2}\mu z\dot{z} - \rho g z \tag{1.28}$$

Equation (1.28), though simple in form, suffers from a fundamental limitation that as $t \to 0$, $M \to 0$ (since $z \to 0$). This gives rise to an ill-posed problem with infinite initial acceleration. The ambiguity may, however, be resolved by introducing the concept of added mass of fluid inducted into the motion initially, also known as the added (or virtual) mass, m_0, which is nothing but the mass of the system lying outside the control volume (capillary) and ready to be inducted into the capillary. This added mass can be approximated by executing a potential flow analysis for an incoming spherical liquid element of radius r that moves with a velocity$\dot{z}$. Executing this analysis, the added mass for a cylindrical tube [33] may be obtained as $m_0 \approx \frac{3\rho\pi r^3}{8}$, which is equivalent to introduction of an additional term in the left hand side of Eq. (1.28), in the form$\frac{3\rho r\ddot{z}}{8}$. This eliminates the prediction of an unrealistic initial burst as $t \to 0^+$.

Equation (1.28), even with the modification of added mass, represents a rather simplified situation. In reality, a number of other complicating factors may be present, resulting in further modifications in the equation of capillary advancement. For example, the capillary walls can adsorb surfactants from the bulk solution, leading to a continuous variation in the surface tension coefficient. The adsorption of the surfactant, Γ_{lv}, can lead to an axial gradient of surface tension, following the Gibbs equation:

$$\gamma = \gamma_0 - \bar{R}T \int_{C=0}^{C(z,t)} \Gamma_{\mathrm{lv}} d \ln C \tag{1.29}$$

where C is the surface concentration of the surfactant. As an illustrative example, an approximate analysis may be carried out by incorporating an exponentially decaying time-dependence of Γ_{lv}, yielding:

$$\gamma = \gamma_0 - \Delta\gamma \phi_B(t) \tag{1.30}$$

where $\Delta\gamma = \gamma_0 - \gamma_\infty$, $\phi_B(t) = 1 - \exp(-\beta t)$, with γ_0 being the initial surface tension, γ_∞ being the equilibrium surface tension, and β being a relaxation constant.

Despite being embedded with certain modifications beyond a fundamental simplistic consideration, the real physical picture of surface tension driven capillary motion appears to be much more complicated than what is depicted by Equation. (1.28). The consideration of fully developed Poiseuillean flow for drag force calculations may be erroneous, since the surface tension-driven flow through a closed conduit is expected to pass through an entrance region of length L_3, followed by a fully developed regime of length L_2 (which is characterized by a so-called Poiseuillean velocity profile), and a meniscus traction regime [34] of length L_1 (see Fig. 1.8). This third regime physically originates due to a deviation of the velocity profile from a classical Poiseuillean velocity profile, on account of strong interplay of adhesion and cohesion forces, which gives rise to a dynamically evolving contact angle between the interface and the channel wall. In reality, the total length of the fluid column (z, say) may vary, as the fluid advances inside the conduit. In case $z<L_1+L_3$, one has $L_2=0$, and hence, the entrance region and the meniscus traction region interact directly. The additional resistances because of the presence of these regimes, in practice, cannot be represented by simple mathematical expressions. However, experimental and/or full-scale computational results can be utilized to a good effect for specification of this enhanced resistance. For the entrance regime, the numerical work of Lew and Fung [35] can be effectively utilized to express the net drag force gradient (dF_{viscous}/dz), in terms of its fully developed counterpart ($dF^*_{\mathrm{viscous}}/dz$) as

$$\frac{dF_{\mathrm{viscous}}}{dz} = \left[1 + f\left(\frac{z}{r_h}\right)\right] \frac{dF^*_{\mathrm{viscous}}}{dz} \tag{1.31}$$

where r_h is the hydraulic radius of the conduit and the function $f(z/r_h)$ takes care of the enhanced flow resistance in the entry region. Details of this fitting function are tabulated in Lew and Fung [35].

In the context of capillary filling dynamics, the dynamics of moving contact lines become an important consideration for appropriate modeling of the flow. For a liquid filling a capillary by displacing a gas, a three phase contact line is established at the tri-junction – solid–liquid, liquid–gas and solid–gas interface. It has been experimentally shown that points on the interfacial lines arrive at the contact line within a finite time span. Therefore, one must pose an effective slip law that relieves a force singularity condition or infinite viscous stress condition, by ensuring that a finite force is necessary to move the contact lines of a fluid, irrespective of the no-slip boundary conditions classically being applied at the channel walls. In order to mathematically ensure this effect, one needs to pose a condition on the contact angle θ, as a function of velocity of the contact line. The apparent dynamic contact angle, θ_a, that the liquid forms with the solid surface is closely described by a universal scaling relationship, known sometimes as Tanner's law, which can be mathematically stated as

$$\theta_a \sim \mathrm{Ca}^{1/3} \tag{1.32}$$

where $\mathrm{Ca} = \frac{\mu u}{\gamma}$ is the capillary number, and u is velocity of the contact line. Equation (1.32) is valid for low capillary numbers, implicating the dominance of surface tension force over viscous forces. The contact line does not abruptly meet the solid, which would have otherwise lead to huge viscous stresses. Instead, the contact line extends in the form of thin film along the channel wall, provided the capillary is prewetted. The meniscus rolls over this thin film, also known as precursor film. However, in order to quantify the proportionality constant in Eq. (1.32), one must resolve the asymptotic behavior of the thin liquid film in vicinity of the microcapillary wall. This thin film region can further be divided into two parts, namely, (a) a lubricating film region followed by (b) a precursor film region. In the precursor film region, one expects the interfacial length scales to approach molecular scales, as intermolecular forces become important. Behind the lubricating film, on the other hand, the length scales are quite large (of the order of r), and a potential challenge remains in devising a quantitative expression for dynamic evolution of θ_a by asymptotic matching from solutions to these regions of widely different length scales. For analysis of the same, one may assume negligible gravitational effects (typically characterized by a low Bond number, $B_0 = (\rho g r^2)/\gamma$), and viscosity of the gas phase to be negligible in comparison to that of the liquid phase so that dynamics of the two phases are essentially decoupled. In that situation, the domain of interest can be divided into two regions, namely (i) the outer region where the lateral and vertical length scales are both $O(1)$ and (ii) the inner lubrication region in which the lateral and vertical length scales are $O(Ca^{1/3})$ and $O(Ca^{2/3})$, respectively, as capillary and viscous forces balance. Further division of the inner region may be necessary when the intermolecular forces become important at very thin films. In the lubricating film region, this effect can be incorporated by invoking an extra term of the same dimension as that of pressure in the overall force balance, which can be described as $A/6\pi z^3$, z being the film thickness. This term is known as the 'disjoining pressure'. The parameter A is called the Hamaker's constant, which is typically negative for wetting fluids. In particular, the surface tension force, which is due to intermolecular forces between the gas and liquid phases, cannot be considered separately from the Van der Waals forces between the gas and the solid phases for very thin films. If one introduces a molecular length scale R_m as: $R_m = \left(\frac{|A|}{6\pi\gamma}\right)^{1/2}$, or its dimensionless counterpart $\lambda = \frac{R_m}{r}$, then the above-mentioned model for lubricating film is valid if its film thickness is much larger than R_m, i.e., $Ca^{2/3} >> \lambda$. Since R_m is typically of the order of a few angstroms, while a typically capillary radius is of the order of 10^{-1} mm, the lower bound on Ca is of the order of 10^{-4}. Beyond this lower bound, the continuum model remains valid for the lubricating film. At relatively larger Ca, the intermolecular forces would turn out to be less effective, so that dependence of Tanner's law on λ becomes progressively weaker. For a significantly large Ca, the meniscus speed may exceed the wetting speed and the meniscus may reverse its curvature as the contact angle passes through 90°. However, the foregoing analysis assumes that such situations do not occur, and accordingly, it remains valid for $\theta<90°$. An asymptotic matching may accordingly be done between the outer region and the precursor film through the lubricating film at the inner region. In front of this region, however, intermolecular forces and a vanishingly small film (spread by fast wetting) stipulate a hyperbolic decay of the film thickness towards zero. Accordingly, a universal relationship for dynamic evolution of the contact angle can be obtained as [36]

$$|\tan\theta_a| = 7.48\mathrm{Ca}^{1/3} - 3.28\lambda^{0.04}\mathrm{Ca}^{0.293} \tag{1.33}$$

It can be noted here that λ is typically of the order of 10^{-8} for wetting fluids flowing through a capillary having hydraulic radius of the order of 1 mm. Physically, for very low values of Ca, the corresponding correction term appearing in Eq. (1.33) may become important, where intermolecular forces come significantly into play and $\tan\theta_a$ diverges slightly away from the asymptotic behavior characterized by Eq. (1.33) towards lower values. Mathematically, Eq. (1.33) has been derived from a matched asymptotic analysis that matches a static outer region to the precursor adhering the wall, through an intermediate lubricating film. The problem, hence, is characterized by three length scales, namely, the capillary radius, film thickness and molecular length scale (λ). This interconnection is captured by Eq. (1.33), and the outcome is very much analogous to the slip model obtained by Hoffman [37], which gives the variation of θ_a as a function of slip length as:

$$g(\theta_a) = g(\theta_a(0)) + \mathrm{Ca}\ln\left(\frac{k}{l_s}\right) \tag{1.34}$$

where the slip length scale is given by: $l_s \sim l_d/\mathrm{Ca}$, l_d being of the order of wall roughness and k being a slip-model dependent constant. Exact form of $g(\theta_a)$ is available in Cox [38], which is given by:

$$g(\theta_a) = \int_0^\theta d\phi\,[f(\phi)]^{-1} \tag{1.35}$$

where the function $f(\phi)$ is given by:

$$f(\phi) = \frac{2\sin\phi\{q^2(\phi^2 - \sin\phi) + 2q[\phi(\pi-\phi) + \sin^2\phi] + (\pi-\phi)^2 - \sin\phi\}}{q(\phi^2 - \sin\phi)[(\pi-\phi) + \sin\phi\cos\phi] + (\phi - sin\phi\cos\phi)[(\pi-\phi)^2 + \sin^2\phi]} \tag{1.36}$$

where q is the ratio of viscosity of two liquids forming the meniscus.

1.4.3.2 Marangoni Effect

A ring of clear liquid is formed on the walls of a glass of wine and it rolls down the glass above the surface of strong wine. This phenomenon is commonly known as 'Tears of Wine', which happens to be the classical example of a phenomenon called Marangoni effect. Alcohol has lower surface tension than that of water. When inhomogeneous mixtures of alcohol and water are prepared, liquid tends to flow away from regions of higher alcohol concentration, on account of a surface tension gradient that is developed because of the alcohol concentration gradients. This phenomenon can also be readily demonstrated by spreading of a thin film of water on a smooth surface and then allowing a drop of alcohol to fall on the centre of the film. The water will tend to move away from the region where the drop of alcohol was put. Wine is a mixture of alcohol and water, with dissolved sugars, acids, colorants and flavours. Glass has a wetting property, as a result of which wine starts climbing as soon as it comes in contact with water. During this process, both alcohol and water evaporate from the rising film, but the alcohol evaporates rapidly, due to its higher vapour pressure and lower boiling point. This change in the composition of the film causes its surface tension to increase – this in turn causes more liquid to be drawn up from the bulk of the wine, which has a lower surface tension because of its higher alcohol content. The wine which moves up the walls of the glass falls back under gravity in the form of droplets. This phenomenon by which fluid motion is induced by the tangential gradients of surface tension is known as the Marangoni effect.

In microfluidics, Marangoni effect may be exploited for flow actuation. One way of achieving that is by creating a temperature gradient in the system, thereby implicitly creating a surface tension gradient by virtue of the fact that surface tension can be a strong function of temperature. For illustration, we may consider a gas bubble of fixed volume [39] in a tube of radius a filled with a liquid of viscosity μ, surface tension γ, density ρ, thermal diffusivity α and temperature coefficient of surface tension $\gamma_T = -d\gamma/dT$. Viscosity and conductivity of the gas phase may be neglected in comparison to those of the liquid phase. We consider the bubble length is taken to be much larger than the radius of the tube. A constant temperature gradient, β is imposed on the tube wall. The bubble starts moving at a steady speed from the cold to the hot region because of the variation in the interfacial tension. A thin layer of fluid (is assumed to be perfectly wetting) engulfing the bubble prevents the bubble to be in direct contact of the wall. The three important non-dimensional numbers charecterizing the bubble motion are the Bond number ($Bo = \rho g a^2/\gamma$), Peclet number ($Pe = Re.Pr$, where Re is the Reynolds number and Pr is the Prandtl number) and capillary number ($Ca = \mu u_b/\gamma$), where u_b is the characteristic axial speed of the bubble. We consider these numbers to be small for mathematical analysis. A small Bond number implies that the gravity effects are negligible; a small Peclet number implies that the advection in the temperature variations are negligible; small Ca and small Re imply that the surface tension forces dominate over the viscous forces and viscous forces in turn dominate over the inertial effects. For small Ca, one can subdivide the flow around the bubble in three distinct regimes, namely, (I) a section parallel to the wall with constant film thickness (say, b), (II) sections of constant curvature hemispherical caps in the front and in the back, and (III) a transitional region connecting the constant film thickness with zero curvature smoothly to the constant curvature region (see Fig. 1.9). Low Re ($Re = \rho_a u_b a/\mu$) consideration allows one to use a lubricating film assumption in the transition regions.

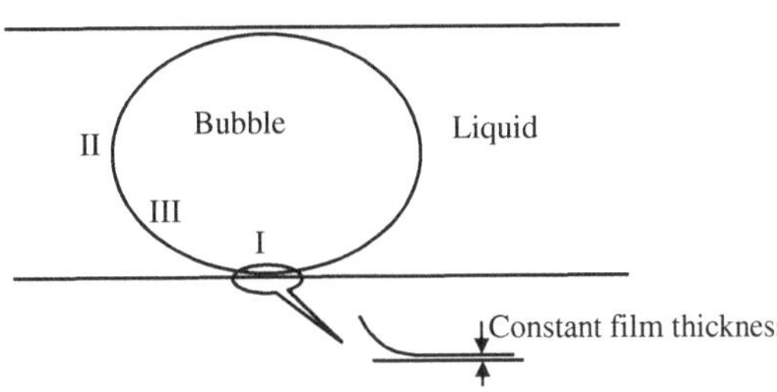

Fig. 1.9 Schematic of a bubble in a liquid with the regions I, II and III shown

For the section of the bubble parallel to the wall (section I), one can obtain the maximum velocity at the edge of the thin film of thickness b, analogous to the situation of a shear-driven Couette flow, as [32]

$$U_T = \frac{\gamma_T \beta b}{\mu} \tag{1.37}$$

where the shear stress, τ, is balanced by the Marangoni stress due to temperature gradient (i.e., $\frac{\partial \gamma}{\partial x} = \frac{\partial \gamma}{\partial T}\frac{\partial T}{\partial x} = -\gamma_T \beta$). The velocity U_T, however, is unknown, since the film thickness (b) is not known a priori. In the transition region, the governing equations of fluid flow, in a reference frame fixed with the bubble, can be written from Eq. (1.17) under low Re approximation as

$$\frac{\partial^2 u}{\partial y^2} = \frac{1}{\mu}\frac{\partial p}{\partial x} \tag{1.38}$$

subject to the following boundary conditions:

$$u = -U_b \text{ at } y = 0 \text{ and } \tau = \mu \frac{\partial u}{\partial y} = -\gamma_T \beta \text{ at } y = h(x) \tag{1.39}$$

where u is the liquid velocity in the moving reference frame, x is the axial direction of motion, y is the transverse direction and $h(x)$ is the film thickness. A solution for Eq. (1.38) can be obtained as:

$$u\left(x,y\right) = \frac{1}{2\mu}\frac{\partial p}{\partial x}y^2 + \frac{1}{\mu}\left(\frac{\partial \gamma}{\partial x} - \frac{\partial p}{\partial x}h\right)y - U_b \tag{1.40}$$

The pressure gradient is related to γ by Young-Laplace equation (given by Eq. (1.22)) and substituting the expressions of radii of curvature in terms of the geometrical parameters of the surface profile (noting that

$$\frac{1}{R_1} = \frac{\frac{d^2h}{dx^2}}{\left[1 + \left(\frac{dh}{dx}\right)^2\right]^{3/2}},$$

where the denominator can be neglected relative to the numerator for small curvature, and $R_2 \to \infty$), one obtains

$$\frac{\partial p}{\partial x} = -\gamma \frac{d^3h}{dx^3} \tag{1.41}$$

The differential equation for the free surface location can be derived by noting that the flux across any cross section needs to be a constant, which mathematically implies $\int_b^{h(x)} u(x,y)dy = 0$, in the moving reference frame. Also noting that $\frac{\partial p}{\partial x} = 0$ at $y = b$, one can utilize Eqs. (1.40) and (1.41) to evaluate the above integral and obtain

$$\frac{\gamma}{3\mu}h^3\frac{d^3h}{dx^3} - \frac{\gamma_T\beta}{2\mu}\left(h^2 - b^2\right) - U_b\left(h - b\right) = 0 \tag{1.42}$$

It is interesting to note that Eq. (1.42) explicitly takes care of the fact that $h \to b$ when the film becomes flat, or in other words, it takes care of an asymptotic matching of the solution in the constant film-thickness region. For an asymptotic matching with the hemispherical cap, one may use a new set of dimensionless coordinates such that $\eta = \frac{h}{b}$, $\zeta = \frac{x}{b(3Ca)^{-1/3}}$. These rescalings cast Eq. (1.42) in a modified Landau–Levich form, as:

$$\eta^3\frac{d^3\eta}{d\zeta^3} - \frac{U^*}{2}(\eta^2 - 1) - (\eta - 1) = 0 \tag{1.43}$$

where $U^* = \frac{U_T}{U_b} = \frac{\gamma_T\beta b}{\mu U_b}$, which is an unknown parameter. However, given a value of U^*, it is possible to integrate Eq. (1.42) numerically, with specified sets of initial conditions. When η is very close to unity, Eq. (1.42) may be linearized to obtain

$$\frac{d^3\eta}{d\zeta^3} = (1 + U^*)(\eta - 1) \tag{1.44}$$

which has the following general solution [39]:

$$\eta = 1 + A\exp\left((1+U^*)\,\zeta\right) + B\exp\left(-\frac{1}{2}\left(1+U^*\right)\zeta\right)\cos\left(\frac{\sqrt{3}}{2}\left(1+U^*\right)\zeta\right) + C\exp\left(-\frac{1}{2}\left(1+U^*\right)\zeta\right)\sin\left(\frac{\sqrt{3}}{2}\left(1+U^*\right)\zeta\right) \tag{1.45}$$

where A, B and C are constants of integration. The asymptotic behaviour of the solution as $\eta \to \infty$ can be used by applying the perturbation theory over Eq. (1.45) to yield [32, 39]

$$\eta = \frac{1}{2}C_0\zeta^2 + C_1\zeta + C_2 - \frac{U^*}{C_0}\left(\zeta\ln|\zeta| - \zeta\right) + \frac{U^{*2}}{2C_0^3}\ln^2|\zeta| + \left(\frac{U^{*3}}{2C_0^3} - \frac{C_1U^*}{C_0^2}\right)\ln|\zeta| + O\left(\frac{\ln^2|\zeta|}{\zeta}\right) \tag{1.46}$$

The coefficients C_0, C_1 and C_2 are determined from the numerical solution for large values of ζ. The constant C_0 (related to the curvature of the end caps) can be determined as the limiting value of the second derivative for large η. When $U^* = 0$, the asymptotic form simplifies to $C_0 = 0.643$ [39]. In general, C_0 is a monotonically increasing function of U^*, as can be obtained numerically. Another matching condition can be obtained by noting that the mean curvature of the caps equals $2/a$ for the leading order approximation for small Ca, which implies

$$h_{xx} \approx \frac{1}{a} \text{ or } \frac{(3Ca)^{2/3}}{b}\eta_{\zeta\zeta} \approx \frac{1}{a} \tag{1.47}$$

From the above considerations, one gets

$$\eta_{\zeta\zeta} \approx \frac{b}{a(3Ca)^{2/3}} \text{ or } C_0(U^*) \approx \frac{b}{a(3Ca)^{2/3}} \tag{1.48}$$

Further, from the considerations of global mass balance, one may note that the rate of volume of fluid pumped by the Marangoni stress, Q_1, must be same as the rate of volume displaced by the moving bubble, Q_2. Since $b<<a$, one can neglect $O(b/a)$ terms and approximate Q_1 as

$$Q_1 = 2\pi \int_{a-b}^{a} U_T(a-r)\, r dr \approx \frac{-\pi \gamma_T \beta a}{\mu} \frac{b^2}{a^2} \tag{1.49}$$

Similarly,

$$Q_2 = \pi U_b (a-b)^2 \approx \pi U_b a^2 \tag{1.50}$$

Since Q_1=–Q_2, one may write

$$Ca = \frac{\mu U_b}{\gamma} = \frac{\gamma_T \beta a}{\gamma} \frac{b^2}{a^2} \tag{1.51}$$

Equation (1.51) can be used with Eq. (48) to express $C_o(U^*)$ as a sole function of *b*/*a*, in the asymptotic matching condition. With $C_o(U^*)$ already being obtained numerically, the ratio of *b*/*a* can therefore be obtained from Eq. (1.44), for a given value of $\Delta\gamma^* = \frac{\Delta\gamma}{\gamma} = \frac{\gamma_T \beta a}{\gamma}$ (which is the ratio of the thermocapillary force to the mean surface tension force). The numerical results can be fitted with the following power law expression [39]:

$$\frac{b}{a} = 13.591 \Delta\gamma^{*^2} \tag{1.52a}$$

and

$$Ca = 184.715 \Delta\gamma^{*^5} \tag{1.52b}$$

One can observe from Eq. (1.52b) that higher temperature gradients may cause the bubble to move faster. The solutions presented above, however, are rather approximate and are valid for vanishingly small Capillary numbers only.

1.4.3.3 Electrocapillary Effects

Surface tension driven flows may be modulated by exploiting several effects, including the effects of variations in the apparent contact angle with applied electrical voltage. In 1875, Lippman first experimentally demonstrated the interfacial tension modulation by electrical effects through a capillary rise phenomenon, which was later termed as Electrocapillarity. Electrocapillary principle is based on the fact that the surface tension occurs to be a strong function of the electric potential acting across an interface.

For illustration, let us consider a discrete liquid droplet in a microchannel, as depicted in Fig. 1.10(a). On application of an electrical voltage, the interfacial tension gets modified, leading to an asymmetric deformation of the meniscus at the two ends and thus a motion of droplet can be actuated. This principle of actuation of fluid motion, commonly known as continuous electrowetting (CEW), happens to be a fundamental principle of operation of different types of optical switches, micromotors and micropumps. However, the high electrical conductance of a fluidic

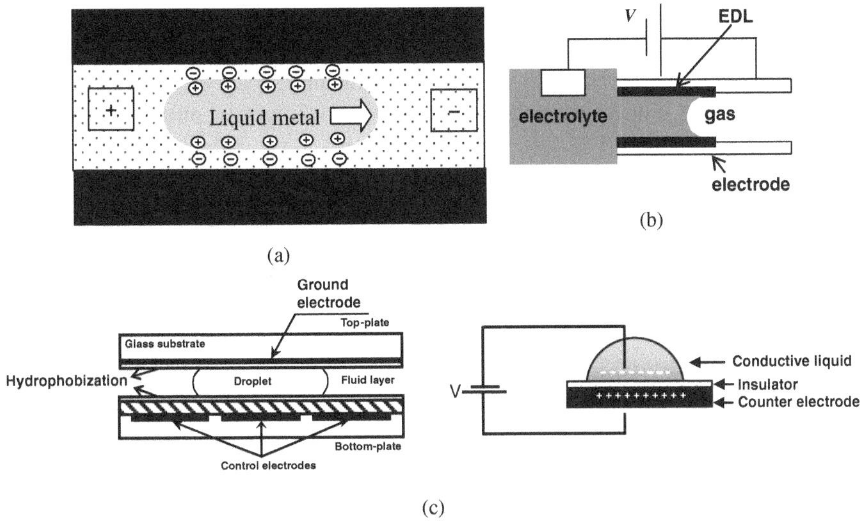

Fig. 1.10 (**a**) An electrically actuated liquid droplet in a microchannel (**b**) An advancing liquid meniscus actuated by electrowetting (**c**) The arrangement for electrowetting on dielectrics of a liquid droplet. The *right hand* diagram is a simpler schematic illustration

system makes the metal-electrolyte systems not-suited for microchannel networks. Moreover, because of large density differences, metal-electrolyte systems are often very sensitive to gravitational forces.

Two other types of electrocapillary principles, namely electrowetting and electrowetting on dielectric (EWOD) are more commonly employed in practice in microfluidic applications, to manipulate liquid droplets without necessitating another liquid medium [32]. Electrowetting (EW) refers to the control of wetting properties of a liquid (commonly, an electrolyte) on a solid (typically, electrode surface), by the modification of electric charges present at the solid–liquid interface. An electrically charged interfacial layer, also known as the electric double layer (EDL), is formed between the electrode and the aqueous solution, typically of a few nm in thickness (for details of EDL morphology, see Section 1.4.5.1). On application of a voltage between the liquid and the electrode, the effective solid–liquid interfacial energy is lowered, on account of the fact that the EDL essentially acts like a parallel plate capacitor that stores electrostatic energy within the same. As a result, the apparent contact angle (θ) gets altered with the applied voltage, resulting in a net change in the driving force that acts on the liquid meniscus, and a consequent advancing or retracting motion of the same (refer to Fig. 1.10b). Large changes in contact angle can be realised by large changes in capacitance ($C \sim \varepsilon/\lambda_D$, λ_D being the Debye screening length, typically of the order of a few nm and ϵ being the permittivity of the ionic medium) of the EDL, even for relatively small applied voltages.

Often, electrochemical reactions take place between the electrode and the aqueous medium, thereby restricting the application of higher voltages. In order to avoid such undesirable reactions, the liquid and the electrodes are separated by a thin dielectric layer (refer to Fig. 1.10c) and such a configuration is commonly known as EWOD or electrowetting on dielectric. Moreover, highly non-wetting surfaces can be designed, which can respond more prominently to the electrowetting actuation mechanisms. For a summary of various types of electrocapillary phenomena, Table 1.2 may be referred to.

Table 1.2 A Comparison of different electrocapillary actuating mechanisms

	CEW	EW	EWOD
Capacitor	EDL	EDL	Dielectric layer
Modulated surface tension, γ	Liquid–liquid interfacial tension	Solid–liquid interfacial tension	Solid–liquid interfacial tension
Contact angle	Always 180°, as the droplet does not directly contact the solid surfaces	Variable	Variable

In comparison to its thermal counterpart (thermocapillary motion), a significant advantage of electrocapillarity is the speed with which electrical potentials can be applied and regulated, with possible characteristic timescales of even less than a few milliseconds. Power consumption in microactuators (using electrocapillary) are less, as compared to the typical thermocapillary microdevices. Thus, electrocapillary flows are much more energy efficient, with a much faster speed of operation (speeds more than 100 mm/s have been successfully achieved by electrowetting action, in contrast to a typical speed of only about 1 mm/s in thermocapillary flows).

Despite having their distinctive implications, all the electrocapillary actuation mechanisms have a common synergy in a sense that these are all based on the variations in surface tension as a function of the electrical potential acting across an interface. Following this principle, the contact angle, θ, gets modified from its value at zero voltage, θ_0, depicted by a relation commonly known as the Young-Lippmann equation, given by:

$$\cos\theta = \cos\theta_0 + \frac{cV^2}{2\gamma_{\mathrm{lv}}} \tag{1.53}$$

where c is the capacitance of the interface per unit area, and V is the interface potential. A physical basis behind this formula may be delineated as follows. When an initially uncharged droplet is connected to the power source (battery), a charge δQ flows from the battery to the system – droplet, the electrode and the dielectric layer (if present, for example in case of EWOD). The work done on the droplet-electrode-dielectric capacitor (of total capacitance C) is given by $\delta W_{\mathrm{droplet}} = V_Q \delta Q$, where V_Q is the potential of the capacitor on being energized with an instantaneous charge of Q. The total work done is given by $\int \delta W_{\mathrm{droplet}} = \frac{1}{2} C V_B^2$ (using $V_Q = Q/C$), where V_B is the potential of the power supply (battery). The incremental work done on the battery, is given by $\delta W_B = V_B \delta Q_B = V_B(-\delta Q)$. Since the battery voltage is a constant, we have, $\int \delta W_B = -V_B \int \delta Q_B = -C V_B^2$. Thus, the net work done on the battery-electrode-dielectric system is given by $\int \delta W_{\text{droplet}} + \int \delta W_B = -\frac{1}{2}CV_B^2$. It may be interpreted, hence, that this extra amount of work contributes to the modification of the surface energy. Electrowetting decreases the effective contact angle, which is driven by the energy gain upon redistributing the charge from the battery to the droplet. This reduction of apparent contact angle is fundamentally related to the fact that a minimization of the free energy requires a maximization of the capacitance. Applying a potential between the droplet and the electrode, therefore, would tend to spread the droplet as much as possible, in an effort to increase the capacitance.

1.4.3.4 Optical Modulation of Surface Tension

The surface tension can also be modified by optical means in small scale systems, unveiling a promising and fascinating area of opto-microfluidics. As a demonstrative example, we consider a substrate that is coated with a wide band gap metal oxide semiconductor material (typically photosensitive materials like TiO_2).The substrate may be exposed to light of wavelength corresponding to the band gap of the material (for instance, UV light for TiO_2 that has a band gap energy of 3.2 eV). As a result, electron-hole pairs are formed [40] which are capable in reducing or oxidizing the species that are adsorbed on the surface. Based on the relative rates of oxidation and reduction (essentially, electron-hole pair formation), an excess or depletion of charge can be generated at the surface. The consequent change in surface energy due to the direct absorption of photonic energy by the optically tunable film may create local gradients in the surface tension forces (altering the state from a hydrophobic to a hydrophilic one, for example), and hence may actuate a fluidic motion on the microchannel substrate. A desired spatial resolution of fluidic control can be achieved by various patterns of designed TiO_2 on the channel surfaces being exposed to the fluid. This would also allow a local control of the flow velocities, without any fringing effects. Further, a temporal control over the fluid motion may be imposed by switching the light on and off, as per designed conditions.

Figure 1.11 demonstrates the use of the above principle for switching the motion of a liquid in a microfluidic circuit. The fluid at the T-junction of a microchannel essentially waits for a logical decision to be triggered to move either towards the right or towards the left, which may be tuned by shedding UV light on the TiO_2-coated inner surface of the horizontal portion of the channel in a selective manner. In the specific example shown in this figure, the UV-exposed right branch alters the surface characteristic from hydrophobic to hydrophilic, thereby allowing the water to flow along that direction. The movement, in terms of speed and directionality may be controlled selecting by controlling the exposure time pulses and by translation of the optical system.

Optical modification of surface tension may also be thermally tuned. In order to actuate the bulk fluid using light, a particle laden fluid with photosensitive nanoparticles (such as gold nanocrescent particles) may be dispersed. Upon photo-actuation [41], these nanoparticles evaporate, creating a concentration difference near the interface because of 'coffee ring' effect (essentially, a ring like pattern

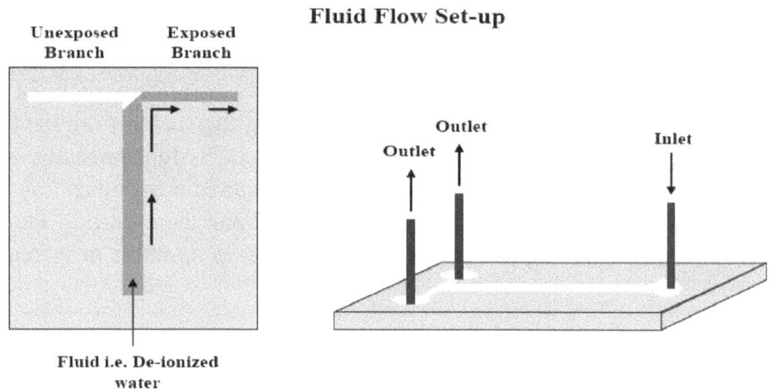

Fig. 1.11 Tunable optofluidic valve

deposited along the perimeter of a spill of coffee after evaporation). The local concentration of the nanoparticles near the liquid–air interface is higher than that of the interior. When a focused light illuminates the nanoparticles near the liquid–air interface, heat is almost instantaneously (within a few nanoseconds) generated and transferred from the nanoparticles to the surrounding liquid. This significantly accelerates the evaporation process at the liquid–air interface forming the meniscus. Simultaneously, the original liquid contact line is pinned and the liquid lost in evaporation is replenished from the interior region. The vapour in the colder air adjacent to the substrate-liquid interface condenses almost immediately after the evaporation. As a consequence, droplets form very close to the liquid–air interface. These droplets then coalesce with each other and grow into the larger ones. These larger droplets eventually merge with the original liquid body and extend its contact line (see Fig. 1.1 in Ref. [41]). The nanoparticles are driven to the modified contact line, by a combined advective–diffusive transport. Coalescence of the droplet can facilitate flow significantly by altering the contact line dynamics to a significant extent, leading to a microcapillary motion.

1.4.4 Rotationally Actuated Microflows

Microflows may also be actuated by rotational (centrifugal and Coriolis) forces, by spinning a disc containing microfluidic networks. The disc in many ways may resemble the Compact Discs (CD) used for external data storage, and hence this type of flow actuation is also known as CD-based microfluidics. CD based microfluidics has gained considerable attention owing to its utility in bio-microfluidic analysis. It can act as a relatively inexpensive platform for chemical analysis and biomedical (pathological) diagnostics, exploiting the advantageous features of portability and rapidity of the analytic platform. Its prime advantages lie in handling wide variety of sample types, the ability to gate the flow of liquids, simple rotational motor requirements, economized fabrication methods, large ranges of flow rates attainable, and the possibility of performing simultaneous and identical fluidic operations. These advantages render the CD an attractive platform for multiple parallel assays [42, 43], despite the apparent constraint that the rotational force is essentially a volumetric force that scales with the cube of a characteristic length scale (which may otherwise not sound to be so attractive over the micro-domain, as compared to the surface forces that scale more favourably with linear dimensions).

A number of research investigations have been reported in the literature [42, 43] on several distinctive aspects of centrifugally-aided microchannel flows on CD-based platforms, leading to the common consensus that rotational effects induce an artificial gravity to pump the fluid in the radial direction without pulsation. For a steady flow under low rotation speeds (hence, ignoring Coriolis effects), the governing equation for rotating frame may be described by using an effective pressure gradient in Eq. (1.17) as:

$$\frac{\partial P}{\partial x} = \frac{\partial p}{\partial x} - \rho \omega^2 x \tag{1.54}$$

where ω is the angular speed of rotation and x is the radial distance from the centre of rotation. It may be noted in this respect, that although the force is a function of the radial coordinate, the flow attains a fully developed condition and the final velocity profile is parabolic in nature. The flow rate (Q) may be obtained by integrating the velocity profile over the area of cross-sectional area (A) and is given by:

$$Q = \frac{\rho \omega^2 \bar{r} \Delta r A D_\mathrm{h}^2}{32 \mu L} \tag{1.55}$$

where D_h is the hydraulic diameter of the channel (define by $4A/p$, where p is the wetted perimeter), $\bar{r}$ is the average distance of the liquid from the centre of the CD, Δr is the radial extent of the fluid, and L is the length of the liquid in liquid channel.

The CD based platforms may also be cleverly designed to act as droplet valves. Beyond a critical rotational frequency (also known as the bursting frequency), centrifugal force acting on a droplet may overcome resistive surface tension and viscous forces, so that the droplet starts moving, whereas below the threshold limit the droplet may be designed to be kept at its position relative to the CD. This arrangement, thus, acts like a smart centrifugal valve. The burst frequency for typical channel valve of hydraulic diameter (D_h) with surface tension coefficient (γ) may be obtained by balancing the surface tension and centrifugal force to obtain:

$$f \geq \left(\frac{\gamma \cos \theta}{\rho \pi^2 \bar{r} \Delta r D_\mathrm{h}} \right)^{1/2} \tag{1.56}$$

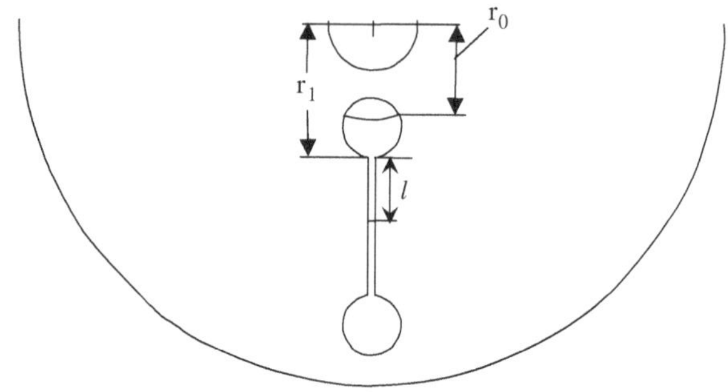

Fig. 1.12 Schematic of one symmetrical half of the rotating CD-based microchannel

where θ is the contact angle at the junction of the valve. The surface tension should resist the centrifugal pumping force, which is directed in radial direction. Hence, the surface should be rendered hydrophobic with contact angle greater than 90°.

For illustration of microcapillary filling principles with centrifugally-aided microfluidics, we focus our attention on a microchannel, which is fabricated in the CD substrate (as shown in fig. 1.12) that may be rotated at various desired angular velocities. Fluid enters the microchannel by the effects of surface tension and outward centrifugal force and advances further with an additional driving influence of the centrifugal effects. The fluid motion is opposed by the viscous resistances, as determined by the different flow regimes instantaneously prevailing within the liquid in the capillary. For mathematical analysis of the above-mentioned physical situation, we consider the assumptions made in Section 1.4.3.1 to be valid for capillary filling analysis. In addition, we neglect Coriolis effects over low ranges of rotational speeds considered in this study. For a microchannel of height h and width w, the axial displacement of the centroid of the capillary meniscus (from the inlet of the microchannel), l, is governed by the following equation in a rotating reference frame (CD) pertinent to a lumped mass system [44]:

$$\frac{d}{dt}\left[(M_{\mathrm{a}} + \rho hlw)\frac{dl}{dt}\right] = P\sigma\cos\theta + F_{\mathrm{C}} - F_{\mathrm{D}} \tag{1.57}$$

where ρ is density of the fluid, P is the wetted perimeter [$P=2(w+h)$], σ is the surface tension coefficient, θ is the contact angle, F_{C} is the centrifugal force and F_{D} is the viscous drag force. The centrifugal force can be estimated as

$$F_{\mathrm{C}} = \int_{r_0}^{l+r_1} \rho\omega^2 rhwdr \tag{1.58}$$

where ω is the rotational speed, r_0 is the positions of the rear end of the meniscus in the inlet reservoir from the centre of rotation and r_1 is the distance of the inlet of the microchannel from the centre of rotation respectively (see Fig. 1.12). The term M_a in Eq. (1.57) is the 'added mass'.

One may first begin with the modeling of the net viscous drag force by noting that it combines the viscous resistances encountered in the entry regime, fully developed regime and the meniscus traction regime. Further, the description of fluid flow in a low Reynolds number regime is governed by the Stokes equation with a centrifugal body force, as

$$\rho \frac{\partial u(y,z,t)}{\partial t} = -\frac{\partial p}{\partial x} + \rho\omega^2(l_0 + x) + \mu\left(\frac{\partial^2 u(y,z,t)}{\partial y^2} + \frac{\partial^2 u(y,z,t)}{\partial z^2}\right) \tag{1.59}$$

where ρ and μ are respectively the density and viscosity of the fluid, u is the velocity component along the axial direction of the channel (x-axis), p is the hydrostatic pressure, t is the time, l_0 is the distance of the channel entrance section from the centre of rotation, ω is the angular velocity of rotation, and y, z are the two perpendicular coordinate axes directed along the height and width of the channel, respectively. Equation. (1.59) may be solved with the boundary conditions of no slip at the walls, and initial condition of zero velocity, to yield:

$$u = \frac{16}{\pi^2\mu}\left(-\frac{\partial p}{\partial x} + \rho\omega^2(l_0 + x)\right)\sum_{m=1}^{\infty}\sum_{n=1}^{\infty}\frac{(-1)^{m+n}}{(2m-1)(2n-1)\lambda_{m,n}}[1 - e^{-\lambda_{m,n}\mu t}]\cos\frac{\beta_m y}{h}\cos\frac{\beta_n z}{b} \tag{1.60}$$

where $\lambda_{m,n} = 4\left(\frac{\beta_m^2}{h^2} + \frac{\beta_n^2}{b^2}\right)$ and $\beta_m = \pi(m - 1/2)$ and $\beta_n = \pi(n - 1/2)$. In case the flow is fully developed, the above-mentioned velocity profile gives a description of the total drag force acting over an axial extent of x, as

$$F_D^* = \mu bhx\frac{dx}{dt}\frac{\sum_{m=1}^{\infty}\sum_{n=1}^{\infty}\frac{1 - e^{-\lambda_{m,n}\mu t}}{(2m-1)^2(2n-1)^2}}{\sum_{m=1}^{\infty}\sum_{n=1}^{\infty}\frac{1 - e^{-\lambda_{m,n}\mu t}}{(2m-1)^2(2n-1)^2\lambda_{m,n}}} \tag{1.61}$$

In reality, the above expression does not give a complete picture of the integrated viscous resistance, since there are additional resistances due to the presence of entrance and the surface traction regimes, as already discussed. From experimental and/or full-scale computational results reported in literature in this regard can be utilized to a good effect for specification of this enhanced resistance (see Eq. (1.31) as well as the pertinent details outlined in [35, 44]). In addition, dynamic nature of the contact angle also needs to be aptly considered, as already outlined in Section 1.4.3.1.

In order to assess the predictive capabilities of the pertinent mathematical models, controlled experiments have also been conducted by researchers [44] on centrifugally-actuated microchannel filling processes. In an effort to delineate detailed insights on the implications of various models vis-à-vis the experimental

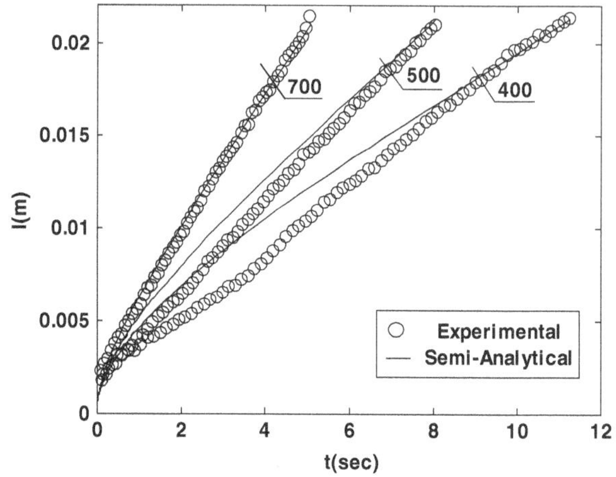

Fig. 1.13 The position of the capillary front (l) versus time (t). Results correspond to three different rpms – 400, 500 and 700, marked in the figure, representing semi-analytical, and experimental results

findings [for details, see Ref. 44], the transients in the capillary meniscus displacement characteristics of Silicone Oil (Fischer Scientific, USA; ρ=963 kg/m^3, σ = 18.83 mN/m, μ= 0.05329 mPa s) are separately depicted in Fig. 1.13, for three different rotational speeds. Interestingly, it is revealed that the deviations between the experimental and theoretical findings appear to be amplified in case of the capillary front advancement predictions, although reasonably good agreements may still be obtained. Partially, the deviations between the experimental and theoretical findings may be attributed to the fact that images from experiments are grabbed from the stroboscopic sequence with an interval of 60/n seconds, where n is the rpm. This implies that when a stroboscopic sequence shows that the capillary front has just advanced up to a distance of $l(t)$, it might in reality have been filled up to that distance at any time instant within a time interval of (t –60/n seconds) to t. Since this error is inversely proportional to the rpm, deviations between the experimental and theoretical findings are expected to be less severe at higher rpms. However, since the error is accumulated in each time interval, the cumulative error tends to get magnified during the later transients, even for higher rpms. Hence, an approximate theoretical model detailed here is the gross manifestation of the underlying consequences in close quantitative proximity with the experimental findings. However, for more detailed descriptions on the pressure distribution and a complete characterization of the meniscus topography, a full-scale numerical model should be resorted to and reader should refer to [44] for detailed analysis.

1.4.5 Electrokinetic Actuation

Electrical forces are used to actuate microflows for various lab-on-a-chip based applications, including pumping, mixing, thermal cycling, dispensing and separating. In addition to the fluid mechanical advantages (to be elucidated later), electrical actuation and control of microflows exploit several advantageous features of micro- and nano-fabrication technology. In fact, with rapid advancements in miniaturized fabrication technology, integration of micro or nano scale electrodes in fluidic device has become a simple procedure.

Electrical actuation and control in microfluidics may be achieved by several means. Several of these techniques strongly depend on the phenomenon of Electrical Double Layer (EDL) formation adhering to an electrically charged substrate. Clubbed altogether, fluid flows influenced by EDL effects are also known as electrokinetic flows.

1.4.5.1 Electrical Double Layer Formation

EDL may form adjacent to the surface either spontaneously (because of electrochemical reactions), or, by virtue of application of surface voltages. In the former case, a solid surface in contact with an electrolyte solution alters the chemical state of the surface. The surface may inherit a charge, because of several reasons, including the following:

(i) ionisation or dissociation of surface groups depending on the nature (pH) of the medium; for example: carboxylic acids dissociates into carboxylic ion (COO^-) and releases proton, common silicate glass, containing SiOH in the presence of H_2O in basic medium, ionizes to produce charged surface groups SiO^- and release of a proton, but ionises to $SiOH_2{}^+$ accepting a proton in acidic medium
(ii) adsorption of ions from a solution onto an uncharged surface; for example: the binding of Ca^{2+} onto a lipid bilayer surface (zwitterionic heads). The adsorption of ions from solution, known as ion exchange, can also occur onto oppositely charged sites; for example: adsorption of cationic Ca^{2+} to anionic COO^- sites vacated by H^+ or Na^+).

Irrespective of the charging mechanism, the surface charge is balanced by a net surplus of oppositely charged ions (counterions) located in a thin fluid layer adhering to the substrate. This charged layer is commonly known as the electrical double layer (EDL) [45, 46]. A schematic depicting the charge and the potential distribution within an EDL is shown in Fig. 1.14. Immediately next to the charged surface, a layer of immobilized counterions is present, which is known as the compact layer or the Stern layer or the Helmholtz layer. This layer is about a few Angstroms in thickness, and the potential distribution within the same is taken almost linear. From this static charge layer to the electrically neutral bulk liquid, the net charge density gradually reduces to zero. The layer of mobile ions beyond the Stern layer is called the Guoy-Chapman layer, or the diffused layer of the EDL. These two layers are separated by a shear plane. The potential at this shear plane is known as the zeta potential (ζ). The thickness of the EDL is known as the Debye length (λ), which is the length from the shear plane over which the EDL potential reduces to $(1/e)$ of ζ. The Debye length, also the characteristic thickness of EDL, depends on the physico-chemical properties of the liquid and not on the surface properties.

When an external electric field is applied across the channel the free charges in the fluid experience a net driving force. Fluid elements in the diffuse layer of the

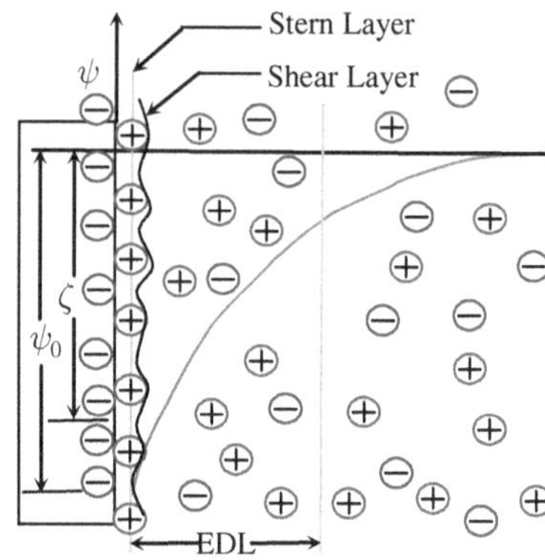

Fig. 1.14 Charge and potential distribution within the EDL

EDL tend to move under the action of electrostatic forces. Due to a cohesive nature of the hydrogen bonding in the polar solvent molecules, the entire buffer solution is pulled, leading to a net electrokinetic body force on the bulk fluid. The resulting flow is known as electroosmotic flow.

Interestingly, EDL effects may be important for fluidic transport even when no external electrical field is applied. For example, in a pure pressure-driven flow occurring through a narrow fluidic channel, the ions in the mobile part of the EDL get transported towards the down-stream end with the liquid motion. This causes an electrical current, known as the streaming current, to flow in the direction of the imposed fluid motion (see Fig. 1.15). However, the resultant accumulation of ions in the downstream section of the channel sets up its own induced electrical field, known as the streaming potential field [47]. This field, in turn, generates a current to flow back against the direction of the imposed pressure-driven flow. This so-called conduction current (fundamentally, electromigration of ions) balances the streaming current at steady state, so that the net electrical current becomes zero, consistent with a pure pressure-driven flow condition. Unlike the cases of electroosmosois and streaming potential that refer to the transport of a fluid relative to a stationary solid

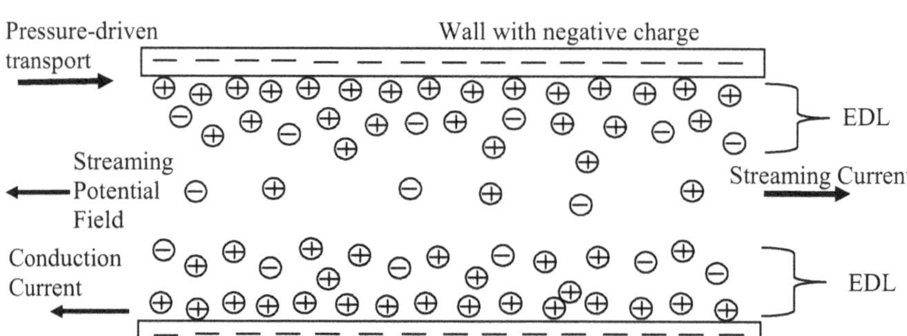

Fig. 1.15 Illustration of the development of streaming potential in a pressure-driven flow

under electrical effects, electrophoresis and Dorn effect refer to the electrical field-driven transport of charged solids relative to a stationary fluid. Electrophoresis is a term to describe the migration of the charged particles in an ionic fluid medium. Under local equilibrium, the force due to electrical field on the particle is balanced by the viscous drag. The velocity of the particle attained under these conditions, per unit electrical field, is also known as its electrophoretic mobility. The converse of electrophoresis is the Dorn effect. It bears the same relation to electrophoresis as streaming potential does to electroosmosis. The Dorn effect produces a potential difference in the solution when charged particles fall through it.

Another variant of electrokinetic effect is known as dielectrophoresis (DEP), in which a force is exerted on a dielectric particle when it is subjected to a non-uniform electrical field (direct or alternating). This force does not require the particle to be charged. Although all particles exhibit dielectrophoretic activity in the presence of electric fields, the strength of the force depends strongly on the medium and particles' electrical properties, on the particles' shape and size, as well as on the frequency of the electric field. As a consequence, fields of a particular frequency can manipulate particles with great selectivity. This has permitted, for example, the separation of cells or the orientation and manipulation of nanoparticles and nanowires [48]. The response of the double layer polarisation in a non-uniform field, leading to DEP forces, depends on the square of the applied field intensity. This is primarily attributable to the fact that polarisation producing the dipole depends upon the field strength to the first power, but the response of the dipole to the field gradient again depends upon the first power of the field strength. The counterpart of the dielectrophoresis is the induction of the polarization of the double layer by relative motion of the particles through the liquid. It presumably arises when suspended particles are accelerated relative to the liquid, when sound waves are impressed upon a colloidal suspension or emulsion.

1.4.5.2 Electro-Mechanics and Thermodynamics Within the EDL

As has been already mentioned, the total charge within the EDL (Fig. 1.14) consists of bound charges and free charges. To calculate the corresponding electric field within the EDL ($\boldsymbol{E}_{\mathrm{EDL}}$), one may invoke Gauss' law (Net electric flux = Total enclosed charge) and write:

$$\varepsilon_0 \oint_S \boldsymbol{E}_{\mathrm{EDL}}\, dS = \int_{\forall} \rho_{\mathrm{e,\,total}}\, d\forall \tag{1.62}$$

where ε_0 is the permittivity of free space, $\rho_{e,\mathrm{total}}$ is the total charge density that consists of charge density due to the free as well as bound charges (i.e., $\rho_{e,\mathrm{total}} = \rho_{e,\mathrm{free}} + \rho_{e,\mathrm{bound}}$; subsequently we call $\rho_{e,\mathrm{free}}$ as ρ_e), S is the surface area, and $\forall$ is the volume. Applying the divergence theorem to convert the surface integral to volume integral and using $\rho_{e,\mathrm{total}} = \rho_{e,\mathrm{free}} + \rho_{e,\mathrm{bound}}$, one may rewrite Eq. (1.62) as:

$$\varepsilon_0 \int_{\forall} (\nabla \cdot \boldsymbol{E}_{EDL})\, d\forall = \int_{\forall} \rho_{e,total} d\forall \tag{1.63}$$

Since the elemental control volume is arbitrary, one may write (considering the notation $\rho_{e,\text{free}} \equiv \rho_e$):

$$\varepsilon_0 (\nabla \cdot \boldsymbol{E}_{\text{EDL}}) = \rho_{\text{e, total}} = \rho_e + \rho_{\text{e, bound}} \tag{1.64}$$

The density of bound charges can be written in terms of the polarization density vector ($\boldsymbol{E}$) as:

$$\rho_{e,\text{bound}} = -\nabla \cdot \boldsymbol{P} \tag{1.65}$$

Polarization density vector is the dipole moment per unit volume that originates due to gradients in local dipole densities within the medium. Hence it can be written as (under the approximation that it varies linearly with the electric field):

$$\boldsymbol{P} = \varepsilon_0 \chi_S \boldsymbol{E}_{\text{EDL}} \tag{1.66}$$

where χ_S is the susceptibility of the medium which can be expressed in terms of the dielectric constant (or relative permittivity) of the medium (ε_r) as

$$\chi_S = \varepsilon_r - 1 \tag{1.67}$$

Using Eqs. (1.64), (1.65), (1.66) and (1.67), it follows:

$$\varepsilon_0 [\nabla \cdot \{(1 + \chi_S) \boldsymbol{E}_{\text{EDL}}\}] = \rho_e \Rightarrow \nabla \cdot (\varepsilon_0 \varepsilon_r \boldsymbol{E}_{\text{EDL}}) = \rho_e \tag{1.68}$$

Expressed in terms of the EDL potential (ψ, i.e., $-\nabla\psi = \boldsymbol{E}_{\text{EDL}}$), Eq. (1.68) reduces to:

$$\nabla \cdot (\varepsilon_0 \varepsilon_r \nabla \psi) = -\rho_e \tag{1.69}$$

Equation (1.69) is also known as the *Poisson Equation* for potential distribution within the EDL.

The electrical description given by Eq. (1.69) is not mathematically closed, since it involves two unknowns, namely ρ_e and ψ. These unknowns, however, may be inter-linked by invoking the pertinent thermodynamic considerations for electrochemical equilibrium of the ionic species. For illustration of this principle, one may consider a single plate or charged surface in contact with an infinite liquid phase. For the system to be in equilibrium, the electrochemical potential ($\bar{\mu}_i$) of the ions need to be constant everywhere, which implies that:

$$\frac{d\bar{\mu}_i}{dy} = 0 \tag{1.70}$$

where y is a direction normal to the plate pointing towards the fluid, and the subscript i indicates type i ions. $\bar{\mu}_i$ can be expressed as the sum of chemical potential (μ_i) and the electric potential ($z_i e\psi$), so that

$$\bar{\mu}_i = \mu_i + ez\psi \tag{1.71}$$

where z_i is the valence of the type i ion and e is the electronic charge. The chemical potential can be expressed (from thermodynamic considerations) by considering an ideal solution of ions, as

$$\mu_i = \mu_i^0 + k_{\mathrm{B}} T \ln(n_i) \tag{1.72}$$

where μ_i^0 is a constant for type i ions, k_{B} is the Boltzmann constant, T is the absolute temperature of the solution and n_i is the number concentration (having units $1/\mathrm{m}^3$) of the type i ions. Using Eqs. (1.70), (1.71) and (1.72), one gets

$$\frac{dn_i}{n_i} = -\frac{z_i e}{k_B T} d\psi \tag{1.73}$$

Equation (1.73) may be integrated to obtain the ionic number density distribution, subjected to the pertinent boundary conditions. For illustration, one may consider the following two cases:

Case 1: At $y = h$, $\psi = \psi_h$, $n_i = n_i^h$, which yields

$$n_i = n_i^h \exp\left(-\frac{ez_i}{k_{\mathrm{B}} T}(\psi - \psi_h)\right) \tag{1.74),}$$

and Case 2: At $y \to \infty$, $\psi = 0$, $n_i = n_i^\infty$, which yields

$$n_i = n_i^\infty \exp\left(-\frac{ez_i\psi}{k_{\mathrm{B}} T}\right) \tag{1.75}$$

Equation (1.74) is the so-called Boltzmann distribution of ionic species near a charged surface. There are a number of important assumptions implicit with the derivation of this distribution. Some of the major assumptions include: (i) ions are uncorrelated point charges, (ii) the system is in equilibrium, with no macroscopic advection or diffusion of ions, (iii) the solid surface is microscopically homogeneous, (iv) the charged surface is in contact with an infinitely large liquid medium, (v) the strength (intensity) of the EDL field significantly overweighs the strength of any imposed electric field, close to the interface, and (vi) the far-stream boundary condition is applicable.

To obtain the EDL potential distribution, one may first express the net charge density of free ions, ρ_e, in the EDL, as

$$\rho_e = e \sum_i z_i n_i \tag{1.76}$$

Substituting the above in the Poisson equation (Eq. 1.69), it follows:

$$\nabla \cdot (\varepsilon_0 \varepsilon_r \nabla \psi) = -e \sum_i z_i n_i \tag{1.77}$$

Considering that the ionic distribution obeys the Boltzmann distribution Eqs. (1.75) and (1.46) maybe re-written as

$$\nabla \cdot (\varepsilon_0 \varepsilon_r \nabla \psi) = -e \sum_i z_i n_i^{\infty} \exp\left(-\frac{e z_i \psi}{k_{\mathrm{B}} T}\right) \tag{1.78}$$

Equation (1.78) is also known as the *Poisson–Boltzmann Equation (PBE)* for describing EDL potential. Under further assumptions that the relative permittivity of the medium, ε_r, is constant, the electrolyte is binary and symmetric ($z_+ = -z_- = z$) and the bulk value (or far stream value) of the number densities of cations and anions are identical ($n_+^{\infty} = n_-^{\infty} = n^{\infty}$), one may simplify Eq. (1.78) as:

$$\nabla^2 \psi = \frac{z e n^{\infty}}{\varepsilon_0 \varepsilon_r} \left[\exp\left(\frac{e z \psi}{k_{\mathrm{B}} T}\right) - \exp\left(-\frac{e z \psi}{k_{\mathrm{B}} T}\right)\right] \Rightarrow \nabla^2 \psi = \frac{2 z e n^{\infty}}{\varepsilon_0 \varepsilon_r} \sinh\left(\frac{e z \psi}{k_{\mathrm{B}} T}\right) \tag{1.79}$$

Equation (1.79) may be solved analytically under different boundary, geometric and physical conditions. This aspect is illustrated in the examples below. In these examples, the potential at the fluid–solid interface is approximated as the ζ (zeta) potential, with the consideration that the shear plane essentially represents the interface between the stationary and the mobile fluid and may be considered as an important reference for describing the hydrodynamics within the EDL. We will consider few examples to illustrate various limits and boundary conditions to obtain the analytical expression for the potential distribution.

Example 1. A single charged plate (located at $y = 0$) with specified ζ and subjected to far-stream boundary conditions (as $y \to \infty$, $\psi = 0$) with sufficiently small ζ such that $\left|\frac{e z \zeta}{k_{\mathrm{B}} T}\right| << 1$: Under these conditions, one may write $\sinh\left(\frac{e z \psi}{k_{\mathrm{B}} T}\right) \approx \frac{e z \psi}{k_{\mathrm{B}} T}$ (This is known as Debye–Hückel Linearization). Accordingly, Eq. (1.79) may be expressed as

$$\frac{d^2 \psi}{d y^2} = \frac{2 z^2 e^2 n^{\infty}}{\varepsilon_0 \varepsilon_r k_{\mathrm{B}} T} \psi \tag{1.80}$$

Here $\sqrt{\frac{2 z^2 e^2 n^{\infty}}{\varepsilon_0 \varepsilon_r k_{\mathrm{B}} T}} = \frac{1}{\lambda}$ represents the inverse of EDL thickness scale (λ being known as the Debye length). Finally, Eq. (1.80) can be integrated twice to yield [49]

$$\psi = \zeta \exp\left(-\frac{y}{\lambda}\right) \tag{1.81}$$

Example 2. A single charged plate with specified ζ and subjected to far-stream boundary conditions (as $y \to \infty$, $\psi = 0$, $d\psi/dy = 0$) without any restrictions on the magnitude of ζ: In this case, multiplying both sides of Eq. (1.80) by $d\psi/dy$ and integrating once (using the condition $d\psi/dy = 0$, $\psi = 0$ as $y \to \infty$ to evaluate the constant of integration), one may obtain

$$\left(\frac{d\psi}{dy}\right)^2 = \frac{4n^{\infty}k_{\mathrm{B}}T}{\varepsilon_0\varepsilon_r}\left[\cosh\left(\frac{ez\psi}{k_{\mathrm{B}}T}\right) - 1\right] \tag{1.82}$$

Using the identity $\cosh(x) - 1 = 2\{\sinh(x/2)\}^2$, Eq. (1.82) can be rewritten as

$$\frac{d\psi}{dy} = \pm 2\sqrt{\frac{2n^{\infty}k_{\mathrm{B}}T}{\varepsilon_0\varepsilon_r}}\sinh\left(\frac{ez\psi}{2k_{\mathrm{B}}T}\right) \tag{1.83}$$

Considering that the positive y-direction points outward from the surface to the liquid in a direction of decreasing magnitude of the EDL potential, one may choose the negative root of Eq. (1.83), and integrate it once more with prescribed ζ to yield [49]

$$\psi = \frac{4k_{\mathrm{B}}T}{ze}\tanh^{-1}\left[\tanh\left(\frac{ez\zeta}{4k_{\mathrm{B}}T}\right)\exp\left(-\frac{y}{\lambda}\right)\right] \tag{1.84}$$

Example 3. Two charged plates (with symmetric ζ prescribed at both) separated by a thin gap between them. The gap is narrow enough for the two EDLs formed at the two plates to interact, but sufficiently large so that the EDLs actually do not overlap: This situation may be addressed by considering a superposition of the potential fields due to the individual confining boundaries, so that one may write

$$\begin{aligned}\psi = \frac{4k_{\mathrm{B}}T}{ze}\Bigg\{&\tanh^{-1}\left[\tanh\left(\frac{ez\zeta}{4k_{\mathrm{B}}T}\right)\exp\left(-\frac{y}{\lambda}\right)\right] \\ &+ \tanh^{-1}\left[\tanh\left(\frac{ez\zeta}{4k_{\mathrm{B}}T}\right)\exp\left(-\frac{2H-y}{\lambda}\right)\right]\Bigg\}\end{aligned} \tag{1.85}$$

Example 4. Same as Example 1, except for the fact that the surface charge density, σ_{c}, is specified instead of a specified ζ: In this case the interfacial boundary condition is replaced by $\varepsilon_0\varepsilon_r\frac{d\psi}{dy} = -\sigma_{\mathrm{c}}$ at $y = 0$. The remaining exercise is straight forward and one finally gets

$$\psi = \left(\frac{\sigma_{\mathrm{c}}\lambda}{\varepsilon_0\varepsilon_r}\right)\exp\left(-\frac{y}{\lambda}\right) \tag{1.86}$$

Example 5. Same as Example 2, except for the fact that the surface charge density, σ_{c}, is specified instead of a specified ζ: Here one may simplify the analysis by prescribing an equivalent ζ from the specified value of σ_{c} by using Grahame equation [50] which may be readily derived by expressing Eq. (1.83) in terms of σ_{c}.

This yields

$$\zeta = \frac{2k_{\mathrm{B}}T}{ze}\sinh^{-1}\left(\frac{\sigma_{\mathrm{c}}ze\lambda}{2\varepsilon_0\varepsilon_r k_{\mathrm{B}}T}\right) \tag{1.87}$$

The remaining procedure is identical to that considered in Example 2.

Example 6. Same as Example 3, except for the fact that the surface charge density, σ_{c}, is specified instead of a specified ζ: With the consideration of Eq. (1.87), this case boils down to the same as that described in Example 3.

Example 7. A single charged plate having infinite liquid medium adjacent to the same. However, the interfacial boundary condition is expressed by a chemical equilibrium condition that may relate the surface charge density with the zeta potential, instead of explicitly specifying either of these: Here one may account for the chemical reactivity of the surface by allowing the charge density σ_{c} to vary due to proton transfer through the following equilibrium [50] condition:

$$\mathrm{SiOH} \rightleftharpoons \mathrm{SiO}^- + \mathrm{H}^+ \tag{1.88}$$

The zeta potential may be related to the charge density σ_{c} and the equilibrium constant (pK) for the dissociation of SiOH, as [50]:

$$\zeta = \frac{k_{\mathrm{B}}T}{ze}\ln\left(\frac{-\sigma_{\mathrm{c}}}{e\Gamma_{\mathrm{S}} + \sigma_{\mathrm{c}}}\right) + \frac{k_{\mathrm{B}}T}{ze}\{\ln(10)\}(pK - \mathrm{pH}) - \frac{\sigma_{\mathrm{c}}}{C_{\mathrm{S}}} \tag{1.89}$$

where Γ_{S} is the surface density of chargeable sites (having units of nm^{-2}) and C_{S} is the per unit area capacitance (having units of F/m^2) of the Stern layer. Equation. (1.24) may be iterated with Eq. (1.89) to obtain an equivalent ζ consistent with the interfacial constraints for this case.

Example 8. Same as Example 3, except for the fact that chemical equilibrium boundary conditions are considered at the two plates: Like the previous case here too one may calculate the equivalent zeta potential at the two surfaces so that the mathematical problem effectively reduces to the case as described in Example 3.

It is instructive to mention here that despite its extensive use, the standard PBE may break down (or needs to be modified) if one or more of the assumptions used for the derivation of Boltzmann equation do not hold. In the foregoing discussions, a deviation in this regard because of finite sized effects of the ionic species (Steric effects) is exemplified. Such deviations turn out to be important for cases in which ionic sizes turn out to be of comparable dimensions as those of the physical system under consideration, as typical to nanochannel transport.

For assessing the consequences of Steric effects in modifying the Poisson-Boltzmann model, it may first be noted that large crowding of counterions at high surface potentials, as predicted by the existing nonlinear PBE, will mean that the ionic concentrations close to the surface exceed the maximal coverage by orders of magnitude. Alternatively, this means an unfeasible situation where the mean ionic spacing at the surface falls well below the ionic radius (~1 Å). This limit of surface potential up to which the nonlinear PBE remains valid, without violating the constraint of minimum inter-ionic spacings, is known as the Steric limit. Typically for $z = 1$ and at room temperature, this value is about 200 mV [51]. Beyond these limits, finite sized effects of the ions need to be aptly considered. For illustration of the underlying consequences, one may consider a symmetric $z{:}z$ electrolyte for which the free energy is assumed to be of the form:

$$F = U - TS \tag{1.90}$$

where

$$U = \int d\mathbf{r}\left(-\frac{\varepsilon_0\varepsilon_r}{2}|\nabla\psi|^2 + zen_+\psi - zen_-\psi\right) \tag{1.91}$$

and

$$-TS = -\frac{k_BT}{a^3}\int d\mathbf{r}\left\{n_+a^3\ln\left(n_+a^3\right) + n_-a^3\ln\left(n_-a^3\right) + \left(1 - n_+a^3 - n_-a^3\right)\ln\left(1 - n_+a^3 - n_-a^3\right)\right\} \tag{1.92}$$

In the right hand side of Eq. (1.91), the first term within parentheses represents the self free energy density of the EDL field and the remaining two terms are the free energy densities due to the electrostatic interaction of the ions with the EDL field. Here, the parameter a signifies the size of the solute ions (taken to be same for anions and cations for simplicity in illustration). The ionic concentration distributions may be derived by assuming that the net electrochemical potential of each type of ions (which one may denote as $\bar{\mu}_\pm = \partial F/\partial n_\pm$) is constant in the system, which implies

$$\bar{\mu}_\pm = \frac{\partial F}{\partial n_\pm} = \int d\mathbf{r}\left[\pm ez\psi + k_BT\ln\left(\frac{n_\pm a^3}{1 - n_+a^3 - n_-a^3}\right)\right] = \text{constant} \tag{1.93}$$

As the differential $d\mathbf{r}$ represents any arbitrary volume in the system, constancy of $\bar{\mu}_\pm$ will mean

$$\pm ez\psi + k_BT\ln\left(\frac{n_\pm a^3}{1 - n_+a^3 - n_-a^3}\right) = \text{constant} \tag{1.94}$$

Thus taking differential on both sides of Eq. (1.94), one gets

$$\frac{dn_\pm}{n_\pm} + \frac{a^3\left(dn_+ + dn_-\right)}{1 - a^3\left(n_+ + n_-\right)} = \mp\frac{ez}{k_BT}d\psi \tag{1.95}$$

Writing $1 - a^3\left(n_+ + n_-\right) = p_1$, Eq. (1.95) may be rewritten as

$$\frac{dn_\pm}{n_\pm} - \frac{dp_1}{p_1} = \mp\frac{ez}{k_BT}d\psi \tag{1.96}$$

Integrating Eq. (1.96), it follows:

$$\ln\left(\frac{n_{\pm}}{p_1}\right) = \mp\frac{ez\psi}{k_{\mathrm{B}}T} + K_1 \tag{1.97}$$

where K_1 is the constant of integration. To evaluate K_1, one may apply the condition that at the bulk (i.e., far away from the charged surface), $n_{\pm} = n^{\infty}$ (hence $p_1 = 1 - 2n^{\infty}a^3 = 1 - \nu$, where $\nu = 2n^{\infty}a^3$ is the Steric or size factor) and $\psi = 0$. This yields: $K_1 = \ln\left(\frac{n^{\infty}}{1-\nu}\right)$. Thus, the ionic concentration distribution equations read:

$$\frac{n_+}{1 - a^3(n_+ + n_-)} = \frac{n^{\infty}}{1-\nu}\exp\left(-\frac{ez\psi}{k_{\mathrm{B}}T}\right) \tag{1.98}$$

and

$$\frac{n_-}{1 - a^3(n_+ + n_-)} = \frac{n^{\infty}}{1-\nu}\exp\left(\frac{ez\psi}{k_{\mathrm{B}}T}\right) \tag{1.99}$$

Evidently, Eqs. (1.98) and (1.99) are not explicit in either n_+ or n_-. To obtain that explicit form, one may divide Eq. (1.98) by Eq. (1.99) to obtain

$$n_+ = n_- \exp\left(-\frac{2ez\psi}{k_{\mathrm{B}}T}\right) \tag{1.100}$$

Using Eq. (1.100) in Eq. (1.99), one finally obtains (with a notation $\frac{ez\psi}{k_{\mathrm{B}}T} = \alpha$; using the identity $\cosh(\alpha) - 1 = 2\{\sinh(\alpha/2)\}^2$):

$$n_- = \frac{n^{\infty}\exp(\alpha)}{1 + \nu(\cosh(\alpha) - 1)} = \frac{n^{\infty}\exp(\alpha)}{1 + 2\nu\sinh^{2}\left(\frac{\alpha}{2}\right)} = \frac{n^{\infty}\exp\left(\frac{ez\psi}{k_{\mathrm{B}}T}\right)}{1 + 2\nu\sinh^{2}\left(\frac{ez\psi}{2k_{B}T}\right)} \tag{1.101}$$

Using Eq. (1.101) in Eq. (1.100), one obtains:

$$n_+ = \frac{n^{\infty}\exp\left(-\frac{ez\psi}{k_{\mathrm{B}}T}\right)}{1 + 2\nu\sinh^{2}\left(\frac{ez\psi}{2k_{\mathrm{B}}T}\right)} \tag{1.102}$$

To obtain the equation governing the potential distribution, one may minimize the free energy expression with respect to the variable ψ, which eventually leads to the Poisson equation. Substituting Eqs. (1.101) and (1.102) in the same, it follows:

$$\nabla^2\psi = -\frac{zen^{\infty}}{\varepsilon_0\varepsilon_r}\frac{\exp\left(-\frac{ez\psi}{k_{\mathrm{B}}T}\right) - \exp\left(\frac{ez\psi}{k_{\mathrm{B}}T}\right)}{1 + 2\nu\sinh^{2}\left(\frac{ez\psi}{2k_{\mathrm{B}}T}\right)} = \frac{zen^{\infty}}{\varepsilon_0\varepsilon_r}\frac{2\sinh\left(\frac{ez\psi}{k_{\mathrm{B}}T}\right)}{1 + 2\nu\sinh^{2}\left(\frac{ez\psi}{2k_{\mathrm{B}}T}\right)} \tag{1.103}$$

Equation (1.103) is essentially a modified form of the PBE considering finite size effects of ions.

The modified PBE derived as above is based on the mean-field approximation that neglects any correlation between the ions. In case ion-ion correlation needs to be accounted for, one may need to postulate an extended free energy description consistent with a statistically based Langevin formalism [52, 53]. Other approaches for incorporating the effects of ion-ion correlation in a modified PB framework include the use of mean spherical approximations [54–56] and statistical density functional theory [57–60]. In principle, any effect beyond the scope of the standard PB description may be incorporated in the mathematical formalism by considering the corresponding additional contribution to the free energy or equivalently the electrochemical potential, and finding out the necessary variations to derive the pertinent equations governing the distribution of the ionic species as well as the EDL potential.

1.4.5.3 Linear DC Electroosmosis

The simplest form of electroosmosis is linear DC electroosmosis that occurs when the applied electric field is relatively small ($\sim 10^4$ V/m) and steady. For illustration, one may consider that such a field has been applied to an aqueous ionic solution enclosed within a narrow parallel plate confinement with charged substrates having constant and symmetric ζ corresponding to each of the confining boundaries. Under such circumstances, the fluid motion is governed by the continuity and momentum equations given as:

$$\nabla \cdot \boldsymbol{V} = 0 \tag{1.104}$$

and

$$\frac{\partial}{\partial t}(\rho \boldsymbol{V}) + (\rho \boldsymbol{V} \cdot \nabla)\,\boldsymbol{V} = \nabla \cdot (\mu \nabla \boldsymbol{V}) + \rho_e \boldsymbol{E} \tag{1.105}$$

where $\boldsymbol{V}$ is the velocity field vector with components (u, v) and $\boldsymbol{E}$ is the electric field. The last term in the right hand side of Eq. (1.105) describes the electrokinetic body force per unit volume. Under the condition of steady, hydrodynamically fully developed flow with constant properties, no pressure gradients, and external electric field applied only in the axial direction, Eqs. (1.104) and (1.105) may be combined into a single equation governing the x-component of the velocity (u), as:

$$\frac{d^2 u}{dy^2} = -\frac{\rho_e}{\mu} E_x \tag{1.106}$$

The net electrical charge density being expressed as $\rho_e = -\varepsilon_0 \varepsilon_r \frac{d^2 \psi}{dy^2}$ (see Eq. 1.69), Eq. (1.106) gets reduced to:

$$\frac{d^2u}{dy^2} = \frac{\varepsilon_0 \varepsilon_r E_x}{\mu} \frac{d^2\psi}{dy^2} \tag{1.107}$$

Considering symmetry conditions ($du/dy = 0$ and $d\psi/dy = 0$) at the channel centreline ($y = H$) and no-slip boundary condition at the walls, Eq. (1.107) may be integrated twice to yield:

$$u = -\frac{\varepsilon_0 \varepsilon_r E_x \zeta}{\mu}\left(1 - \frac{\psi}{\zeta}\right) = u_{\mathrm{HS}}\left(1 - \frac{\psi}{\zeta}\right) \tag{1.108}$$

where $u_{\mathrm{HS}} = -\frac{\varepsilon_0 \varepsilon_r E_x \zeta}{\mu}$ is known as the Helmholtz–Smoluchowski velocity. In case of combined pressure driven electroosmotic flows, the velocity field may be obtained as a linear superposition of the electroosmotic and pressure driven components in a hydrodynamically fully developed limit, to yield

$$u = -\frac{1}{2\mu}\frac{dp}{dx}\left(2Hy - y^2\right) - \frac{\varepsilon_0 \varepsilon_r E_x \zeta}{\mu}\left(1 - \frac{\psi}{\zeta}\right) \tag{1.109}$$

In Eq. (1.109), dp/dx represents the imposed pressure gradient and $y = 0, 2H$ represent the locations of the confining boundaries.

The special case with zero pressure gradient is interesting, a typical velocity profile corresponding to which is depicted in Fig 1.16, pertaining to the case of relatively thin EDL with respect to the channel height (κH=100). From the figure, it is evident that the velocity profile is virtually uniform across the channel section. This has a couple of interesting consequences. First, there is negligible dispersion in

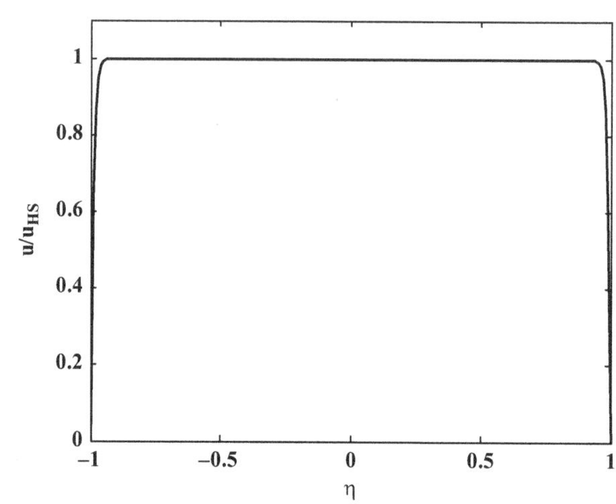

Fig. 1.16 Dimensionless velocity profile corresponding to electroosmotic flow in a parallel plate channel, for κH=100. Here $u_{\mathrm{HS}} = -\frac{\varepsilon_0 \varepsilon_r E_x \zeta}{\mu}$, $\eta = \frac{y}{H} - 1$

a sample that is transmitted with the flow, because of uniformity in the velocity profile. Secondly, one may predict the flow rates with negligible errors by considering the uniform velocity profile prevailing throughout the channel section (plug flow), and introducing an artificial slip velocity of magnitude u_{HS} at the channel walls, instead of considering the underlying body force effect in the momentum conservation equation. Although this simplifies the analysis significantly, by effectively decoupling the hydrodynamics from the electrostatics, it needs to be cautioned that this approach would not work when the thickness of the EDL as compared to the channel height may not be trivially neglected.

1.4.5.4 AC Electroosmosis

AC electroosmosis is a particular frequency dependant electrosomotic phenomenon in which the velocity field responds to the amplitude and frequency of the applied electric field. DC electroosmosis is a special case of frequency dependent electroosmosis with frequency and phase are zero. One special case of AC electroosmosis confers to the use of time periodic unidirectional applied electric field, which we exemplify subsequently. For that purpose, we consider sinusoidally driven time periodic pure electroosmotic flows (no applied pressure gradients) to obtain the governing equation for fluid flow as [61]:

$$\rho \frac{\partial u}{\partial t} = \mu \frac{\partial^2 u}{\partial y^2} + \rho_e E_x \sin(\Omega t) \tag{1.110}$$

where E_x is the magnitude and Ω is frequency of the time periodic applied external AC field, u is the axial velocity and μ is the dynamic viscosity of the fluid. Using Eq. (1.110), the above equation may be rewritten as:

$$\rho \frac{\partial u}{\partial t} = \mu \frac{\partial^2 u}{\partial y^2} + \frac{\mu u_{\mathrm{HS}}}{\alpha \lambda^2} \sinh(\alpha \psi^*) \sin(\Omega t) \tag{1.111}$$

where $\alpha = \frac{ez\zeta}{k_B T}$ is the ionic energy parameter and $\psi^* = \frac{\psi}{\zeta}$. Equation (1.111) is non-dimensionalised with length scale (λ) and time scale ($1/\Omega$) to yield:

$$\frac{\partial U}{\partial \theta} = \frac{1}{\kappa^2}\left(\frac{\partial^2 U}{\partial \chi^2} + \sinh(\alpha\psi)\frac{\sin\theta}{\alpha}\right) \tag{1.112}$$

where the nondimensional time is defined as $\theta = \omega t$, nondimensional distance as $\chi = y/\lambda$, nondimensional velocity as $U = u/u_{\mathrm{HS}}$ and κ is the ratio of the Debye length scale and the diffusion length scale and is given by:$\sqrt{\Omega\lambda^2/\nu}$ with ν is the kinematic viscosity. The differential equation may be solved with no slip boundary condition at the walls. For that purpose, one may use the separation of variables approach by substituting $U(\chi,\theta) = \mathrm{Im}\left(e^{i\theta}F(\chi)\right)$and seeking the imaginary part of the final solution with the consideration that $e^{i\theta}$ is nonzero at all times, to yield:

$$\frac{d^2F(\chi)}{d\chi^2} = i\kappa^2 F(\chi) - \frac{\sinh(\alpha\psi^*)}{\alpha} \tag{1.113}$$

On solving Eq. (1.113), we obtain the nondimensional velocity as a function of the nondimensional time and space coordinates, as [61]:

$$U(\chi,\theta) = \operatorname{Im}\left\{e^{i\theta}\left(Ae^{\sqrt{i\kappa}\chi} + Be^{-\sqrt{i\kappa}\chi}\right)\right\}$$
$$+\operatorname{Im}\left\{\frac{e^{i\theta}}{2\kappa\alpha\sqrt{i}}\left(e^{-\sqrt{i\kappa}\chi}\int_0^{\chi} e^{\sqrt{i\kappa}\chi}\sinh(\alpha\psi^*)d\chi - e^{\sqrt{i\kappa}\chi}\int_0^{\chi} e^{-\sqrt{i\kappa}\chi}\sinh(\alpha\psi^*)d\chi\right)\right\} \tag{1.114}$$

The constants A and B may be evaluated from the no-slip condition $F(\chi = 0) = 0$ and symmetry at the channel centreline, to yield:

$$A = -B = \frac{1}{4\sqrt{i\kappa}\alpha\cosh(\sqrt{i\kappa}\alpha)}\left(e^{-\sqrt{i\kappa}\chi}\int_0^{\infty} e^{\sqrt{i\kappa}\chi}\sinh(\alpha\psi^*)d\chi\right)$$
$$+\frac{1}{4\sqrt{i\kappa}\alpha\cosh(\sqrt{i\kappa}\alpha)}\left(e^{\sqrt{i\kappa}\chi}\int_0^{\infty} e^{-\sqrt{i\kappa}\chi}\sinh(\alpha\psi^*)d\chi\right) \tag{1.115}$$

On analysing Eq. (1.114), it may be observed that for κ=0.01, there is vorticity in the *entire* bulk flow region. The vorticity is mathematically defined as the curl of the velocity and representing fluid rotation. For $\kappa = 0.01$ and $\kappa = 0.1$, it may be observed that the time-periodic electroosmotic flows are not irrotational, when the diffusion length scale is comparable to or less than the channel half-height. The vorticity on the walls alternates its magnitude by time because of the external field. An important difference between steady and time-periodic electroosmotic flows is that the vorticity diffuses deeper into the channel, while its value on the wall is alternating. The velocity profiles in thin EDL limits resemble the classical solution of flat plate oscillating in a semi-infinite flow domain, also known as the Stokes' second problem. Unlike the time periodic electroosmotic flow, the fluid is driven by an oscillating plate with a velocity $u_w = u_{HS}\sin(\Omega t)$, where Ω is the frequency of the plate oscillations and u_{HS} is the amplitude of the plate velocity.

1.4.5.5 Electrophoresis

Electrophoresis refers to the motion of dispersed particles in a fluid medium relative to a static fluid under the influence of spatially uniform electric field. It was first observed in clay particles when they were separated in water. Particles dispersed in water migrate under influence of an applied electric field. The electric field exerts Coulombic force on the particles bearing surface charge. However, recent molecular dynamics simulations have suggested that surface charge is not always necessary for electrophoresis and that even neutral particles can show electrophoresis due to the specific molecular structure of water at the interface [62].

A charged EDL surrounds a charged particle in a fluid medium. The double layer in nature is same as described in electroosmosis, consisting of two layers of charged ions – Stern layer and diffuse layer. It bears a charge equal in magnitude and opposite in sign to that of the charged surface. The Coulomb force exerted on a charged particle is screened by an opposing force of electrostatic in nature, because of the presence of EDL. The electric field induces force on the diffuse layer as well as on the charged surface. Only a part of the force is applied on the particle and mostly it is applied to the ions in the diffuse layer. These ions, located at some distance from the particle surface, experience force due to applied external electric field and transfer part of this electrostatic force (as reaction force) to the particle. While the body starts to move, the particle experiences hydrodynamic resistance in the form of viscous drag. If the Debye layer is large compared to the radius of the particle, then the particle is assumed to be a point charge. In the Stokes flow regime, the balance of the Coulombic force and the drag force gives:

$$q_s E = 6\pi a \mu U \tag{1.116}$$

where q_s is the surface charge on the particle, E is the applied electric field, a is the radius of the particle, μ is the viscosity of the surrounding medium and U is the velocity of the particle in the fluid medium. It may be noted that the charge on the surface of the particle has to be equal to that of the total charge of the Debye layer. Further, invoking the Poisson equation with the limits $\phi \to 0$ as $r \to \infty$ and $\phi = \zeta$ at $r = a$, along with Eq. (1.116) under the approximation of $a \ll \lambda_D$ [63], it follows:

$$U = \frac{2}{3}\frac{\varepsilon E \zeta}{\mu} \tag{1.117}$$

Alternatively, if the Debye layer is thin comparable to the particle characteristic dimension, the curvature effects of the EDL may be neglected, rendering the particle surface as locally planar in an effective sense. The tangential component of velocity and the tangential component of the electric field may be related by Eq. (1.117) in the form: $U_t = \frac{\varepsilon E_t \zeta}{\mu}$. The electric field has to follow: $\boldsymbol{E} = -\nabla\phi$ and ϕ has to satisfy the Laplace equation: $\nabla^2\phi = 0$, subjected to the boundary conditions – the normal component of the current density vanishes at the surface and far from the particle, and the potential approaches the value corresponding to the uniform applied field E_x. Within thin EDLs, the velocity slips at the wall, going from u_t to zero discontinuously. Irrotational flow corresponds to the velocity field satisfying the velocity potential to satisfy the Laplace equation, and also boundary conditions – no penetration and far from the surface, the velocity approaches the value corresponding to the uniform velocity U. Finally, from the similarity and transformation of reference frame in which the particle is moving relative to the reference frame, the velocity may be obtained as:

$$U = \frac{\varepsilon E_x \zeta}{\mu} \tag{1.118}$$

Equation (1.118) is just the Helmholtz–Smoluchowski equation, implying the electrophoresis is just the complimentary of electroosmosis. Equation (1.118) differs from Eq. (1.117) by a factor 2/3. If the applied electric field is low, it may be observed that the speed varies linearly with electric field, and the proportionality constant is known as electrophoretic mobility, μ_e. Equation (1.118) works for any arbitrary shape and concentration. From the consideration of finite EDL thicknesses, three effects may alter the electrophoretic velocity, which are not mutually exclusive, namely, electrophoretic retardation, surface conductance, and relaxation. Electrophoretic retardation results from the fact that the ions in the EDL will have net movement opposite to that of the particle. Local electroosmotic flow of the solvent is created through drag forces that oppose the motion of the particle. Surface conductance refers to a region of the flow near the surface in which charge neutrality is absent and the counterions are excess in comparison to the bulk electrolyte because of the finite double layer. The excess counterion concentration gives rise to regions of higher conductivity in which the applied electric field is reduced. The third effect of relaxation stems from the fact that with motion of the ions because of the electrosmotic transport, the EDL distorts from sphericity leading to asymmetric behaviour, and the centre of the EDL lags behind the centre of the particle. Finite time is required for the system to relax and the EDL to adjust to the original symmetry by electromigration and diffusion in the moving system.

1.4.5.6 Dielectrophoresis

An interesting variant of electrophoresis is known as dielectrophoresis. When a polarizable particle is subjected to non-uniform electric field, a dipole is induced in the particle. If the electric field is diverging, the particle experiences a force that can move it towards the high or low electric field region (see Fig. 1.17), depending on the particle polarizability as compared with the suspending medium. This type of actuation of motion is termed as dielectrophoresis (DEP). If the polarizability of the particle is higher than the medium, the force acts towards the high field strength region (positive DEP). In the other case, the force acts towards the lower field region (negative DEP). DEP is commonly used for generating translational motion of different objects. It is possible to discriminate different cell types according to their properties, by utilizing DEP.

It is helpful in this context to summarize the important differences between electrophoresis and dielectrophoresis, as follows [64]:

(i) Direction of the dielectrophoretic motion is independent of the sign of the electric field with either a.c. or d.c. sources. Electrophoresis produces a motion of the suspended particles in a path depending on the sign of the electric field and charge on the particles.

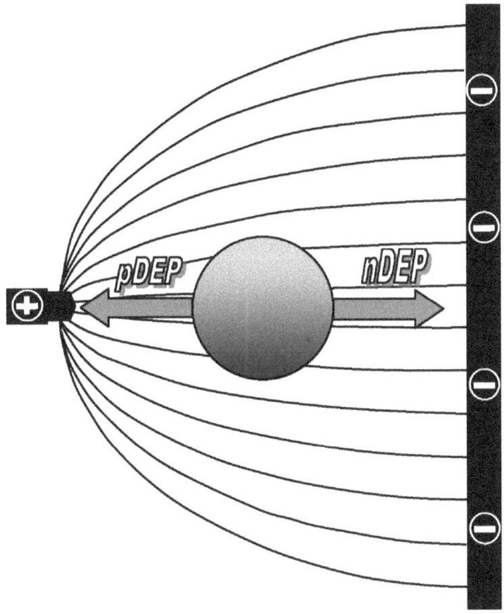

Fig. 1.17 Polarisation of a neutral particle in non-uniform field

(ii) Electrophoretic force is independent of the size of the particles, whereas DEP is dependant on the volume of the polarizable particles. Consequently, DEP is observed more easily on coarse particles and not commonly at the molecular level.

(iii) DEP requires divergent fields for strong effects. Electrophoresis operates in both uniform and divergent fields.

(iv) DEP necessitates high electric field strengths and substantial differences in the relative permittivities of the particles and the surrounding medium. Typically, for low dielectric constants (with typical values ranging from 2 to 7), the strength of electric field required is 10^4 V/m, whereas for high dielectric constants (such as 80, for water as example), a lower electric field of 500 V/m may suffice. Electrophoresis requires relatively low fields to drive the previously charged particles.

In order to quantify DEP forces, the electrical force acting on a small neutral body under static electrical field in equilibrium is given by [64]:

$$F = (\mathbf{p}.\nabla)\, E \tag{1.119}$$

where E is the applied external electric field, $\boldsymbol{p}$ is the constant dipole moment vector. For a dielectric which is isotropically, linearly and homogeneously polarizable, the dipole moment vector is given by:

$$\mathbf{p} = \alpha V E \tag{1.120}$$

where α is the polarizability tensor or the dipole moment per unit volume in unit field and V is the volume of the body. Substituting Eq. (1.120) in Eq. (1.119), we obtain the expression for force acting on the particle as:

$$F = 1/2\alpha V\nabla |E|^2 \tag{1.121}$$

The induced dipole moment of a polarizable sphere of permittivity ε_2 and volume V in an infinite medium of permittivity ε_1 present is a uniform electric field (uniform before the introduction of the sphere) is given as: $p = 3V\varepsilon_1 \frac{\varepsilon_1 - \varepsilon_2}{\varepsilon_2 + 2\varepsilon_1} E$, which upon comparing with Eq. (1.120), the above gives the polarizabilty α as:

$$\alpha = 3\varepsilon_1 \frac{\varepsilon_1 - \varepsilon_2}{\varepsilon_2 + 2\varepsilon_1} \tag{1.122}$$

Substituting, Eq. (1.122) in Eq. (1.121), one may obtain the dielectrophoretic force acting on a small spherical neutral particle, as

$$F = 2\pi r^3 \varepsilon_1 \frac{\varepsilon_1 - \varepsilon_2}{\varepsilon_2 + 2\varepsilon_1} \nabla |E|^2 \tag{1.123}$$

where r is radius of the spherical particle. The term 'small sphere' is mentioned to implicate the approximation that the field does not vary strongly across the particle so as to appreciably alter the degree of polarisation, although the non-uniformity is enough to appreciably produce different forces on the positive and negative portions of the induced charges. Equation (1.123) may be cast in a more general form [64], as:

$$F_{\mathrm{DEP}} = \frac{1}{2} Re\left[m(\omega) \cdot \nabla E^*\right] \tag{1.124}$$

where E^* is the complex conjugate of the electric field and $m(\omega)$is the dipole moment that can be written as:

$$m(\omega) = 4\pi \varepsilon_{\mathrm{m}} r^3\, K(\omega) E \qquad (1.125),$$

for a spherical particle of radius r and a solution of permittivity ε_m, with ω being the angular field frequency (rad/s). The parameter $K(\omega)$ is the Clausius–Mossotti factor, which is given by [64]:

$$K(\omega) = \frac{\tilde{\varepsilon}_{\mathrm{p}} - \tilde{\varepsilon}_{\mathrm{m}}}{\tilde{\varepsilon}_{\mathrm{p}} + \tilde{\varepsilon}_{2\mathrm{m}}} \tag{1.126}$$

where $\tilde{\varepsilon}_{\mathrm{p}}$ and $\tilde{\varepsilon}_{\mathrm{m}}$ are the permittivities of the particle and the medium respectively. For an isotropic homogeneous dielectric, the complex permittivity is represented as:

$$\tilde{\varepsilon} = \varepsilon - \frac{k_{\mathrm{el}}}{\omega} i \tag{1.127}$$

where ε is the permittivity and k_{el} is the conductivity of the dielectric and i is the square root of -1.

1.4.5.7 Streaming Potential

As depicted in Fig. 1.15, ions in the mobile part of the EDL are transported downstream along with the liquid motion in a pure pressure driven flow. An electrical current known as streaming current is developed because of this motion of the ions, which flows in the direction of the imposed fluid motion. The accumulation of ions downstream induces an electric field known as the streaming potential [65]. This field, in turn, generates a current to flow back against the direction of the pressure-driven flow.

An important consequence of the EDL-induced counteracting ionic migration mechanism in pure-pressure driven flows through narrow fluidic confinements is manifested through an enhanced effective resistance, so as to oppose the very cause to which the forward motion of the ionic charges is due. If the reduced flow rate is compared with the flow rate predicted by conventional fluid dynamics without considering the presence of the EDL, it appears that the liquid would have an enhanced effective viscosity. This is usually referred to as the electroviscous effect. Theoretically, this effect is commonly analyzed with the assumption of an open-circuit channel, where the steady-state streaming potential is determined by equating the streaming current with the conduction current. The ratio of the apparent viscosity to the actual viscosity of the fluid, may be as high as 1.3 [66], depending on the channel size, the zeta potential, and the ionic concentration, etc. Several other studies have also been reported [47, 67–72] depicting a complex interplay between the developed EDL and the resistive hydrodynamics within the same under various simplified assumptions.

Towards quantifying the streaming potential, it may first be noted that the total ionic current may be estimated as:

$$I_{\text{ionic}} = e\int\limits_0^{2H} (z_+u_+n_+ + z_-u_-n_-)\,dy \tag{1.128}$$

Here u_+ (u_-) refer to the axial velocities of the cations (anions), expressed as: $u_\pm = u + \dfrac{z_\pm eE}{f_\pm}$. Under the assumption of symmetric electrolyte ($z_+ = -z_- = z$), identical values of cationic/anionic friction coefficient of charge $f_\pm$ ($f_+ = f_- = f$) and under non-overlapped EDL conditions and certain other simplifying assumptions (as described earlier), the ion concentration n_+ and n_- may be expressed through the Boltzmann distribution, as: $n_\pm = n_0 exp\left(-\frac{ez_\pm\psi}{k_BT}\right)$. The above expression for I_{ionic} simplifies to:

$$I_{\text{ionic}} = ez \int_0^{2H} (n_+ - n_-)\, u dy + \frac{z^2 e^2 E}{f} \int_0^{2H} (n_+ + n_-)\, dy \tag{1.129}$$

For pure pressure-driven transport, I_{ionic} becomes identically zero at steady state; the corresponding value of E being known as the streaming potential field (E_S).

Consistent calculation of the net ionic current necessitates an appropriate substitution of the velocity field, as evident from the above expression. In addition to the pressure-driven component, the induced streaming potential is also likely to introduce an additional convective transport of the fluid medium, opposing the driving convective influences of the imposed pressure gradient. Accordingly, the velocity field may be expressed as

$$u = u_p + u_{E_S} = -\frac{1}{2\mu}\frac{dp}{dx}\left(2Hy - y^2\right) - \frac{\varepsilon\zeta E_S}{\mu}\left(1 - \frac{\psi}{\zeta}\right) \tag{1.130}$$

where ζ is the potential at the plane of zero shear (also known as the zeta potential), and E_S is the induced streaming potential field. Setting $I_{ionic} = 0$ at steady state for pure pressure-driven transport, one may obtain an expression for the streaming current, as

$$E_S = \frac{n_0 ez \int_0^{2H} u_P \sinh\left(\frac{ez\psi}{k_B T}\right) dy}{\sigma \int_0^{2H} \cosh\left(\frac{ez\psi}{k_B T}\right) dy + \frac{n_0 ez\varepsilon\zeta}{\mu} \int_0^{2H} \left(1 - \frac{\psi}{\zeta}\right) \sinh\left(\frac{ez\psi}{k_B T}\right) dy}. \tag{1.131}$$

In the above expressions, σ $\left(\sigma = \frac{n_0 e^2 z^2}{f}\right)$ is the electrical conductivity of the fluid (this ionic friction coefficient of charge f can be related to the ionic mobility Λ, Faradays constant F and Avogadro number N_A as $f = \frac{e^2 N_A}{F^2 \Lambda}$), and n_0 is the bulk ionic number density that can be expressed in terms of the Debye layer thickness, λ, as $n_0 = \frac{\varepsilon k_B T}{2\lambda^2 e^2 z^2}$. Researchers have recently demonstrated that significant errors may be incurred in the streaming potential predictions in case the electrically-driven advective transport of the ionic species is neglected in the pertinent calculations, as often done traditionally. This error appears to be magnified for an intermediate range of surface charge densities between 'low' and 'high' surface charge limits [65], more emphatically under overlapped EDL conditions.

The streaming potential can be utilised to convert hydrostatic pressure difference into useful electrical energy by directly exploiting micro/nanoscale flow physics. The results are more favourably scalable for nanochannels than microchannels, because of the fact that the extent of zone in which a net volumetric charge density exists in a nanochannel is much more than that in a microchannel, when normalized with respect to their respective hydraulic radii. This can be further used to characterise interfacial charge of organic thin films, to measure wall charge inversion in presence of multivalent ions in nanochannel, to analyze ion transport through nanoporous membranes, to design efficient nanofluidic batteries etc, as some of the technologically relevant applications. The energy conversion efficiency has been found to be very poor (about 5%), with the loses primarily attributable to huge flow-pumping power necessary to overcome strong frictional resistances against pressure-driven microflows and nanoflows. High energy-conversion efficiency and high output power are the fundamental requirements for such devices to be practical and capable of solving some burning global problems in the field of energy science and engineering. Exploring the interfacial phenomenon by modification of the surface roughness may play a key role in an attempt to reduce the losses due to fluid friction. However, technologically it is difficult to produce controlled surface roughness nanochannels. Exploration of this nascent and novel technology for generating electrical energy directly from a hydraulic form (instead of going through other mechanical/ coupling losses) in miniaturized devices without necessitating any fuels, thus, appears to be very much pertinent and appealing in the global energy crisis scenario, and indeed a very attractive proposition.

1.4.6 Electrothermal Effects

Electrothermal flows originate from temperature gradients in the medium, generated as a consequence of Joule heating effects. This temperature gradient induces local gradients in conductivity, permittivity, density and viscosity, which in turn, give rise to net forces acting on the liquid. For instance, conductivity gradients produce free volumetric charges and Coulombic forces, whereas permittivity gradients produce dielectric forces. To begin with, one may express the electrical force per unit volume in the following general form [73, 32]:

$$\vec{f}_e = \rho_e \vec{E} - \frac{1}{2}\vec{E}^2 \nabla \varepsilon + \frac{1}{2}\nabla\left(\rho \frac{\partial \varepsilon}{\partial \rho}\vec{E}^2\right) \tag{1.132}$$

where ρ_e is the volumetric charge density, $\vec{E}$ is the electric field strength, ε is the permittivity and ρ is the density of the medium. The first and second terms in the right hand side of Eq. (1.132) are the Coulombic and dielectric forces, respectively. The last term is known as electrostriction, which, being the gradient of a scalar, can be combined with the pressure gradient term in the Navier Stokes equation. For an incompressible fluid, this has no additional effect in the fluid dynamics, and accordingly, this term will be neglected in the subsequent analysis. The relative contributions of the first two terms in Eq. (1.132) can be assessed by adding a small perturbation field, $\vec{E}_1$, over and above the applied electric field, $\vec{E}_0$, such that $\vec{E} = \vec{E}_0 + \vec{E}_1$, with $|\vec{E}_1| << |\vec{E}_0|$. Noting that the charge density is given by the Poisson equation

$$\nabla \cdot \left(\varepsilon \vec{E}\right) = \rho_e \tag{1.133}$$

one can write

$$\nabla \cdot \left[\varepsilon\left(\vec{E}_0 + \vec{E}_1\right)\right] = \rho_e, \text{or equivalently}$$
$$\nabla\varepsilon \cdot \vec{E}_0 + \nabla\varepsilon \cdot \vec{E}_1 + \varepsilon\nabla \cdot \vec{E}_0 + \varepsilon\nabla \cdot \vec{E}_1 = \rho_e \tag{1.134}$$

Since $\nabla \cdot \vec{E}_0 = 0$ and $\left|\vec{E}_1\right| << \left|\vec{E}_0\right|$, Eq. (1.134) reduces to the following approximate form:

$$\nabla\varepsilon \cdot \vec{E}_0 + \varepsilon\nabla \cdot \vec{E}_1 = \rho_e \tag{1.134a}$$

Using Eq. (1.134a) in Eq. (1.132), one gets

$$\vec{f}_e = \left(\nabla\varepsilon \cdot \vec{E}_0 + \varepsilon\nabla \cdot \vec{E}_1\right)\vec{E}_0 - \frac{1}{2}\vec{E}_0^2\nabla\varepsilon \tag{1.135}$$

The charge conservation equation can be written in the following form:

$$\nabla \cdot \vec{J} + \frac{\partial\rho_e}{\partial t} = 0 \tag{1.136}$$

where

$$\vec{J} = \underbrace{\sigma\vec{E}}_{\text{conduction current}} + \underbrace{\rho_e\vec{v}}_{\text{convection current}} \tag{1.136a}$$

where σ is the electrical conductivity. The relative contributions of conduction and convection currents in Eq. (1.136) can be assessed by comparing the divergence of the convection charge with the divergence of the Ohmic current, as

$$\frac{\left|\nabla \cdot (\rho_e\vec{v})\right|}{\left|\nabla \cdot \left(\sigma\vec{E}\right)\right|} \approx \frac{\left|\rho_e\vec{v}\right|}{\left|\sigma\vec{E}\right|} = \frac{\left|\nabla \cdot \left(\varepsilon\vec{E}\right)\vec{v}\right|}{\left|\sigma\vec{E}\right|} \approx \frac{\varepsilon v}{\sigma l} \tag{1.137}$$

In expression (1.137), l and v are the length scales and the velocity scales pertinent to the physical problem under consideration. Equation (1.137) can also be interpreted as the ratio of two time-scales, namely, $t_{\text{relax}} = \varepsilon/\sigma$ and $t_{\text{advection}} = l/v$, where t_{relax} is the charge relaxation time scale and $t_{\text{advection}}$ is the advection time scale. With a typical value of $l \sim 10$ μm and $v \sim 200$ μms^{-1}, $t_{\text{advection}}$ comes out to be of the order of 0.1s, which is several orders of magnitude higher than the typical values of t_{relax} in an aqueous solution. Thus, $\nabla\cdot(\rho_e\vec{v})$can be neglected in comparison to $\nabla \cdot \left(\sigma\vec{E}\right)$, in Eq. (1.136). Accordingly, with the help of Eqs. (1.134a), and (1.136) can be re-written as (noting that $\left|\vec{E}_1\right| << \left|\vec{E}_0\right|$)

$$\nabla\sigma \cdot \vec{E}_0 + \sigma\nabla \cdot \vec{E}_1 + \frac{\partial\left(\nabla\varepsilon \cdot \vec{E}_0 + \varepsilon\nabla \cdot \vec{E}_1\right)}{\partial t} = 0 \tag{1.138}$$

Assuming a time-varying electric field of the form

$$\vec{E}_0(t) = \text{Re}\left(\vec{E}_0 \exp(j\omega t)\right) \tag{1.139}$$

where $\vec{E}_0$ is the time-independent component (amplitude) of the unperturbed electric field, one gets, from Eq. (1.138)

$$\nabla\sigma \cdot \vec{E}_0 + j\omega\nabla\varepsilon \cdot \vec{E}_0 + \sigma\nabla \cdot \vec{E}_1 + j\omega\varepsilon\nabla \cdot \vec{E}_1 = 0 \tag{1.140}$$

From Eq. (1.140), the divergence of the perturbation field can be obtained as

$$\nabla \cdot \vec{E}_1 = \frac{-(\nabla\sigma + j\omega\nabla\varepsilon) \cdot \vec{E}_0}{\sigma + j\omega\varepsilon} \tag{1.141}$$

Assuming that the liquid does not respond to the instantaneous value of force at the frequency of the electric field, one can time-average Eq. (1.135) and obtain

$$\left\langle \vec{f}_e \right\rangle = \frac{1}{2} Re \left[\left(\nabla\varepsilon \cdot \vec{E}_0 + \varepsilon\nabla \cdot \vec{E}_1 \right) \vec{E}_0^* - \frac{1}{2} \vec{E}_0^2 \nabla\varepsilon \right] \tag{1.142}$$

where $\vec{E}_0^*$ is the complex conjugate of $\vec{E}_0$. Substituting Eq. (1.141) in Eq. (1.142), one gets

$$\left\langle \vec{f}_e \right\rangle = \frac{1}{2} Re \left[\left\{ \frac{(\sigma\nabla\varepsilon - \varepsilon\nabla\sigma) \cdot \vec{E}_0}{\sigma + j\omega\varepsilon} \right\} \vec{E}_0^* - \frac{1}{2} \left| \vec{E}_0 \right|^2 \nabla\varepsilon \right] \tag{1.143}$$

For real $\vec{E}_0$, Eq. (1.143) can be simplified as [74]

$$\left\langle \vec{f}_e \right\rangle = -\frac{1}{2} \left[\underbrace{\left(\frac{\nabla\sigma}{\sigma} - \frac{\nabla\varepsilon}{\varepsilon} \right) \cdot \vec{E}_0 \frac{\varepsilon\vec{E}_0}{1 + (\omega\tau)^2}}_{\text{Coulomb force}} + \underbrace{\frac{1}{2} \left| \vec{E}_0 \right|^2 \nabla\varepsilon}_{\text{Dielectric force}} \right] \tag{1.144}$$

where $\tau = \varepsilon/\sigma$ is the charge relaxation time ($=t_{\text{relax}}$). For certain ranges of frequencies, the Coulombic force dominates, whereas for other ranges the dielectric force dominates. At a certain frequency, say f_c, these two forces are of equal strength. This frequency can be estimated from Eq. (1.144) as (noting that $\nabla\sigma = \frac{\partial\sigma}{\partial T}\nabla T$, $\nabla\varepsilon = \frac{\partial\varepsilon}{\partial T}\nabla T$), to yield

$$\omega_c = 2\pi f_c \approx \frac{1}{\tau} \left(\frac{2\left| \frac{1}{\sigma}\frac{\partial\sigma}{\partial T} \right|}{\left| \frac{1}{\varepsilon}\frac{\partial\varepsilon}{\partial T} \right|} \right)^{1/2} \tag{1.145}$$

For water, $\frac{1}{\sigma}\frac{\partial\sigma}{\partial T} = 2\%$ per K and $\frac{1}{\varepsilon}\frac{\partial\varepsilon}{\partial T} = -0.4\%$ per K [74], for which $2\pi f_c\tau \approx \sqrt{10}$. In an order of magnitude sense, thus, $f_c \sim 1\tau$, or $f_c = \sigma/\varepsilon$. For $f >> \sigma/\varepsilon$ (typically high frequencies), the dielectric force dominates, whereas for $f << \sigma/\varepsilon$ (low frequencies) the Coulombic force dominates.

Fig. 1.18 Two parallel plates with a small inter-electrode gap, being covered in a dielectric. A potential of 'V' is applied across the gap, with the field direction as shown in the figure

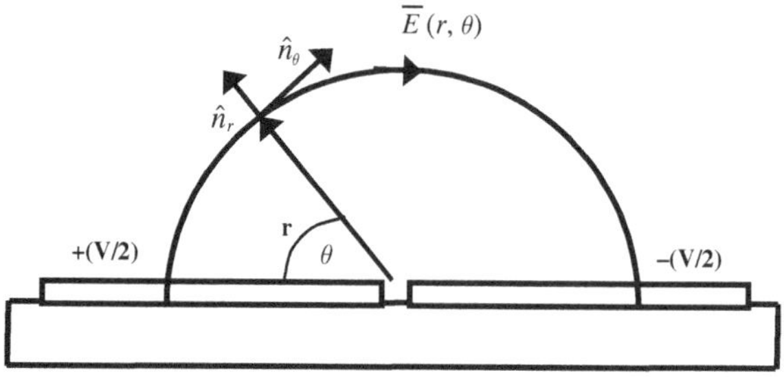

The derivation of a comprehensive mathematical model on electrothermal effects is presented in details in the review of Ramos et al. [74]. A brief description of the pertinent mathematical model is outlined here, for the sake of completeness. For more details, the work of Ramos et al. [74] needs to be referred. Following their analysis, simple analytical estimates of electrothermal forces on a liquid can be obtained by considering two thin parallel metallic plates, with a very small inter-electrode gap. The plates are covered in a dielectric liquid and are subjected to a potential difference of V across the gap, which sets up an electric field $\vec{E}(r,\theta)$ (see Fig. 1.18). Neglecting end effects, this electric field can be expressed as

$$\vec{E} = \frac{V}{\pi r}\hat{n}_\theta \tag{1.146}$$

The corresponding energy dissipation per unit volume (σE^2) can be introduced into the governing equation of energy conservation as a volumetric source term, to yield

$$\frac{k}{r}\frac{\partial}{\partial r}\left(r\frac{\partial T}{\partial r}\right) + \frac{k}{r^2}\frac{\partial^2 T}{\partial \theta^2} + \frac{\sigma V^2}{\pi^2 r^2} = 0 \tag{1.147}$$

where k is the thermal conductivity of the liquid. Writing $T^* = T + \frac{\sigma V^2 \theta^2}{2\pi^2 k}$, one can express Eq. (1.147) in the form

$$\frac{k}{r}\frac{\partial}{\partial r}\left(r\frac{\partial T^*}{\partial r}\right) + \frac{k}{r^2}\frac{\partial^2 T^*}{\partial \theta^2} = 0 \tag{1.147a}$$

Assuming that the electrodes act as thermal baths, i.e., T =0 (a suitably chosen reference temperature) for $\theta = 0, \pi$, one can conclude that T^* is independent of r, which implies

$$\frac{\partial^2 T^*}{\partial \theta^2} = 0 \tag{1.147b}$$

A general solution Eq. (1.147b) can be obtained in the form

$$T^* = a_0\theta + b_0 \tag{1.148}$$

where a_0 and b_0 are obtained from the boundary conditions at $\theta = 0,\ \pi$, as $a_0 = \frac{\sigma V^2}{2\pi k}$ and $b_0 = 0$. Thus,

$$T = \frac{\sigma V^2 \theta}{2\pi k} - \frac{\sigma V^2 \theta^2}{2\pi^2 k} \tag{1.149}$$

The maximum temperature occurs at $\theta = \frac{\pi}{2}$, as $T_{\max} = \frac{\sigma V^2}{8k}$. It is important to note here that for the calculation of electrothermal forces, the most critical factor is not merely the temperature itself but also the gradient of the same. The later can be estimated from Eq. (1.149) as

$$\nabla T = \frac{\sigma V^2}{2\pi k}\left(1 - \frac{2\theta}{\pi}\right)\frac{1}{r}\hat{n}_\theta \tag{1.150}$$

Equation (1.150) can be substituted in Eq. (1.144), with $\nabla\sigma = \frac{\partial\sigma}{\partial T}\nabla T$, $\nabla\varepsilon = \frac{\partial\varepsilon}{\partial T}\nabla T$, to yield

$$\langle f_{\mathrm{e}}\rangle = -M(\omega,\ T)\frac{\sigma\varepsilon V_{\mathrm{rms}}^4}{2k\pi^3 r^3 T}\left(1 - \frac{2\theta}{\pi}\right)\hat{n}_\theta \tag{1.151}$$

Here,

$$M(\omega, T) = \left[\frac{\frac{T}{\sigma}\frac{\partial\sigma}{\partial T} - \frac{T}{\varepsilon}\frac{\partial\varepsilon}{\partial T}}{1 + (\omega\tau)^2} + \frac{1}{2}\frac{T}{\varepsilon}\frac{\partial\varepsilon}{\partial T}\right] \tag{1.152}$$

is a dimensionless factor describing the variation of the time-averaged electrothermal force as a function of the applied frequency and the prevailing temperature. A plot of $M(\omega, T)$as a function of frequency is presented in Fig. 1.2, with T=300 K. The parameters considered for plotting this figure are as follows: $\frac{1}{\sigma}\frac{\partial\sigma}{\partial T}$=2% per K, $\frac{1}{\varepsilon}\frac{\partial\varepsilon}{\partial T}= -0.4\%$ per K, $\sigma = 0.01\ \mathrm{Sm}^{-1}$, $\varepsilon = 640\times10^{-12}\ \mathrm{C^2/N\,m^2}$. It can be observed from the figure that M is positive for low frequencies, whereas it is negative for higher frequencies. For low frequencies, the force is dominated by the space charge, and consequently, the flow takes place in a manner schematically represented in Fig. 1.3. At frequencies greater than the critical frequency, on the other hand, the flow direction gets reversed and the liquid streams up in the centre of the gap between the two electrodes. These conclusions, however, are only valid if the inter-electrode gap is infinitesimally narrow. For finite inter-electrode spacings, a numerical solution of the fluid flow equations may be necessary, to reveal the underlying flow picture.

In order to obtain analytical expressions for the velocity field pertinent to the physical problem described as above, one may begin with the equation of motion (Navier Stokes equation) in the steady state as

$$-\nabla p + \eta\nabla^2 v + \langle f_{\mathrm{e}}\rangle = 0 \tag{1.153}$$

where η is the fluid viscosity and $\langle f_e \rangle$ is given by Eq. (1.151). One can eliminate the pressure gradient term from Eq. (1.153) by differentiating the r-components and θ-components of that equation partially with respect to θ and r, respectively, and subtracting the two resultant equations, to obtain

$$-\eta \nabla^4 \psi + \left(\nabla \times \left\langle \vec{f}_e \right\rangle \right) \cdot \hat{n}_z = 0 \tag{1.154}$$

where $\hat{n}_z$ is a unit vector in the z-direction and $\vec{v} = \nabla \times \left(\psi \hat{n}_z \right)$. Incorporating the expression for $\left\langle \vec{f}_e \right\rangle$ from Eq. (1.151) in Eq. (1.154), one can write

$$\left[\frac{1}{r} \frac{\partial}{\partial r} \left(r \frac{\partial}{\partial r} \right) + \frac{1}{r^2} \frac{\partial^2}{\partial \theta^2} \right]^2 \psi = \frac{2C}{r^4} \left(1 - \frac{2\theta}{\pi} \right) \tag{1.155}$$

where $C = M(\omega, T) \dfrac{\sigma \varepsilon V_{\text{rms}}^4}{2k\pi^3 T \eta}$. Similar to the considerations made for the solution of the energy equation, here also one may note that $\psi = \psi(\theta)$ only. This leads to the following general solution of Eq. (1.155):

$$\psi = -\frac{C}{2} \left(\frac{\theta^2}{2} - \frac{\theta^3}{3\pi} \right) + A_1 \sin 2\theta + A_2 \cos 2\theta + A_3 \theta + A_4 \tag{1.156}$$

where A_i are arbitrary independent constants of integration. Their values can be obtained by imposing the following boundary conditions:

$$\begin{aligned} &\psi = 0 && \text{at } \theta = 0, \pi \\ &\frac{\partial^2 \psi}{\partial \theta^2} = 0 && \text{at } \theta = 0, \pi/2 \text{ (from symmetry)} \end{aligned} \tag{1.157}$$

This gives the final solution of ψ as

$$\psi = -\frac{C}{2} \left(\frac{\theta^2}{2} - \frac{\theta^3}{3\pi} \right) - \frac{C\pi}{24} (\sin 2\theta - 2\theta) \tag{1.158}$$

The velocity distribution can be obtained from the above expression, as follows:

$$v_r = \frac{1}{r} \frac{\partial \psi}{\partial \theta} = \frac{C}{r} \left[-\frac{1}{2} \left(\theta - \frac{\theta^2}{\pi} \right) - \frac{\pi}{12} (\cos 2\theta - 1) \right] \tag{1.159}$$

It can be observed from Eq. (1.159) that $v_r = v_r|_{\max} = \frac{\pi C}{24 r}$ at $\theta = \pi/2$. Also, $v_r = 0$ at $\theta = 0.286\pi$ and $\theta = 0.714\pi$.

It is interesting to compare the electrothermal flow velocity (v_{fluid}) with the dielectrophoretic velocity (v_{DEP})of sub-micron sized charged particles suspended in the aqueous medium. Noting that the time-averaged dielectrophoretic force on a particle of radius R can be described as $\left\langle \vec{F}_{\text{DEP}}(t) \right\rangle = 2\pi \varepsilon_m R^3 Re\left[K(\omega) \right] \nabla \left| \vec{E}_{\text{rms}} \right|^2$

(see Eq. 1.124), where ϵ_m is the permittivity of the dielectric medium, K(ω) is the Clausius–Mossotti factor, given by $K(\omega) = \frac{\tilde{\varepsilon}_p - \tilde{\varepsilon}_m}{\tilde{\varepsilon}_p + 2\tilde{\varepsilon}_m}$, $\tilde{\varepsilon}_p$ and $\tilde{\varepsilon}_m$ being the complex permittivities of the particle and the medium, respectively $(\tilde{\varepsilon} = \varepsilon - j\frac{\sigma}{\omega})$, and $\nabla \left|\vec{E}_{rms}\right|^2$ is the gradient of the square of the r.m.s electric field. Based on this consideration, one may compare the electrothermal and dieleletrophoretic flow velocities as

$$\frac{v_{fluid}}{v_{DEP}} \propto \frac{M(\omega, T)}{\mathrm{Re}\left[K(\omega)\right]} \frac{\sigma V_{rms}^2 r^2}{kTR^2} \tag{1.160}$$

The above implies that the electrothermal flow velocity increases in proportion with the electrical conductivity of the medium. Not only that, the dielectrophoretic force dominates near to the electrode edges ($r \rightarrow 0$). On the other hand, the influence of electrothermal forces becomes progressively stronger for larger-sized particles.

The electrothermal effects in microflows can also be modulated by illuminating the electrodes with fluorescent light, as observed by Green et al. [75]. In their work, the importance of the illumination in generating the fluid flow was described, and the flow was found to be dependent on both the intensity of illumination and the applied electric field. The authors also compared the theoretical estimates of the electrothermally induced flow velocities with their experimental observations. A reasonable agreement was found between the experiments and the theory, with the light generating temperature gradients, and therefore gradients in the fluid permittivity and conductivity, and the electric field being responsible for the motive force. Although there might be several mechanisms through which the light could produce heat in the system, some of these mechanisms could be precluded under the experimental conditions. For example, because of employment of an infrared filter, possibilities of infrared heating were eliminated. Further, although the fluorescent particles could gain energy through the absorption/re-emission process and could generate heat, no global temperature gradients could be generated in this process, since the particles were uniformly distributed. For the laboratory experiments conducted in the above-mentioned study, one of the major reasons behind the electrothermal flow could be attributed to the heating of the electrodes by the light, since a percentage of the light got absorbed by the metal during reflection. With the thin electrodes being more thermally resistant than water, the heat could easily get radiated away through the electrolyte, thereby creating temperature, conductivity and permittivity gradients. The electrothermal fluid flow, therefore, could be attributed to the interaction between the electric field and these gradients.

Sinton et al. [76], in a recent work, have demonstrated that an axially nonuniform temperature distributions can induce a pressure disturbance, which in turn can give rise to velocity gradients. This can lead to velocity profiles that are significantly different from the ideal plug-like electroosmotic flow velocity profiles in microchannels. In their experiments, the axial thermal gradients were induced passively by increased dissipation of Joule heat through the optical infrastructure of a viewing window surrounding a capillary. When large fields were applied, their temperature measurements indicated that fluid in the viewed region was as much as 30°C lower in temperature than in the remainder of the capillary. Despite an increase in viscosity because of this local cooling effect, this also resulted in a locally increased electroosmotic wall velocity, which induced a concave velocity profile in the viewed portion and a convex velocity profile elsewhere.

The electrothermal effects can also be exploited to design efficient micro-actuators, as demonstrated through the recent work of Li and Uttamchandani [77]. In this work, the authors presented a modified design to generate large deflection and to control the peak temperature of the hot beam of a two-beam asymmetric thermal microactuator. Their analysis revealed that by changing the dimensions of a section of the hot beam, it was possible to achieve a higher average temperature but a lower peak temperature within the same. Their analysis also demonstrated the effect of the hot beam geometry on the temperature distribution, and possible methods of optimization so as to avoid local hot spots which lead to thermal failure.

1.4.7 Electro-magneto-hydrodynamic Actuation

Earlier in this section, electrical effects have been outlined as one of the preferred means of flow actuation over microscopic length scales. Off late, it has been also appreciated that combined electro-magnetohydrodynamic (EMHD) [78] effects can potentially be utilized to enhance the liquid flow rates in microchannels. It has also been experimentally established [79] that the average flow rates in micropumps can be substantially augmented by employing low-magnitude magnetic fields. We will subsequently present a brief overview of EMHD actuation in presence of EDL interactions.

Let us consider an electromagnetically driven fluidic transport through a parallel plate microchannel of height $2a$, length l and width W, with $W>>2a$. The fluid flow is simultaneously acted upon by an electric field of strength$\vec{E}$ and a magnetic field of strength $\vec{B}$, as shown in Fig. 1.19. The component of $\vec{E}$ along the axis of the channel provides the necessary driving force for an electrokinetic flow to take place, subjected to the influence of an EDL that forms near the liquid-wall interface and interacts with the externally applied electric field. Further, the magnetic field also interacts with the applied electric field, and eventually acts as an additional influencing parameter that governs the overall microflow. This electroosmotic force is a combined function of the charge density distribution and the imposed electrical field. Superimposed on this effect is the influence of an electromagnetic field, which gets established on account of the interactions between the applied electrical and the magnetic fields.

The body force $\overrightarrow{F}$ that acts on the fluid is essentially contributed by the electrical and magnetic effects imposed on the system. Accordingly, one may write

$$\overrightarrow{F} = \rho_e \vec{E} + \vec{J} \times \vec{B} \tag{1.161}$$

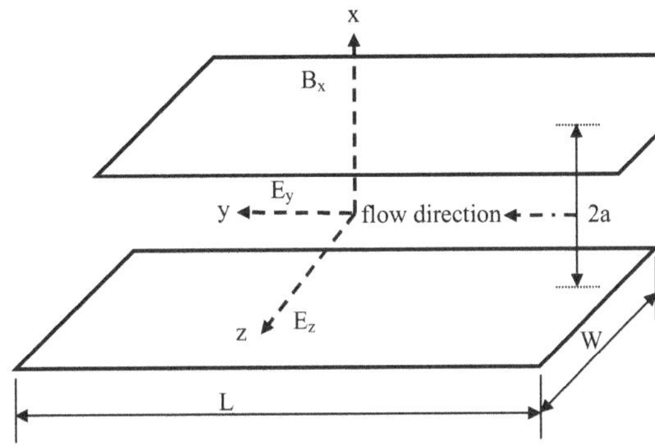

Fig. 1.19 A schematic depicting the physical problem of flow actuation under EMHD influences

where

$$\vec{J} = \sigma_e \left(\vec{E} + \vec{u} \times \vec{B}\right) \tag{1.162}$$

Here ρ_e is the charge density in the EDL and σ_e is the electrical conductivity of the medium. The charge density (ρ_e) in the EDL is described in detail in Section 1.4.5.2 and with the nondimensionalisation as: $\overline{x} = \frac{x}{a}$, $\overline{\psi} = \frac{ze\psi}{k_B T}$, so as to obtain:

$$\rho(\overline{x}) = -\frac{\kappa^2 \varepsilon \zeta}{a^2} \frac{\cosh(k\overline{x})}{\cosh(k)} \tag{1.163}$$

where $\kappa = (a/\lambda_D)$ with λ_D is the characteristic thickness of the EDL and ζ is the wall zeta potential. One important assumption that goes with the present analysis is that the magnetic Reynolds number ($Re_m = \sigma_e \mu_e v_{ref}/l_{ref}$), where v_{ref} is the reference velocity scale, l_{ref} is the reference length scale, and μ_e is the magnetic permeability) is small, which is common to typical microchannel flows. This, in turn, implies that the magnetic field is independent of the flow velocities, by drawing an analogy between the magnetic field equation and the vorticity transport equation [79]. This leads to the following simplified form of the momentum conservation equation of Newtonian fluid along the y-direction, assuming a hydrodynamically fully developed flow in steady state:

$$-\frac{dp}{dy} + \mu \frac{d^2 v}{dx^2} - \sigma_e B_x^2 v + \sigma_e E_z B_x + \rho E_y = 0 \tag{1.164}$$

Equation (1.164) is subjected to the following boundary conditions:

$$v(x = a) = 0, \frac{dv}{dy}(x = 0) = 0 \tag{1.165}$$

Assuming $\frac{dp}{dy} = -\frac{\Delta p}{L}$ (which is a constant for a fully developed flow), Eq. (1.164) can be analytically solved, to obtain

$$\frac{v}{v_{\mathrm{HS}}} = \frac{\left(SHa\frac{v_{\mathrm{HS}}}{v_{\mathrm{p}}}+1\right)}{Ha^2}\frac{v_{\mathrm{p}}}{v_{\mathrm{HS}}}\left(1-\frac{\cosh(Ha\bar{x})}{\cosh Ha}\right) - \frac{k^2}{k^2-Ha^2}\left(\frac{\cosh(k\bar{x})}{\cosh(k)}-\frac{\cosh(Ha\bar{x})}{\cosh Ha}\right) \tag{1.166}$$

where $Ha = \sqrt{\frac{\sigma_{\mathrm{e}}B_x^2}{\mu/a^2}}$ is the Hartmann number, $v_{\mathrm{p}} = \frac{\Delta pa^2}{\mu L}$ is a reference pressure-driven flow velocity, $v_{\mathrm{HS}} = \frac{-\varepsilon\xi E_y}{\mu}$ is a reference electroosmotic velocity, and $S = \frac{E_z a}{v_{\mathrm{HS}}}\sqrt{\frac{\sigma_{\mathrm{e}}}{\mu}}$. In the limit as $Ha \to 0$, Eq. (1.166) yields the classical combined pressure-driven and electroosmotic velocity profile, as

$$\frac{v|_{Ha\to 0}}{v_{HS}} = \frac{v_p}{2v_{HS}}\left(1-\bar{x}^2\right)+\left(1-\frac{\cosh(k\bar{x})}{\cosh k}\right) \tag{1.167}$$

Equations (1.166) and (1..167) can be integrated across the microchannel section, in order to obtain a ratio of the flow rates in presence and absence of the magnetic field, as

$$\frac{Q}{Q|_{Ha=0}} = \frac{\frac{\left(SHa\frac{v_{\mathrm{HS}}}{v_{\mathrm{p}}}+1\right)}{Ha^2}\frac{v_{\mathrm{p}}}{v_{\mathrm{HS}}}\left(1-\frac{\tanh(Ha)}{Ha}\right)-\frac{k^2}{k^2-Ha^2}\left(\frac{\tanh(k)}{k}-\frac{\tanh(Ha)}{Ha}\right)}{\frac{v_{\mathrm{p}}}{3v_{\mathrm{HS}}}+\left(1-\frac{\tanh(k)}{k}\right)} \tag{1.168}$$

It is important to note here that for an effective amplification of the flow rate because of EMHD actuation, the ratio appearing in Eq. (1.168) needs to have a magnitude greater than unity. Based on this, one can define an amplification factor, A, positive values of which represent an effective enhancement in the flow rates. On the other hand, negative values of A would indicate an effective decrement in the net volume flow rate, despite the combined EMHD influences. With these considerations, the amplification factor can be defined as

$$A = \frac{\tanh(k\alpha)}{k\alpha}\left[\frac{1}{1-\alpha^2}-\frac{S}{k\alpha}\right]-\frac{\tanh k}{k}\left[\frac{1}{1-\alpha^2}+\frac{S}{k\alpha}-1\right]-1 \tag{1.169}$$

where $\alpha = Ha/k$.

Another interesting aspect of the electromagnetohydrodynamic microfluidic transport is the net current in the flow, which can be calculated as the sum of the axial conduction current ($i_1 = 2\int_0^a \rho v dx$) and the convection current ($i_2 = 2a\sigma_{\mathrm{e}}E_y$). An interesting situation arises when the sum total of conduction and convection current equates to zero, the corresponding axial potential being known as the streaming potential, as discussed earlier. This condition leads to the following expression for v_{HS}:

$$v_{HS}|_{\text{streaming}} = v_p \left(HaS\frac{v_{HS}}{v_p} + 1 \right) f \tag{1.170}$$

where

$$f = \frac{f_1}{\left(\frac{k^2}{k^2-Ha^2}f_2 - \frac{R}{k^2}\right)Ha^2} \text{ with}$$

$$f_1 = \frac{\tanh k}{k} - \frac{1}{2\cosh k \cosh Ha}\left(\frac{\sinh(k+Ha)}{k+Ha} + \frac{\sinh(k-Ha)}{k-Ha}\right),$$

$$f_2 = \frac{1}{2}\left(\frac{\tanh k}{k} + \text{sech}^2\, k\right) - \frac{1}{2\cosh k \cosh Ha}\left(\frac{\sinh(k+Ha)}{k+Ha} + \frac{\sinh(k-Ha)}{k-Ha}\right)$$

and the parameter R is defined as follows: $R = \dfrac{\sigma_e a^2 \mu}{\varepsilon^2 \zeta^2}$.

It is also interesting to note here that when the liquid is forced through the microfluidic

channel, one can define an apparent (or effective) viscosity, which, under the sole presence of a pressure-driven flow, would have resulted in the same volumetric flow rates as obtained under the influences of all the body forces acting in tandem. For the present situation, this leads to the following expression for the apparent viscosity in the streaming potential limit:

$$\frac{\mu_{app}}{\mu} = \frac{1}{3F\left(SHa\frac{v_{HS}}{v_p} + 1\right)} \tag{1.171}$$

where

$$F = \frac{1}{Ha^2}\left(1 - \frac{\tanh(Ha)}{Ha}\right) - \frac{fk^2}{k^2 - Ha^2}\left(\frac{\tanh(k)}{k} - \frac{\tanh(Ha)}{Ha}\right).$$

With the aid of the above analysis, it becomes possible to quantitatively estimate the net electro-viscous effect influencing the microflow. The apparent viscosity, in a physical sense, gives an estimation of the relative retarding or enhancing effects of the other forces acting on the flow field, as compared to the pressure forces. A value of μ_{app}/μ greater than unity indicates a net retarding effect of the external forces, whereas a value less than unity indicates an overall enhancing effect of the influencing forces.

It is interesting to note that, the amplification factor, A is found to be a strong function of the non-dimensional parameter, S. In fact, for low values of S (typically, of the order of unity or less), the values of A are observed to be negative. This implies that instead of obtaining a gain in the flow rates, there could, in fact, be reductions in the rate of volumetric transport on application of the magnetic field, for relatively low strengths of the transverse electric field (noting that S is directly proportional to E_Z). This, in turn, suggests that for obtaining real benefits associated with the employment of the magnetic field, the transverse electric field strength also needs to be judiciously chosen beyond certain threshold limits. Beyond the threshold limit, any further enhancement in E_Z is expected to result in progressive augmentations in the microflow rates, for low magnetic field strengths (more precisely, low values of Ha). However, as mentioned earlier, this strategy cannot be employed indiscriminately to derive too high a flow rate augmentation, since, the transverse electric field needs to be also constrained within a threshold limit, so as to avoid the adverse effects of Joule heating.

1.4.8 Acoustic Streaming

Acoustic streaming originates from the word *acoustics* or actuation based on sound, although, it may be a misnomer. The interaction caused by an external high-frequency oscillation driven by sound or any other obstacles present in the fluid flow may to this phenomenon. When the fluid experiences a high-frequency oscillation given by an ultrasound source, a progressive wave is established in the air. Due to the attenuation of the wave, a nonzero time averaged Reynolds stress is built in the region close to the sound source, and this stress pushes the fluid in the direction of wave propagation. The resultant wind is called a quartz wind. Eulerian streaming refers to the flow driven by the time averaged Reynolds stress term. When a fluid within a duct receives a standing wave, a nonzero time average of the Reynolds stress is built inside the duct. Due to the interaction between the air and the duct wall, a steady recirculating flow takes place within the duct. The net effect is that dust or particles accumulate at nodes. This is commonly referred to as Kundt's dust pattern. The Stokes drift flow is purely kinematic and hence fundamentally different from the Eulerian streaming flow. However, the mass-transport effect given by the Stokes drift flow is not weaker than the Eulerian streaming. For a progressive wave, the Stokes drift flow plays a dominant role (e.g., application of flexural plate waves to pumping and mixing in microfluidics). In this Section, we will briefly discuss the phenomenon of acoustic streaming experienced by both compressible and incompressible flow conditions.

1.4.8.1 One-Dimensional Compressible Flow Model (Quartz Wind)

Let us consider the case when the sound is generated from a source and travels along a certain direction, say the x^*-direction, in a space. The governing equation for the motion of the compressible fluid in 1-dimension is given by [32, 80]:

$$\frac{\partial \rho^*}{\partial t^*} + \frac{\partial (\rho^* u^*)}{\partial x^*} = 0 \tag{1.172a}$$

$$\rho^* \left(\frac{\partial u^*}{\partial t^*} + u^* \frac{\partial u^*}{\partial x^*} \right) = -\frac{\partial p^*}{\partial x^*} + \frac{4}{3}\mu \frac{\partial^2 u^*}{\partial x^{*2}} \tag{1.172b}$$

where u^* is the flow velocity, p^* the pressure, t^* the time, ρ^* the density of the fluid and μ is the dynamic viscosity of the fluid. The fluid particle is assumed to oscillate back and forth with amplitude A_0, frequency ω, and wavelength λ. We consider $1/\omega$, $\lambda/2\pi$, $A_0\omega$, and $\rho^*{}_0 A_0\omega^2(\lambda/2\pi)$ as the reference quantities for the time, length, velocity, and pressure, respectively. Further, the density and the dimensionless velocity (u) and pressure (p) are expanded as:

$$\rho^* = \rho_0^*(1 + \varepsilon\rho_0 + \varepsilon^2\rho_1 + \cdots) \tag{1.173a}$$

$$u = u_0^* + \varepsilon u_1 + \varepsilon^2 u_2 + \cdots \tag{1.173b}$$

$$p = p_0^* + \varepsilon p_1 + \varepsilon^2 p_2 + \cdots \tag{1.173c}$$

where ρ_0^* corresponds to the undisturbed fluid density and the two dimensionless parameters

$$\varepsilon = \frac{A_0}{\lambda/2\pi} \text{ and } \delta = \frac{\mu}{\rho\omega\,(\lambda/2\pi)^2}$$

represents dimensionless amplitude of

the flow motion and the inverse of the Reynolds number, respectively. These parameters are considered to be small. The leading-order equations obtained from Eqs. (1.172a) and (1.172b), for small values of the parameter ε, become:

$$\frac{\partial \rho_0}{\partial t} + \frac{\partial u_0}{\partial x} = 0 \tag{1.174a}$$

$$\frac{\partial u_0}{\partial t} = -\frac{\partial p_0}{\partial x} + \frac{4}{3}\delta \frac{\partial^2 u_0}{\partial x^2} \tag{1.174b}$$

Using the relation: $\frac{\partial p^*}{\partial \rho^*} = c^{*2} = \omega^2 \left(\frac{\lambda}{2\pi}\right)^2$, the perturbed density in Eq. (1.174a) and the pressure in Eq. (1.174b) may be eliminated to yield:

$$\frac{\partial^2 u_0}{\partial t^2} - \frac{\partial^2 u_0}{\partial x^2} - \frac{4}{3}\delta \frac{\partial^3 u_0}{\partial t \partial x^2} = 0 \tag{1.175}$$

The solution of Eq. (1.175) is a form of the progressive wave and is given by:

$$u_0 = e^{-2/3\delta x} \cos(x - t) \tag{1.176}$$

The next higher order equation obtained from Eq. (1.172b) is:

$$\frac{\partial u_1}{\partial t} + \frac{\partial p_1}{\partial x} - \frac{4}{3}\delta \frac{\partial^2 u_1}{\partial x^2} = -u_0 \frac{\partial u_0}{\partial x} \tag{1.177}$$

The time average of Eq. (1.177) results in:

$$\frac{\partial \bar{p}_1}{\partial x} - \frac{4}{3}\delta\frac{\partial^2 \bar{u}_1}{\partial x^2} = -\left\langle u_0 \frac{\partial u_0}{\partial x}\right\rangle \tag{1.178}$$

where the term $\left\langle u_0 \frac{\partial u_0}{\partial x}\right\rangle$ represents a dimensionless body force per unit mass with u_0 obtained from Eq. (1.176). Lighthill [81] presented solutions of this system of equations for the case in which this force acts as a point source in an infinite space. It is observed that when the viscosity is low enough (at high streaming Reynolds numbers), the flow from the source acts like a jet. This flow is sometimes called the quartz wind. The term acoustic streaming refers to this flow in the case of ultrasound. As can be seen from Eq. (1.178), the driving force for this current vanishes when there is no attenuation, i.e., when $\delta = 0$. Therefore we can say that the acoustic streaming for the case of ultrasound in a compressible fluid is attributed to the attenuation of sound. On the other hand, we can also calculate the Stokes drift velocity from the solution of Eq. (1.176). We can follow the path of the particle's path $x(x_0, t)$, where x_0 is a reference point independent of the time. From the definition of the flow velocity we can write

$$\frac{dx}{dt} = \varepsilon u \tag{1.179}$$

Substituting Eq. (1.176) into Eq. (1.179) and expanding in terms of $x = x_0 + \varepsilon x_1$, we obtain:

$$x_1 = e^{-2/3\delta x_0}\cos(x_0 - t) \tag{1.180a}$$

$$\text{and } u_0 = e^{-2/3\delta x_0}\cos(x_0 - t) + \left.\frac{\partial u_0}{\partial x}\right|_{x=x_0} \varepsilon x_1 \tag{1.180b}$$

The last term contains the steady component and it becomes

$$\varepsilon u_d = \frac{1}{2}\varepsilon e^{-\frac{4}{3}\delta x_0} \tag{1.181}$$

which is known as the Stokes drift velocity. The steady streaming flow $\varepsilon \bar{u}_1$, as given by the solution of Eq. (1.178), is called the Eulerian streaming flow. Comparing with the Eulerian streaming flow, the Stokes drift flow is confined in a region close to the sound source, i.e., within the region $x = O(1/\delta)$. Actually this region corresponds to the one that the force $\left\langle u_0 \frac{\partial u_0}{\partial x}\right\rangle$ acts upon. In this region, the steady flow velocity is composed of the Eulerian streaming velocity and the Stokes drift velocity, as:

$$u_L = \varepsilon \bar{u}_1 + \varepsilon u_d \tag{1.182}$$

which is known as the Lagrangian velocity. Beyond this region the Stokes drift flow vanishes and only the Eulerian streaming exists. In many cases the Stokes drift velocity is ignored, but when the location of interest is not far from the sound source it should be considered. In general, the Stokes flow is expected to occur only when the primary wave is progressive, not when it is standing.

1.4.8.2 Eulerian Streaming

Even the incompressible fluid can also give rise to Eulerian streaming flow when a solid obstacle is in contact with the oscillating fluid. Consider a two-dimensional incompressible flow around a solid body governed by the following dimensionless equations:

$$\nabla.\mathbf{u} = 0 \tag{1.183a}$$

$$\frac{\partial \mathbf{u}}{\partial t} + \varepsilon \mathbf{u}.\nabla \mathbf{u} = -\nabla p + \frac{1}{Re}\nabla^2 \mathbf{u} \tag{1.183b}$$

where u $= (u, v)$ and ∇ is the two-dimensional gradient operator. We use $1/\omega$, L, $A_0\omega$ and $\rho^* A_0 \omega^2 L$ as the reference quantities for the time, length, velocity and pressure, respectively. Note that the reference length L used here represents a typical dimension of the obstacle. The Reynolds number Re is defined as $Re = \frac{\rho \omega L^2}{\mu}$ is assumed to be large. For the small value of ε, we expand

$$(u, p) = (u_0, p_0) + \varepsilon(u_1, p_1) + \ldots \tag{1.184}$$

The leading-order equation of Eq. (1.183b) represents a potential flow equation given as:

$$\frac{\partial u_0}{\partial t} = -\nabla p_0 \tag{1.185}$$

The solutions can take a separable form, as:

$$u_0 = f(x, y)e^{it} \tag{1.186}$$

where the complex functions f must satisfy the continuity Eq. (1.183a) and i is the square root of unity. Note that the inviscid region governed by these equations comprises most of the flow domain except the thin layer near the solid boundary.

Since the potential flow solution (Eq. 1.186) does not satisfy the no-slip condition on the solid surface, we must consider a thin layer (called the Stokes layer) adjacent to the surface $n = 0$, where n refers to the local coordinate normal to the wall. In this thin layer, we use the stretched coordinate Y defined as: $n = \sqrt{2/Re}Y$. We also use the velocity components U and V along the local coordinates s and n, respectively (the coordinate s is along the surface). Then the boundary layer equation becomes:

$$\frac{\partial U}{\partial t} - \frac{\partial u_e}{\partial t} = \frac{1}{2}\frac{\partial^2 U}{\partial Y^2} - \varepsilon\left(U\frac{\partial U}{\partial s} + V\frac{\partial V}{\partial n}\right) \tag{1.187}$$

where u_e denotes the tangential component of the potential flow velocity evaluated at the solid surface. The velocity components may be expanded as:

$$(U,V) = (U_0,\sqrt{2/Re}V_0 + \varepsilon(U_1,\sqrt{2/Re}V_1) + \ldots \tag{1.188}$$

The leading-order solution for Eq. (1.187) is of the form:

$$U_0 = u_{0e}(s)\left[1 - e^{-(1+i)Y}\right]e^{it} \tag{1.189}$$

The normal component V_0 can also be obtained from the continuity equation. The next order velocity, O(ε), in Eq. (1.187) then becomes

$$\frac{\partial U_1}{\partial t} - \frac{\partial u_{1e}}{\partial t} = \frac{1}{2}\frac{\partial^2 U_1}{\partial Y^2} - \left(U_0\frac{\partial U_0}{\partial s} + V_0\frac{\partial V_0}{\partial n}\right) \tag{1.190}$$

In this context, we are interested in the time independent flow. Taking the time average of the above equation over one period of oscillation, one gets:

$$\frac{1}{2}\frac{\partial^2 \bar{U}_1}{\partial Y^2} = \left\langle U_0\frac{\partial U_0}{\partial s} + V_0\frac{\partial V_0}{\partial n}\right\rangle \tag{1.191}$$

The solution of this equation yields the streaming velocity at the edge of the boundary layer,

$$\bar{U}_{1\infty} = -\frac{3}{8}\left[(1-i)\,\tilde{u}_{oe}\frac{du_{0e}}{ds} + (1+i)\,u_{oe}\frac{d\tilde{u}_{0e}}{ds}\right] \tag{1.192}$$

where $\tilde{u}_{oe}$ denotes the complex conjugate. This velocity then acts as a boundary condition for the exterior bulk region. The governing equation of the steady streaming flow takes the following form:

$$(\bar{\mathbf{u}} + \bar{\mathbf{u}}_d)\cdot\nabla\bar{\mathbf{u}} = -\nabla\bar{p} + \frac{1}{\mathrm{Re}_s}\nabla^2\bar{\mathbf{u}} \tag{1.193}$$

where Re_s is the streaming Reynolds number based on the streaming velocity at the edge of the Stokes layer; $Re_s = \varepsilon^2 Re$. This equation looks very much similar to the Navier–Stokes equation, but here the convective velocity is replaced by the Lagrangian velocity $\bar{\mathbf{u}}_L = \bar{\mathbf{u}} + \bar{\mathbf{u}}_d$

The above formulation is effective and suitable when the streaming Reynolds number is large so that the Reynolds stress action is confined within the thin Stokes layer. In the microfluidic application, however, Re_s is usually small. Accordingly, the Reynolds stress term may be added to the streaming-flow Eq. (1.193), so that

$$(\bar{\mathbf{u}} + \bar{\mathbf{u}}_d)\cdot\nabla\bar{\mathbf{u}} = -\nabla\bar{p} + \frac{1}{\mathrm{Re}_s}\nabla^2\bar{\mathbf{u}} - \langle(\mathbf{U_0}\cdot\nabla)\,\mathbf{U_0} + \mathbf{U_0}\,(\nabla\cdot\mathbf{U_0})\rangle \tag{1.194}$$

where the second term within <> on the right-hand side is nonzero for the compressible fluid case; for the compressible fluid case Eq. (1.194) itself must be modified. This means that for the case with a low streaming Reynolds number, Eq. (1.194) must be solved over the whole domain including the Stokes layer. Another important point for the incompressible flow is that the Reynolds stress vanishes when the primary oscillating flow is of the progressive wave type. On the other hand, it needs to be noted that in the microfluidic area no literature has taken into account the Stokes drift flow in the convective velocity in the numerical simulation of the streaming flow.

We can experience a typical example of the Eulerian streaming flow around a circular cylinder [82]. Here, the fluid surrounding the cylinder oscillates with high frequency; or the cylinder may oscillate in the otherwise quiescent fluid without fundamental difference in the results. The steady flow within the Stokes layer at high streaming Reynolds numbers shows a four-cell structure around the circular cylinder. There are two streams coming out of the cylinder from both the sides in the direction of oscillation. Kundt's dust pattern manifests another simple example of the Eulerian streaming flow given by the two-dimensional standing wave in a duct. When an acoustic standing wave is established in the duct with a compressible fluid, the steady streaming reveals four-cell structure over a half wavelength (or over the space between two neighboring nodes). Near the duct wall, the steady streaming is toward the nodes, and near the duct center it is coming out of the nodes. Therefore dust within the duct should cluster near the nodal points of the standing wave. The detailed solution for this case has been given by Riley [80].

1.5 Microfluidics of Droplets

Till this point, we have discussed flow actuation and control considering a continuous liquid motion. However, another interesting facet of microfluidics lies in the handling of discrete droplets for several interesting applications, including bio-chemical and bio-medical analysis. In this Section, we will briefly elucidate some of the fundamental principles concerning droplet-based microfluidics.

Fundamental considerations on droplet motion in a confined microsystem may be simplistically illustrated by considering a liquid slug of mass m and volume $\forall$ in a capillary tube of radius R (see Fig. 1.20). The caps at the ends of the droplet meet the wall at angles θ_1 and θ_2, corresponding to the contact angles of receding and advancing ends respectively. Let x_c be the centre of mass of the slug (refer to Fig. 1.20) and V_{r1} and V_{r2} be the contact line velocities of the two end caps. As the flow is associated with low Reynolds number, the inertial effects may be neglected and the equation of motion of the slug can be written as [32]:

$$m\ddot{x}_c = (\cos\theta_2 - \cos\theta_1)\,2\pi R - c\frac{V_{r1} + V_{r2}}{2} \tag{1.195}$$

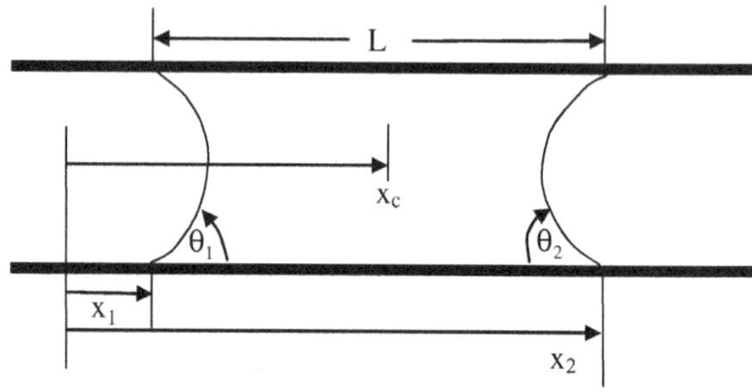

Fig. 1.20 A deformable slug (droplet) in a micro-tube

where c is an equivalent frictional coefficient for the slug ($c = 8\pi\mu L$, for a fully developed flow, for example).

The volume enclosed by the spherical caps and the plate that includes the contact line is given by:

$$\forall_{\text{sph}}(\theta) = \frac{\pi R^3}{3}\frac{\cos\theta\,(2+\sin\theta)}{(1+\sin\theta)^2} \tag{1.196}$$

The centroid of the volume of the end cap, as measured from the plane of the contact line, can be obtained as

$$h(\theta) = \frac{3 - 2\sin\theta - \sin^2\theta}{4\cos\theta\,(2+\sin\theta)}R \tag{1.197}$$

In order to maintain the total volume of the slug as constant, the instantaneous slug length, $L(\theta_1, \theta_2)$, is determined by the initial slug length L_s and the static contact angle θ_s as

$$\forall = \pi R^2 L_s - 2\forall_{\text{sph}}(\theta_s) = \pi R^2 L(\theta_1, \theta_2) - \forall_{\text{sph}}(\theta_1) - \forall_{\text{sph}}(\theta_2) \tag{1.198}$$

This gives the instantaneous length as:

$$L(\theta_1, \theta_2) = L_s + \frac{1}{\pi R^2}\left[\forall_{\text{sph}}(\theta_1) + \forall_{\text{sph}}(\theta_2) - 2\forall_{\text{sph}}(\theta_s)\right] \tag{1.199}$$

The contact line velocities at the two end caps (V_{r1} and V_{r2}) are related to the corresponding instantaneous positions as

$$V_{r1} = \frac{dx_1}{dt} \tag{1.200a}$$

$$V_{r2} = \frac{dx_2}{dt} \tag{1.200b}$$

where the positions x_1 and x_2 are given in terms of the centre of mass by:

$$x_1 = x_c - \frac{1}{\forall}\left(\frac{\pi R^2 L(\theta_1,\theta_2)^2}{2} - \forall_{sph}(\theta_1)\,h(\theta_1) - \forall_{sph}(\theta_2)\left[L(\theta_1,\theta_2) - h(\theta_2)\right]\right) \tag{1.201a}$$

and

$$x_2 = x_c + \frac{1}{\forall}\left(\frac{\pi R^2 L(\theta_1,\theta_2)^2}{2} - \forall_{sph}(\theta_2)\,h(\theta_2) - \forall_{sph}(\theta_1)\left[L(\theta_1,\theta_2) - h(\theta_1)\right]\right) \tag{1.201b}$$

Equations (1.195), (1.198), (1.200a), (1.200b), (1.201a), (1.201b) and the relation $\dot{x}_c = \frac{dx_c}{dt}$ are seven independent equations with nine unknowns, namely, θ_1, θ_2, $x_c, \dot{x}_c$, V_{r1}, V_{r2}, x_1, x_2 and L. To match the number of independent equations with the number of unknowns, two additional independent constraining equations are necessary. These additional equations are available from the considerations of contact line slip, which can be generally described in the form of

$$\theta_{1,2} = f\,(V_r, Ca)_{1,2} \tag{1.202}$$

where Ca is the Capillary number ($Ca = \eta V/\gamma_{lv}$). Detailed expressions for Eq. (1.202) depend on the particular slip models chosen as given by Eqs. (1.33), (1.34), (1.35) and (1.36).

The above analysis represents a lumped model – a one dimensional droplet transport model which is somewhat generic in nature, and can take special forms depending on the specific modes of droplet motion actuation. For example, one may consider the thermocapillary driven droplet motion in a cylindrical capillary, in which the surface tension varies as a function of the local temperature. For small temperature variations, this dependence is approximately linear, and can be described as

$$\gamma_{lv} = \gamma_{lv,0} - a\,(T - T_0) \tag{1.203}$$

where $\gamma_{lv,0}$ and a are constants, depending on the specific fluid being transported. Under these conditions, a simple analytical solution of Eq. (1.195) can be obtained, by neglecting the droplet deformation and contact line slip, as

$$\begin{aligned}\dot{x}_c = &\frac{R}{4\mu L}\left(1 - \exp\left(-\frac{8\mu t}{\rho R^2}\right)\right)\left[\left\{\gamma_{lv,0} - a\,(T - T_0)\cos\theta\right\}_2\right.\\ &\left.- \left\{\gamma_{lv,0} - a\,(T - T_0)\cos\theta\right\}_1\right]\end{aligned} \tag{1.204}$$

The above discussion has been restricted to single straight channels. Droplet dynamics in other microchannel geometries have interesting features which are not usually apparent with flows in single straight microchannels. As an example, one may cite the case of droplet motion in a microchannel geometry that is characterized with a sudden contraction in the cross sectional area. Rosengarten et al. [83] have recently presented an interesting study on this aspect, especially dealing with the cases in which the droplet size is larger than the contraction but is smaller than the original microchannel section. The droplet transport can be of a filament type or a slug type, depending on the contraction Capillary number (Ca_{con}) and the Reynolds number (Re) of flow. The average extensional strain rate due to a contraction can be approximated as $\frac{\bar{u}_2-\bar{u}_1}{L}$, where $\bar{u}_2 - \bar{u}_1$ is the difference in mean velocity between the upstream and the downstream channels and L is the distance over which this change takes place. If $\beta = d_1/d_2$ be the contraction ratio of the two capillaries then it can be inferred that the entry length and the upstream vortex attachment length for the flow through the contraction are $0.25d_2$ and $0.17d_1$, respectively [69]. Assuming $L \approx 0.25d_2 + 0.17d_1$, accordingly, the extensional strain rate becomes

$$\dot{\varepsilon} = \frac{\bar{u}_2 - \bar{u}_1}{0.25d_2 + 0.17d_1} \tag{1.205}$$

For an axisymmetric contraction, $\bar{u}_2 = \beta^2 \bar{u}_1$ (from continuity), which implies

$$\dot{\varepsilon} = \frac{\bar{u}_1(\beta^2 - 1)}{0.25d_2 + 0.17d_1} \tag{1.205a}$$

Hence, the contraction Capillary number can be described as

$$Ca_{\mathrm{con}} = \frac{\mu_c \dot{\varepsilon} R_0}{\gamma} \tag{1.206}$$

where R_0 is the original radius of the initially-spherical droplet and $\dot{\varepsilon}$ is given by Eq. (1.205). Beyond a threshold value of Ca_{con}, there occurs a transition of the droplet morphology in the contraction, from the slug shape to the filament shape. The transitional value of Ca_{con} decreases as the flow Reynolds number is increases. For details of the droplet shape and contact angle evolution under these circumstances, one may refer to the work of Rosengarten et al. [83].

In T-junctions, it has been observed that the droplets can be sheared, extended and split. Let us consider a non-deformed droplet (with viscosity μ_{b}) of radius R in a fluid of viscosity μ_{c}, with a local velocity gradient of magnitude G and surface tension γ. The stresses scale as $\mu_{\mathrm{c}} G$ and the drop extension time-scale is as $\dfrac{1 + \mu_b/\mu_c}{G}$. The surface tension tends to relax a deformed droplet back to its spherical shape by the virtue of energy minimising. The capillary velocity scale obtained from the scaling analysis is as γ/μ_{c}, the drop relaxation time-scale is as $\dfrac{(1 + \mu_b/\mu_c)\,\eta_c R}{\gamma}$, and the capillary stresses scale as γ/R. The corresponding extensional flow Capillary number representing the ratio of the viscous to the capillary stresses, or equivalently, inverse of the ratio of the corresponding time-scales is given as $\mathrm{Ca}_{\mathrm{ext}} = \dfrac{\mu_c G R}{\gamma}$. For the onset of droplet break-up, $\mathrm{Ca} = O(1)$. Since $G \sim Q_0/R_i^3$, where R_{i} is the hydraulic radius of the microchannel and Q_0 is the volumetric flow rate, the above condition implies

$$R \sim \frac{\gamma R_i^3}{\mu_c Q_0} \tag{1.207}$$

Expression (1.207) reveals that larger the flow rate, smaller is the droplet size. It is important to mention here that this expression is applicable only when $R < R_i$, i.e., when the droplets are small enough so that the hydrodynamic forces exerted by the channel walls are rather inconsequential and the breakup solely relies on the straining of the imposed flow. Identical considerations can be corroborated by the trends of the droplet breakup phenomena that can be observed across the microchannel T-junctions [84]. For an extensional flow in the neighbourhood of a T-junction, $G \sim u/w_0$, where w_0 is the channel width. One may note here that $Ca_{\text{ext}} = \frac{\mu_c G R}{\gamma} = Ca\frac{R}{w_0}$, where $Ca = \frac{\mu_c u}{\gamma}$ is the standard deformation Capillary number. Extensional stress at the stagnation point, being much larger than the shear stresses upstream of the junction, change the droplet length from l_0 to l_e. For the critical condition at which the droplet just breaks, one may write

$$(l_e - l_0) \propto (G - G_0)^{1/2} \tag{1.208}$$

where G_0 is the upstream shear rate. When $G_0 << G$,

$$Ca_{ext} \propto \left(\frac{l_e - l_0}{R}\right)^2 \tag{1.208a}$$

At the stability limit of the droplet,

$$\frac{l_e}{\pi w_w} \sim 1 \tag{1.208b}$$

where w_e is the width of the stretched droplet. This criterion is consistent with the classical Rayleigh-Plateau instability, in which a cylindrical liquid thread can reduce its total surface area by breaking when its length (l_e) exceeds its circumference (πw_e). Further, from the considerations of volume conservation of the incompressible droplet, one may write

$$l_0 w_0^2 = l_e w_e^2 \tag{1.208c}$$

Using (1.208b) and (1.208c) in (1.208a) and denoting the initial extension as $\varepsilon_0 = l_0/\pi w_0$, a critical capillary number of droplet breakage at the T-junction can be estimated as

$$Ca_{\text{cr}} = \alpha \varepsilon_0 \left(\frac{1}{\varepsilon_0^{2/3}} - 1\right)^2 \tag{1.209}$$

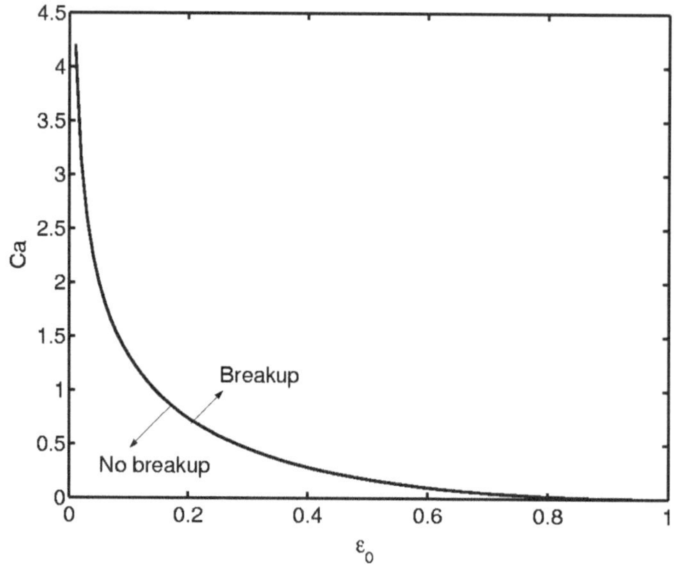

Fig. 1.21 Critical conditions for droplet breaking at T-junctions

where α is a dimensionless constant, which is a function of η_b/η_c and the geometry of the channel.

For symmetrical T-junctions, $\alpha\sim 1$ has been found to excellently match with the experimental observations corresponding to the critical conditions for the breaking of droplets [84]. It is also important to note here that with $\varepsilon_0 > 1$, droplets are always found to break in the T-junction, even for the lowest achievable values of *Ca* (see Fig. 1.21). On the other hand, they never break up upstream in the straight portion of the channel preceeding the T-junction, despite satisfying the breaking criterion mathematically. This can be attributed to the fact that before the onset of breaking, the droplet may be perturbed with associated crests and troughs. The portions of the greatest lateral extensions (i.e., the crests) are characterized with pressures that are higher than the pressures at the regions of greatest lateral contractions (troughs). To restore the equlibrium, liquid would flow from higher pressure to lower pressure, thereby stabilizing the droplet. Physically, the confinement of the channel walls imparts the stabilizing effect to the droplet at the upstream locations, preventing the droplet breakup. At the T-junction, however, the droplet is offered with a provision of being free of any confinement over at least one portion of its lateral faces, allowing the perturbations to grow and the droplet to break up.

The development of fundamental models on droplet dynamics through microchannels has lead to a number of analytical and numerical studies reported in the literature in the recent past. Scheeizer and Bonnecaze [85] employed the boundary-integral method to numerically simulate the displacement of a two-dimensional droplet attached to a solid surface when the inertial and gravitational forces are negligible. These authors showed that as the capillary number was increased, the deformation of the droplet increased until a critical value was reached. Beyond this limit, no steady droplet shapes were observed. Increasing the droplet size resulted in an increase in the deformation of the interface at a given capillary number. The deformation of the droplet was found to be more severe with higher viscosity ratios. Both shear and pressure-driven flow regimes were examined and were shown to have similar values of the critical capillary number. Further, addition of surfactants was shown to reduce the deformation of the pinned droplets, as the surface Peclet number was increased.

Digital microfluidics has emerged as an efficient microscale liquid manipulation technique, in which droplets of picolitre size are manipulated on arrays of electrodes. This technique is used to miniaturize a wide range of applications, with the advantages of reduced sample size, fast heat transfer and reaction rates, and integration capacity (i.e., the lab-on-chip concept). It often exploits the principle of electrocapillarity. Lab-on-a-chip applications of electrowetting have been primarily directed towards moving, merging, mixing and splitting of droplets. A sandwich design consisting of a droplet confined between two parallel substrates (typically, separated by a distance of ~100 μm) has become a standard practice in this respect. A droplet containing samples or reagents can be dispensed from reservoirs, moved, merged, and split into smaller droplets, each independently from the others. The primary advantage may be attributed in handling of individual droplets precisely. A second difference is reagent isolation – droplets serve as discrete microvessels, in which reactions can be carried out without cross-talk between samples or reagents; this stands in contrast to microchannels, which are prone to undesirable hydrostatic and capillary flows. Digital microfluidics provides an array-based applications, added advantage of dynamic device formats and fabrication, the physics of droplet actuation, and a sampling of the myriad applications to which the technology is being applied, which may be broadly classify as biological and nonbiological applications. In context of the later, it may be pertinent to mention that microscale fluid flows have often been employed for electronics cooling applications, and have been shown to be capable of achieving cooling rates as high as 100 W/cm^2. However, such capacity is not sufficient to cool local hot spots on integrated circuits (300 W/cm^2). Controlling droplets seem to be well suited for this application, which may be moved directly to hot spots, bypassing the regions not requiring cooling.

1.6 Summary and Outlook

Many microfluidic devices have been developed in the past few years, based on the ability to pattern substrates, implement the lab-on-a-chip concept, control and enhance chemical reactions and heat transfer, manipulate particle position, orientation and transport rates, and develop mixing and separation processes. These systems are progressively being applied in the biomedical, pharmaceutical, printing and chip-cooling applications, to name just a few. It is natural to think that these systems will be integrated with smart materials and fully automated devices for more efficient operations, offering both research and engineering opportunities for the future. The importance of scaling down these devices, as well as characterizing and understanding the interplay of fluid flow, surface forces, and potentially statistical and molecular interactions, are among the research questions that will need to be addressed in more details in the years to come, so that micro and nano systems technology sheds its dogma from a research hype into useful products that are simple, compact, affordable, multi-functional, and evolutionary in all respects. Understanding the underlying science, more than hitting the technology through trial and error processes, will perhaps hold the key in maturing the research on microfluidics towards achieving this important objective.

References

1. Whitesides GM (2006) The origins and the future of microfluidics. Nature 442:368–373.
2. Batchelor GK (2000) An Introduction to Fluid Dynamics, Cambridge Mathematical Library, London.
3. Maxwell JC (1879) On stresses in rarefied gases arising from temperature inequalities. Phil. Trans. R. Soc. Part 1 170:231.
4. Karniadakis GE, Beskok A, and Aluru N. (2005) Microflows – Fundamentals and Simulations. Springer-Verlag, New York.
5. Brenner H (2005) Navier–Stokes revisited. Physica A 349:60–132.
6. Brenner H (2006) Fluid mechanics revisited. Physica A 370:190–224.
7. Brenner H and Bielenberg JR (2005) A continuum approach to phoretic motions: Thermophoresis. Physica A 355:251–273.
8. Chakraborty S and Durst F (2007) Derivations of extended Navier–Stokes equations from molecular transport considerations for compressible ideal gas flows: Towards extended constitutive forms. Phys. Fluids 19:088104.
9. Kennard EH (1938) Kinetic theory of gases with an introduction to statistical mechanics. McGraw-Hill, New York.
10. Thompson PA and Troian SM (1997) A general boundary condition for liquid flow at solid surfaces. Nature 389:360–362.
11. Peiyi W and Little WA (1983) Measurement of friction factors for the flow of gases in very fine channels used for microminiature Joule-Thomson refrigerators. Cryogenics 23:273–277.
12. Wu PY and Little WA (1984) Measurement of the heat transfer characteristics of gas flow in fine channel heat exchangers used for microminiature refrigerators. Cryogenics 24:415–420.
13. Yang CY and Webb RL (1996) Condensation of R-12 in small hydraulic diameter extruded aluminum tubes with and without micro-fins. Int. J. Heat Mass Transfer 39:791–800.
14. Qu W, Mala GM, and Li D (2000) Pressure-driven water flows in trapezoidal silicon microchannels. Int. J. Heat Mass Transfer 43:353–364.
15. Mala GM and Li D (1999) Flow characteristics of water in microtubes. Int. J. Heat Fluid Flow 20:142–148.
16. Kleinstreuer C and Koo J (2004) Computational analysis of wall roughness effects for liquid flows in micro-conduits. J. Fluids Eng. 126:1–13.
17. Wang BX and Peng XF (1994) Experimental investigation on liquid forced-convection heat transfer through microchannels. Int. J. Heat Mass Tran. 37:73–82.
18. Peng XF and Peterson GP (1996) Convective heat transfer and flow friction for water flow in microchannel structures. Int. J. Heat Mass Tran. 39:2599–2688.
19. Zhu Y and Granick S (2002) Limits of the hydrodynamic no-slip boundary condition. Phys. Rev. Lett. 88:106102–106104.
20. Majumder M, Chopra N, Andrews R, and Hinds BJ (2005) Nanoscale hydrodynamics: Enhanced flow in carbon nanotubes. Nature 438:4.

21. Jirage KB, Hulteen JC, and Martin CR (1997) Nanotubule-Based Molecular-Filtration Membranes. Science 278:655–658.
22. Hummer G, Rasaih JC, and Noworyta JP (2001) Water conduction through the hydrophobic channel of a carbon nanotube. Nature 414:188–190.
23. Sokhan VP, Nicholson D, and Quirke N (2002) Fluid flow in nanopores: Accurate boundary conditions for carbon nanotubes. J. Chem. Phys. 117:8531–8539.
24. Skoulidas AI, Ackerman DM, Johnson JK, and Sholl DS (2002) Rapid transport of gases in carbon nanotubes. Phys. Rev. Lett. 89:185901.
25. Chakraborty S (2007) Order parameter modelling of fluid dynamics in narrow confinements subjected to hydrophobic interactions. Phys. Rev. Lett. 99:094504–094504.
26. Chakraborty S (2007) Towards a generalized representation of surface effects on pressure-driven liquid flow in microchannels. Appl. Phys. Lett. 90:034108-3.
27. Chakraborty S, Das T, and Chattoraj S (2007) A generalized model for probing frictional characteristics of pressure-driven liquid microflows. J. Appl. Phys. 102: 104907–104911.
28. Steinberger A (2007) High friction on a bubble mattress. Nat. Mat. 6:665–668.
29. Chakraborty S and Anand KD (2008) Implications of hydrophobic interactions and consequent apparent slip phenomenon on the entrance region transport of liquids through microchannels. Phys. Fluids 20:043602-9.
30. Fox RW, McDonald AT, and Pritchard PJ (2000) Introduction to Fluid Mechanics, John Wiley & Sons Inc., New York.
31. White FM (2005) Viscous Fluid Flow, McGraw Hill Education, New York.
32. Li D (2008) Encyclopedia of Microfluidics and Nanofluidics, Springer, New York.
33. Yih CH (1995) Kinetic-energy mass, momentum mass, and drift mass in steady irrotational subsonic flow. J. Fluid Mech. 297:29–36.
34. Huang W, Bhullar RS, and Fung YC (2001) The surface-tension-driven flow of blood from a droplet into a capillary tube. ASME J. Biomed. Eng. 123:446–454.
35. Lew HS and Fung YC (1970) Entry flow into blood vessels at arbitrary Reynolds number. J. Biomech. 3:23–38.
36. Kalliadasis S and Chang HC (1994) Apparent dynamic contact angle of an advancing gas-liquid meniscus. Phys. Fluids 6:12–23.
37. Hoffman RL (1975) A study of the advancing interface I. Interface shape in liquid–gas systems. J. Colloid Interface Sci. 50:228–235.
38. Cox RJ (1986) The dynamics of the spreading of liquids on a solid surface. Part 1. Viscous flow. J. Fluid Mech. 168:169–194.
39. Mazouchi A and Homsy GM (2000) Thermocapillary migration of long bubbles in cylindrical capillary tubes. Phys. Fluids 12:542–549.
40. Moorthy J, Khouri C, Moore JS, and Beebe DJ (2002) Active control of electroosmotic flow in microchannels using light. J. Sensors Actuators A 75:223–229.
41. Liu GL, Kim J, Lu Y, and Lee LP (2006) Optofluidic Control via Photothermal Nanoparticles. Nature Mat. 5:27–32.
42. Madou M, Zoval J, Jia G, Kido H, Kim J, and Kim N (2006) Lab on a CD. Annu. Rev. Biomed. Eng. 8:601–628.
43. Ducree J, Haeberle S, Lutz S, Pausch S, Stetten FV, and Zengerle R (2007) The centrifugal microfluidic Bio-Disk platform. J. Micromech. Microeng. 17:S103–S115.
44. Chakraborty D, Gorkin R, Madou M, Kulinsky L, and Chakraborty S (2009) Capillary filling in centrifugally actuated microfluidic devices with dynamically evolving contact line motion. J. Appl. Phys. 105:084904–084910.
45. Li D (2004) Electrokinetics in Microfluidics. Elsevier, Amsterdam.
46. Israelachvili JN (1992) Intermolecular and Surface Forces. Academic Press, New York.
47. van der Heyden FH, Stein D, and Dekker C (2005) Streaming Currents in a Single Nanofluidic Channel. Phys. Rev. Lett. 95:116104.

48. Rajaramana S, Noha H, Hesketha PJ, and Gottfriedb DS (2006) Rapid, low cost microfabrication technologies toward realization of devices for dielectrophoretic manipulation of particles and nanowires. Sens. Actuators B Chem. 114:392–401.
49. Hunter RJ (1981) Zeta Potential in Colloid Science, Academic Press, London.
50. van der Heyden FHJ, Stein D, and Dekker C (2005) Streaming currents in a single nanofluidic channel. Phys. Rev. Lett. 95:116104.
51. Kilic MS, Bazant MZ, and Ajdari A (2007) Steric effects in the dynamics of electrolytes at large applied voltages. I. Double-layer charging. Phys. Rev. E 75:021503–16.
52. Schuss Z, Nadler B, Singer A, and Eisenberg RS (2001) Derivation of Poisson and Nernst-Planck equations in a bath and channel from a molecular model. Phys. Rev. E 64:036116–036114.
53. Nadler B, Schuss Z, Singer A, and Eisenberg RS (2004) Ionic diffusion through confined geometries: from Langevin equations to partial differential equations. J. Phys. Condens. Matter 16:S2153–S2165.
54. Levine S and Outhwaite CW (1978) Comparison of theories of the aqueous electric double layer at a charged plane interface. J. Chem. Soc. Faraday Trans. 2, 74:1670–1689.
55. Outhwaite CW, Bhuiyan LB, and Levine S (1980) Theory of the electric double layer using a modified Poisson–Boltzmann equation. J. Chem. Soc. Faraday Trans. 2, 76:1388–1408.
56. Bhuiyan LB, Outhwaite CW, and Henderson D (2005) A modified Poisson–Boltzmann analysis of the capacitance behaviour of the electric double layer at low temperatures. J. Chem. Phys. 123:034704-5.
57. Tang Z, Scriven LE, and Davis HT (1992) Structure of a dipolar hard sphere fluid near a neutral hard wall. J. Chem. Phys. 96:4639–4645.
58. Tang Z, Scriven LE, and Davis HT (1992) A three-component model of the electrical double layer. J. Chem. Phys. 97:494–503.
59. Tang Z, Scriven LE, and Davis HT (1994) Effects of solvent exclusion on the force between charged surfaces in electrolyte solution. J. Chem. Phys. 100:4527–4530.
60. Zhang L, Davis HT, and White HS (1993) Simulations of solvent effects on confined electrolytes. J. Chem. Phys. 98:5793–5799.
61. Dutta P and Beskok A (2001) Analytical solution of time periodic electroosmotic flows: Analogies to stokes' second problem. Anal. Chem. 73:5097–5102.
62. Knecht V, Risselada HJ, Mark AE, and Marrink SJ (2008) Electrophoretic mobility does not always reflect the charge on an oil droplet. J. Colloid Interface Sci. 318:477–486.
63. Probstein RF (2003) Physicochemical Hydrodynamics–An introduction. Willey Inter-Science.
64. Pohl HA (1978) Dielectrophoresis-The behavior of neutral mater in non-uniform electric fields, Cambridge University Press, Cambridge.
65. Chakraborty S and Das S (2008) Streaming-field-induced convective transport and its influence on the electroviscous effects in narrow fluidic confinement beyond the Debye–Hückel limit. Phys. Rev. E 77:037303.
66. Levine S, Marriott JR, Neale G, and Epstein N (1975) Theory of electrokinetic flow in fine cylindrical capillaries at high zeta-potentials. J. Coll. Inter. Sci. 52:136–149.
67. Yang C and Li D (1997) Electrokinetic effects on pressure-driven liquid flows in rectangular microchannels. J. Coll. Inter. Sci. 194:95–107.
68. Ren CL and Li D. (2004) Electroviscous effects on pressure-driven flow of dilute electrolyte solutions in small microchannels. J. Coll. Inter. Sci. 274:319–330.
69. Li D (2001) Electro-viscous effects on pressure-driven liquid flow in microchannels. Phys. Eng. Asp. 195:35–57.
70. Vainshtein P and Gutfinger C (2002) On electroviscous effects in microchannels. J. Micromech. Microeng. 12:252–256.
71. Chun MS, Lee TS, and Choi NW (2005) Microfluidic analysis of electrokinetic streaming potential induced by microflows of monovalent electrolyte solution. J. Micromech. Microeng. 15:710–719.

72. Brunet E and Ajdari A (2007) Stability and transient dynamics of thin liquid films flowing over locally heated surfaces. Phys. Rev. E 76:056306.
73. Stratton JA (1941) Electromagnetic Theory. McGrawHill, New York.
74. Ramos A, Morgan H, Green NG, and Castellanos A (1998) AC electrokinetics: a review of forces in microelectrode structures. J. Phys. D 31:2338–2353.
75. Green NG, Ramos A, González A, castellanos A, and Morgan H (2000) Electric field induced fluid flow on microelectrodes: the effect of illumination. J. Phys. D 33:L13–L17.
76. Sinton D, Xuan X, and Li D (2004) Thermally induced velocity gradients in electroosmotic microchannel flows: the cooling influence of optical infrastructure. Exp. Fluid. 37:872–882.
77. Li L and Uttamchandani D (2004) Modified asymmetric micro-electrothermal actuator: analysis and experimentation. J. Micromech. Microeng. 14:1734–1741.
78. Chakraborty S and Pal D (2006) Microchannel flow control through a combined magneto-electro-hydrodynamic transport. J. Phys. D 39:5364–5371.
79. Jang J and Lee SS (2000) Theoretical and experimental study of MHD (magnetohydrodynamic) micropump. Sens. Actuators A. 80:84–89
80. Riley N (2001) Steady streaming. Ann. Rev. Fluid Mech. 33:43–65.
81. Lighthill J (1978) Waves in Fluids. Cambridge University Press, Cambridge.
82. Schlichting H and Gersten K (2003) Boundary Layer Theory. Springer, Berlin.
83. Rosengarten G, Harrie D, and Cooper-White J (2005) Using computational fluid dynamics to study the effect of contact angle on microdroplet deformation. ANZIAM J. 46:C304–C319.
84. Link DR, Anna SL, Weitz DA, and Stone HA (2004). Geometrically mediated breakup of drops in microfluidic devices. Phys. Rev. Lett. 92:054503(1–4).
85. Schleizer A and Bonnecaze RT (1999) Dispacement of a two-dimensional immiscible droplet adhering to the wall in shear and pressure-driven flows. J. Fluid Mech. 383:29–54.

Chapter 2
Mathematical Modeling of Electrokinetic Effects in Micro and Nano Fluidics

Sandip Ghosal

Abstract In engineering applications familiar from everyday experience fluid flow is almost always pressure driven. The scaling law for the pressure head needed to drive a fixed flux through a circular capillary, is according to the celebrated Poiseuille formula, inversely proportional to the fourth power of capillary diameter. Thus, when it comes to small scales: microns and below; the electrokinetic method of transporting fluids appear increasingly attractive. Here the Voltage drop needed to maintain a flux increases inversely as only the second power of the capillary diameter. In this brief introduction to the subject of electrokinetic flows, the foundations are developed assuming that the reader has only minimal prior knowledge in the area. First, the laws of incompressible hydrodynamics are summarized followed by a review of electrostatics. These two streams are then interwoven with the laws of ionic transport to explore electroosmotic flow and its effect on transport of solutes. The focus of this exposition is on practical applications in microfluidic systems such as capillary electrophoresis.

Keywords Electroosmosis · Electrophoresis · Debye layer · Reynolds number · Stokes flow · Gauss law · Nerst-Planck equation · Poisson-Boltzmann equation · Gouy-Chapman model · Debye-Huckel approximation · Lubrication approximation

2.1 Introduction and Historical Overview

Electroosmotic flow (EOF) or Electroosmosis was first reported by F.F. Reuss in 1809 in a paper entitled 'Sur un nouvel effet de lé électricité galvanique' that appeared in the *Proceedings of the Imperial Society of Naturalists of Moscow* [25].

S. Ghosal (✉)
Department of Mechanical Engineering, Northwestern University, Evanston, IL, USA
e-mail: s-ghosal@northwestern.edu

S. Chakraborty (ed.), *Microfluidics and Microfabrication*,
DOI 10.1007/978-1-4419-1543-6_2, © Springer Science+Business Media, LLC 2010

In that paper, Reuss demonstrated that water could be made to percolate through porous clay diaphragms by the application of an electric field. Notwithstanding the name, the mechanism behind the phenomena is unrelated to osmosis, which is, the flow of water across a semipermeable membrane driven by a solute concentration gradient. Instead, the mechanism behind this observed mobility of water is as follows. Particles of clay and other silicate materials such as glass or silicon, acquire a surface charge when in contact with an electrolyte such as water due to chemical dissociation of surface ionic groups. The amount of this surface charge is determined by various factors but a typical value for silica–water interfaces is in the range of -4 to -60 milli-Coulomb (mC) per square meter. These fixed charges on the substrate attract the free charges in solution of unlike sign and repel those of like sign (the free charges in solution are ionic dissociation products of the water molecule itself as well as salt ions that even relatively pure water contains in copious amounts). This results in the formation of a thin (1–10 nm is typical) charged region in the solute next to the substrate boundary known as the Debye layer. In the presence of an external electric field, the fluid in this Debye layer experiences a body force thereby acquiring momentum which is then transmitted to adjacent layers of fluid due to viscosity. Obviously the effect causes relative motion between solute and substrate, this may mean a resultant liquid flow (if the solid phase is immobile) or particle transport (if the liquid phase is immobile) or the motion of both phases. Figure 2.1 illustrates the mechanism. A number of physical effects are closely related to EOF and are collectively known as 'electrokinetic effects'. Some of the important ones are: electrophoresis, streaming potential, sedimentation potential, the electroviscous effect, the seismoelectric effect, and dielectrophoresis. The reader may refer to a standard text on physico-chemical fluids [22] for an elementary discussion or to more specialized papers and monographs for detailed investigations related to these phenomena. One of the earliest applications of EOF is in Civil Engineering as a method for drying soil. The application of a strong electric field to a porous media such as clay drives out water by electroosmosis. The electrically driven flow of water can also be used to leach out contaminants in the soil in land reclamation projects [24] or in the desalination of salt water [23]. These early applications have now been completely eclipsed in importance by modern applications in microfluidics.

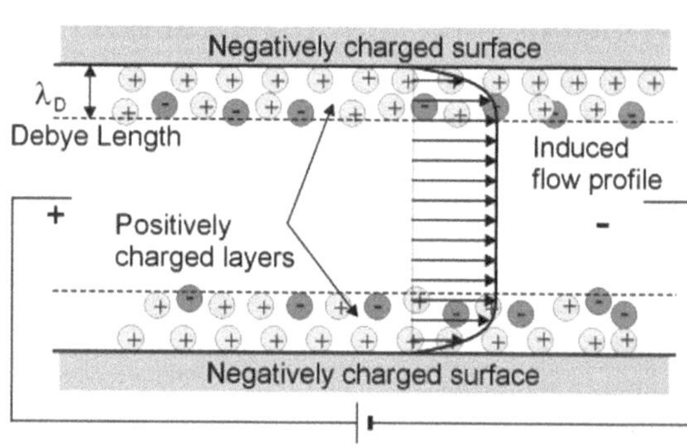

Fig. 2.1 A sketch showing the Debye layers formed at the walls of a parallel channel and the resultant electroosmotic flow in response to an applied Voltage (Image: courtesy of Prof. H. Bruus of the Technical University of Denmark)

Microfluidics is the science of manipulating fluids on spatial scales anywhere between one to a hundred micron. Thus, microfluidics involves engineering structures for manipulating fluids on scales that are microscopic in comparison to human dimensions but that are nevertheless much larger than atomic dimensions so that the systems can still be treated in the continuum approximation. The primary driving force behind the development of microfluidics has come from Molecular Biology and the Life Sciences where progress has become dependent upon the ability to perform analytical chemistry at heretofore unprecedented speeds. According to some studies, the global market for microfluidic based technologies was at 2.9 billion dollars in 1995 and expected to grow at an average annual rate of 14.5% to a projected 6.2 billion dollars in 2011. The distinction between the words 'microfluidics' and 'nanofluidics' is not always very clear. As an operational definition we will use nanofluidics to refer to flow through nanochannels: a conduit between two reservoirs of fluid with a characteristic internal diameter of roughly one to several tens of nanometers. The phrase 'one to several tens of nanometers' in the above operational definition may seem both arbitrary and vague, but it is not so. Since the subject of discussion is the flow of water through narrow channels, they must be wide enough to allow the passage of a molecule of water, which is about 0.1 nm. On the other hand, if they are much wider than this size, let us say more than a 100 nm or 0.1 μm we are in the domain of microfluidics, where water can safely be regarded as a continuous material as we do in everyday engineering practice. Thus, the specification of 'one to several tens' derives from the scale provided by the water molecule itself, that is, the subject of discussion are channels that are neither too narrow nor too wide in comparison to the size of a water molecule. These definitions however are based on common usage and are necessarily imprecise; there is no exact boundary as to where nanofluidics ends and microfluidics begins! Problems involving nanochannels usually bring in either or both of the following characteristics compared to microchannels (a) the Debye layer thickness becomes comparable to the channel diameter (b) single molecule effects become important, that is, the continuum approximation can no longer be applied with confidence. In this article, though we will on occasion refer to the situation (a), we will not discuss at all effects related to (b). Thus, the mathematical modeling described here apply primarily to microchannels and only to a limited extent to nano channels.

Before concluding this introductory discussion, it is worthwhile to point out that though man made microfluidic and nanofluidic devices some involving electrokinetic effects are of recent vintage, such devices are common place in the world of living things. For example, the narrowest capillaries in the human circulatory system are of the order of 5–10 μm and the diameter of a swimming bacteria may be about 10μm. The African bombardier beetle (*Stenaptinus insignis*) is armed with a spectacular microfluidic weapon system [11]. Its abdomen contains a reaction chamber where a propellant (hydroquinone) and an oxidizer (hydrogen peroxide) from separate storage organs are mixed in the presence of catalysts. The result of the highly exothermic reaction is a boiling mixture of steam and corrosive liquid which the beetle is able to deliver on target through a fully steerable micro-nozzle and an accompanying loud bang! Less spectacular but no less amazing applications of microfluidic technologies are fluid flow in the proboscis of insects such as butterflies and mosquitos, the flagellar propulsion of bacteria such as *Ecoli*, propulsion by a single waving flagellum in spermatozoa, the use of cillia by numerous small animals for moving fluids along and the innovative use of surface tension forces by insects and larva that live on the air water interface in stagnant pools [9]. Nanochannels and ion transport across nanochannels is a very common feature in living cells [1]. For example, nerve cells contain channels that are selectively permeable to Sodium, Pottasium or Calcium ions the collective action of which is responsible for neuronal activity in all living things. Another nanochannel in living systems help regulate the water content of cells; these are called aquaporins. They play an important role for example in the functioning of the kidney.

In the following sections we discuss the mathematical modeling of some selected phenomena in micro and nanofluidics pertaining to electrokinetic effects. The choice of material is obviously biased by the author's own research expertise and do not necessarily correspond to the relative importance of the subject matter.

2.2 Review of Underlying Physical Principles

Electrokinetic effects involve the movement of ions in response to an electric field as well as due to molecular diffusion and physical or bulk transport by the movement of water. Thus, our subject lies at the confluence of three physical phenomena (a) Fluid mechanics (b) Electrostatics (c) Ion Transport. In this section we briefly review the relevant part of each of these subjects.

2.2.1 Fluid Mechanics

In the applications considered here, relevant physical dimensions are sufficiently large in comparison to atomic scales that it is permissible to treat the fluid as if it were a continuum. Thus, the fluid velocity $\mathbf{u}$ and pressure p are regarded as continuous functions of position $\mathbf{x}$ and time t, and they obey the incompressible Navier-Stokes equations with a volume density of external forces $\mathbf{f}_e$

$$\rho_0(\partial_t \mathbf{u} + \mathbf{u} \cdot \nabla \mathbf{u}) = -\nabla p + \mu \nabla^2 \mathbf{u} + \mathbf{f}_e. \tag{2.1}$$

This is supplemented by the continuity equation which takes into account the fact that in a liquid the density changes are slight, even for large changes in pressure:

$$\nabla \cdot \mathbf{u} = 0. \tag{2.2}$$

In the above, ρ_0 and μ are the (constant) density and viscosity of the fluid, p is the pressure and $\mathbf{u}$ is the flow velocity.

The relative size of the term on the left of Eq. (2.1) (due to fluid inertia) and the second term on the right (due to viscosity) is characterized by the Reynolds number

$$\mathrm{Re} = \frac{UL\rho_0}{\mu} \tag{2.3}$$

where U and L denote a characteristic velocity and length for the flow. In most applications of microfluidics, $\mathrm{Re} \ll 1$. In some applications, $\mathrm{Re} \sim 1$. By contrast, in large scale flows (aircraft engines, geophysical flows etc.) as a rule, $\mathrm{Re} \gg 1$. Because of the smallness of the Reynolds number, Re, in microfluidics, most well known flow instabilities leading to period doubling, chaos and finally turbulence are absent. Furthermore, the left hand side of Eq. (2.1) which corresponds to fluid inertia can either be neglected, or treated as a small perturbation. In the former case, we arrive at the Stoke's equation:

$$-\nabla p + \mu \nabla^2 \mathbf{u} + \mathbf{f}_e = 0 \tag{2.4}$$

which is often said to describe 'slow', 'creeping' or 'highly viscous' flow. All of these terms mean the same thing, namely $\mathrm{Re} = (UL\rho_0)/\mu \ll 1$. The unknown scalar field p in Eq. (2.4) is determined by the constraint provided by Eq. (2.2).

When solving Eq. (2.1) or (2.4) in a domain with boundaries, 'boundary conditions' must be specified. If the boundaries are rigid, the 'no-slip' boundary conditions are used:

$$\mathbf{u}_{fluid}(P) = \mathbf{u}_{solid}(P), \tag{2.5}$$

that is, at any point P on the solid–fluid interface, the velocity of the fluid must match the velocity of the solid at that point.

2.2.2 Electrostatics

The electric field $\mathbf{E}$ (the electric force experienced by a unit charge) in the neighborhood of a fixed distribution of charges is determined by the equations

$$\nabla \times \mathbf{E} = 0 \tag{2.6}$$

$$\kappa \epsilon_0 \nabla \cdot \mathbf{E} = \rho_e \tag{2.7}$$

where ρ_e is the density of electric charges, κ is the (dimensionless) dielectric constant of the medium and $\epsilon_0 = 1/(4\pi)$ in CGS units and 8.854×10^{-12} Farads per meter in SI units. The first of these equations indicate that the electrical force is a conservative force and the second is the differential form of Gauss's law of electrostatics. Using the familiar transformations of vector analysis, these equations may be equivalently expressed in integral form:

$$\int_C \mathbf{E} \cdot d\mathbf{s} = 0 \tag{2.8}$$

$$\kappa\epsilon_0 \int_S \mathbf{E} \cdot d\mathbf{S} = \int_V \rho_e \, dV \tag{2.9}$$

where C is an arbitrary closed contour and S is a simple closed surface enclosing a volume V. The familiar 'Coulomb's Law' of electrostatics is contained in this more general formulation. To show this, consider an isolated charge Q. Due to symmetry, the electric field must be in the radial direction and its magnitude $E(r)$ is a function of a single variable, r, the radial distance from the charge. Applying Gauss's law in its integral form, Eq. (2.9) to a spherical surface of radius 'r' centered around the charge,

$$\kappa\epsilon_0(4\pi r^2)E(r) = Q \tag{2.10}$$

so that

$$E(r) = \frac{Q}{4\pi\kappa\epsilon_0 r^2} \tag{2.11}$$

Therefore, the force on a second charge q placed at a distance r from the first charge is

$$F = qE = \frac{Qq}{4\pi\kappa\epsilon_0 r^2} \tag{2.12}$$

along the line joining them, which is Coulomb's law.

Since the electric field $\mathbf{E}$ is irrotational, it admits a potential ϕ called the electric potential defined by

$$\mathbf{E} = -\nabla\phi. \tag{2.13}$$

If this is substituted in Eq. (2.7) we get

$$\kappa\epsilon_0\nabla^2\phi = -\rho_e \tag{2.14}$$

which is known as the Poisson's equation of electrostatics.

Equations (2.6) and (2.7) are a reduced form of the more general Maxwell's equations describing electromagnetic phenomena when electric charges are stationary. These equations remain approximately true even when charges are in motion, provided the characteristic velocity of the charges, $\upsilon \ll c$ where c is the speed of light in the medium. In the applications of concern to us, ions move about primarily due to thermal fluctuations, the characteristic speed of which may be estimated as $\upsilon \sim \sqrt{3k_B T/m}$ where k_B is Boltzman's constant, T is the absolute temperature and m is the ionic mass. At room temperature, for a potassium ion, we have

$\upsilon \sim 7.1 \times 10^2$ m/s which is six orders of magnitude less than the speed of light in water, $c \approx 2.3 \times 10^8$ m/s.

2.2.3 Ion Transport in Solvents

If the electrolyte contains N species of ions with charges ez_k (e is the magnitude of the electronic charge) and concentration n_k ($k = 1, \ldots, N$) then $\rho_e = \sum_{k=1}^{N} ez_k n_k$. Each ion species obeys a conservation equation

$$\frac{\partial n_k}{\partial t} + \nabla \cdot \mathbf{j}_k = 0. \tag{2.15}$$

Here $\mathbf{j}_k$, the flux vector for the species k can be modeled by the Nernst-Planck equation for ion transport [19]

$$\mathbf{j}_k = -\nu_k z_k e n_k \nabla \phi - D_k \nabla n_k + n_k \mathbf{u}. \tag{2.16}$$

In Eq. (2.16) ν_k is the ion mobility: the velocity acquired by the ion when acted upon by a unit of external force. The diffusivity of the kth species is D_k and $\mathbf{u}$ is the fluid velocity.

The boundary conditions are those of no slip at the wall for the velocity, Eq. (2.5) and no ion flux normal to the wall $\mathbf{j}_k \cdot \hat{\mathbf{n}} = 0$ ($\hat{\mathbf{n}}$ is the unit normal directed into the fluid). In the absence of external electric fields, the chemistry at the electrolyte substrate interface leads to the establishment of a potential, $\phi = \zeta$. This so called ζ-potential at an interface depends on a number of factors including the nature of the substrate and ionic composition of the electrolyte, the presence of impurities, the temperature and the buffer pH. Methods of determining the ζ-potential and measured values for a wide variety of surfaces used in microfluidic technology have been reviewed in [17] and [18].

The ion distribution near a planar wall at $z = 0$ with potential $\phi(z)$ is known from statistical thermodynamics: $n_k = n_k(\infty) \exp(-z_k e\phi/k_B T)$ where k_B is the Boltzmann constant and T is the absolute temperature of the solution. In order that this expression be a steady solution of equations (2.15) we must have the Einstein relation $D_k/\nu_k = k_B T$. Therefore Eq. (2.16) can also be written as

$$\mathbf{j}_k = -n_k \nu_k \nabla \psi_k + n_k \mathbf{u} \tag{2.17}$$

where $\psi_k = ez_k \phi + k_B T \ln n_k$ is called the chemical potential for the species k.

2.3 Structure of the Equilibrium Debye Layer

Suppose that the system is in the steady state and that there is no fluid flow or imposed electric fields. Further suppose that the geometry is such that the electrolyte–substrate interface is an iso-surface of ψ_k (the simple semi-infinite half-plane is just an example of such a geometry). Then it readily follows from Eqs. (2.15) and (2.17) and the boundary condition of no flux into the wall that $\nabla\psi_k = 0$ everywhere. Therefore, $n_k = n_k^{(\infty)} \exp(-z_k e\phi/k_B T)$ where $n_k^{(\infty)}$ is the ion concentration where the potential $\phi = 0$; usually chosen as a point very far from the wall. Using the solution for n_k in the charge density ρ_e we get the non-linear Poisson-Boltzmann equation for determining the potential

$$\nabla^2\phi = -\frac{e}{\kappa\epsilon_0}\sum_{k=1}^{N} n_k^{(\infty)} z_k \exp(-z_k e\phi/k_B T)\,. \tag{2.18}$$

with the boundary condition $\phi = \zeta$ on walls. Equation (2.18) was the starting point of a detailed investigation of the structure of the Electric Double Layer (EDL) or 'Debye Layer' by Gouy and Chapman. The description in terms of Eq. (2.18) is therefore known as the Gouy-Chapman model of the EDL.

Equation (2.18) is a nonlinear equation. It can be linearized by expanding the exponential terms on the right hand side in Taylor series and discarding all terms that are quadratic or of higher order in ϕ, which gives

$$\nabla^2\phi - \frac{\phi}{\lambda_D^2} = 0 \tag{2.19}$$

where

$$\lambda_D^{-1} = \left[\sum_{k=1}^{N} \frac{z_k^2 e^2 n_k^{(\infty)}}{\varepsilon_0 \kappa k_B T}\right]^{1/2} \tag{2.20}$$

is a constant determined by the ionic composition of the electrolyte. In arriving at (2.19) we used the condition $\sum_{k=1}^{N} n_k^{(\infty)} z_k = 0$ which expresses the fact that the bulk solution ($\phi = 0$) is free of net charge. It is easily verified that λ_D has units of length. It is called the Debye-Length, and the linearization of (2.18) that led to (2.19) is known as the Debye-Hückel approximation. The solution to (2.19) near a charged plate of potential ζ may be written as $\phi(z) = \zeta \exp(-z/\lambda_D)$ where z is distance normal to the plate. Thus, the potential due to the charged plate is shielded by the free charges in solution and the effect of the charge penetrates a distance of the order of the Debye-length λ_D; which gives a physical meaning to this very important quantity. For 'ordinary' water the Debye length is typically 1–10 nm. The linearization proposed by Debye-Hückel is justified provided that $|z_k\phi| \ll k_B T/e$ uniformly in all space and for all k. At room temperature $k_B T/e \approx 30\,mV$. However, for silica substrates $|\zeta| \sim 50$–$100\,mV$ in typical applications. Thus, the Debye-Hückel approximation is often not strictly valid. Nevertheless, it is a very useful approximation because it enormously simplifies mathematical investigations related to the Debye layer.

Equation (2.19) is linear and can be solved analytically under fairly general conditions. It appears in several other contexts such as in the solution of the diffusion equation and in wave propagation problems with evanescent waves.

Exact solutions to (2.18) are known only for symmetric electrolytes (that is, $N = 2, z_1 = -z_2 = Z$ where Z is positive) in certain special geometries.

2.3.1 Half Plane

In front of an infinite charged flat plate (defined as $z = 0$) charged to a fixed ζ potential, the solution to Eq. (2.18) is

$$\tanh\left[\frac{Ze\phi}{4k_BT}\right] = \tanh\left[\frac{Ze\zeta}{4k_BT}\right]\exp(-z/\lambda_D) \tag{2.21}$$

where $z > 0$. This solution can also be used to describe the Debye layers at the walls of a planar channel provided that the channel walls are sufficiently far apart that their Debye layers do not overlap. Clearly, if ϕ and ζ are both sufficiently small that the hyperbolic tangent terms can be approximated by their respective one term Taylor expansion, $\tanh x = x + \ldots$, then the Debye-Hückel theory $\phi(z) = \zeta\exp(-z/\lambda_D)$ is recovered.

2.3.2 Between Parallel Plates

An exact solution of Eq. (2.18) can be written [4] for a pair of plates separated by a distance $2H$ and each held at a potential ζ. The origin $z = 0$ is chosen on one of the plates so that the second plate is located at $z = 2H$. It is convenient to introduce the intermediate variables $k = \exp(-Ze\zeta/k_BT)$ and Φ defined by $\sin\Phi = k^{-1/2}\exp[-Ze\phi/(2k_BT)]$. Then the solution may be expressed in terms of the elliptic integral $F(\Phi, k)$ defined as

$$F(\Phi, k) = \int_0^{\Phi} \frac{d\theta}{(1 - k^2\sin^2\theta)^{1/2}}. \tag{2.22}$$

This solution (originally due to Langmuir) is

$$z/\lambda_D = 2k^{1/2}[F(\pi/2, k) - F(\Phi, k)]. \tag{2.23}$$

For small potentials (that is, in the Debye-Hückel limit) this can be shown to reduce to

$$\phi(z) = \zeta\frac{\cosh[(H - z)/\lambda_D]}{\cosh(H/\lambda_D)} \tag{2.24}$$

2.3.3 Circular Cylinders

An analytical solution for the region inside infinitely long circular cylinders is available only for Eq. (2.19) and is due to Rice and Whitehead [26]. The potential is

$$\phi(r) = \zeta \frac{I_0(r/\lambda_D)}{I_0(a/\lambda_D)} \tag{2.25}$$

where r is the radial distance, a is the radius of the cylinder and I_0 is the zeroth order modified Bessel function of the first kind.

2.4 Electroosmosis

In the presence of external fields and fluid flow the equilibrium Gouy-Chapman model is generally not applicable and one must proceed from the full electrokinetic equations presented earlier. However, if the external field and fluid velocity are both along the iso-surfaces of the charge density ρ_e then the presence of the flow or the imposed field does not alter the charge density distribution which may still be obtained from the Gouy-Chapman model. Examples where such a situation holds would be

1. A planar uniformly charged substrate at $z = 0$ with an applied electric field E_0 that is tangential to the surface (the x-direction).
2. A uniform infinite cylindrical capillary with an imposed electric field E_0 along the axis (the x-direction).
3. A narrow slit with uniformly charged walls and an imposed constant electric field E_0 along the slit (the x-direction).
4. A thin Debye layer adjacent to an electrically insulating solid; here the equipotential surfaces and iso-surfaces of charge density are almost orthogonal.

For any of the above geometries, the fluid flow equations reduce to (assuming steady state and zero imposed pressure gradient)

$$\mu \nabla^2 u + \rho_e E_0 = \mu \nabla^2 u - \varepsilon_0 \kappa E_0 \nabla^2 \phi^{(EDL)} = 0 \tag{2.26}$$

where u is the axial velocity and $\phi^{(EDL)}$ is the electric potential for the equilibrium problem, that is, without the external field or flow. Therefore,

$$u = \frac{\varepsilon_0 \kappa E_0}{\mu} \phi^{(EDL)} + \chi \tag{2.27}$$

where χ satisfies

$$\nabla^2 \chi = 0 \tag{2.28}$$

and $\chi = -\varepsilon_0 \kappa E_0 \zeta / \mu$ at the boundaries in order to satisfy the 'no-slip' boundary condition. For a homogeneous channel, the quantity $(\varepsilon_0 \kappa E_0 \zeta)/\mu$ is a constant along the wall so that the solution to the Dirichlet problem for χ is

$$\chi = -\varepsilon_0 \kappa \zeta E_0 / \mu \tag{2.29}$$

Thus, the velocity is determined in terms of the potential distribution in the equilibrium EDL as:

$$u = \frac{\varepsilon_0 \kappa E_0}{\mu} \left[\phi^{(EDL)} - \zeta \right] \tag{2.30}$$

If we adopt the Debye-Hückel approximation then

$$\phi^{(EDL)} = \begin{cases} \zeta \exp\left(-z/\lambda_D\right) & \text{for (1) infinite plane} \\ \zeta I_0(r/\lambda_D)/I_0(a/\lambda_D) & \text{for (2) infinite cylindrical capillary} \\ \zeta \cosh\left(z/\lambda_D\right)/\cosh\left(b/\lambda_D\right) & \text{for (3) narrow slit} \end{cases} \tag{2.31}$$

where a is the capillary radius, r the distance from the axis and I_0 is the zero order modified Bessel function of the first kind. In the last formula, $2b$ is the channel width and z is the wall normal co-ordinate with origin on the plane (in Case 1) or origin at a point equidistant between the two walls (in Case 3). Since the fluid flow equation is linear in this limit, clearly a pressure driven flow can be added to the solution (superposition) in the event that both a pressure gradient and an electric field are simultaneously applied. The solution for an infinite capillary was first obtained by Rice and Whitehead [26]. Solutions in a narrow slit were obtained by Burgreen and Nakache [8] in the context of the Debye-Hückel approximation as well as for a 1:1 electrolyte (that is, in our notation $N = 2$ and $z_1 = -z_2$) directly from the full Poisson-Boltzmann equation.

2.5 Limit of Thin Electric Double Layers

Since the characteristic radius of microfluidic channels $\sim$ 10–100 μm, whereas, the Debye length $\lambda_D \sim 1-10$ nm, the thin Debye Layer approximation is usually an excellent one in microfluidic applications. It may fail to be valid in nanochannels with diameters of tens of nanometers or less. In the limit of thin EDL, the Navier-Stokes/Poisson-Boltzmann system described in the last section may be replaced by a simpler set of equations. The EDL then forms a very thin 'boundary layer' at the solid fluid interface where the electrical forces are confined.

At leading order, the dominant balance is between the viscous and the electrical forces in the boundary layer;

$$\mu \frac{\partial^2 u}{\partial x^2} + \rho_e E \cong 0 \tag{2.32}$$

where E is the external electric field which is in the tangential direction. Since rates of change across the boundary layer (z-axis) are much larger than along it (x-axis), Poisson's equation may be written as

$$\rho_e \cong -\varepsilon_0 \kappa \frac{\partial^2 \phi}{\partial z^2}. \tag{2.33}$$

On eliminating ρ_e between (2.32) and (2.33), integrating the resulting differential equation and using the boundary conditions at the inner and outer edges of the EDL, the following jump condition across the EDL is derived

$$\mathbf{u} - \mathbf{u}_{\text{solid}} \equiv \Delta \mathbf{u} = -\frac{\varepsilon_0 \kappa \mathbf{E} \zeta}{\mu}. \tag{2.34}$$

where in Eq. (2.34), $\mathbf{u}_{\text{solid}}$ is the velocity of the solid at a point on the solid-fluid interface and $\mathbf{u}$ is the velocity of the fluid at the correponding point, just above the (infinitely thin) Debye layer. A formal asymptotic derivation of the result in terms of the small parameter λ_D/a_0 (where a_0 is a characteristic radius) has been presented by Anderson [4]. Equation (2.34) is known as the 'Helmholtz-Smoluchowski (HS) slip boundary condition' after the pioneering work of Helmholtz [16] & Smoluchowski [28]. Thus, in the limit of thin Debye layers the term $-\rho_e \nabla \phi$ may be dropped from (2.1), instead, at the boundary, the 'no slip' boundary condition is replaced by (2.34). Since the external field $\mathbf{E}$ is tangential to the interface, (2.34) implies that the normal component of the velocity is continuous.

With this approximation, the equations of fluid flow inside the capillary become exactly identical to the classical fluid flow equations, the coupling to the electrical problem is only felt through the boundary condition Eq. (2.34). Figure 2.2 which shows the variation of the flow profile in a cylindrical capillary for different ratios of capillary radius to Debye length (a/λ_D) illustrates the concept of the thin EDL limit. It is seen that when a/λ_D is of order unity or smaller (as is typical in nanochannels) the flow profile is close to parabolic as in classical Poiseuille flow. However, when a/λ_D exceeds about 10 (as is often the case in micro-channels), the flow profile increasingly shown a 'boundary layer' type structure with the profile corresponding to a uniform flow or 'plug flow' over most of the channel cross-section and dropping to zero over a thin sheath like region next to the wall. The latter corresponds to the 'thin EDL' limit.

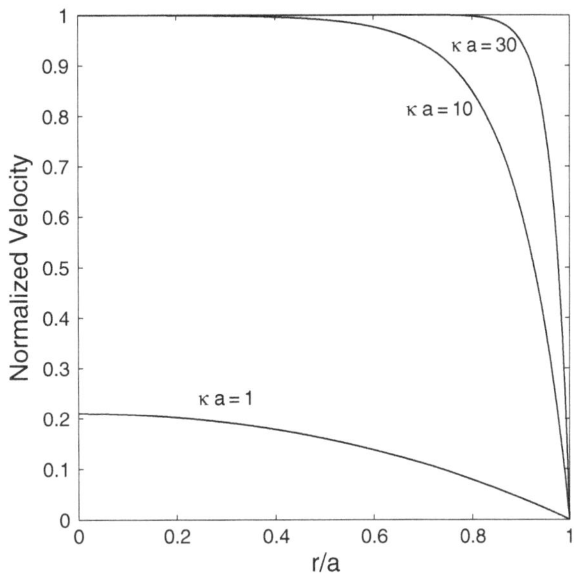

Fig. 2.2 Velocity profiles (normalized by the HS slip velocity) as a function of radial distance (r) in a cylindrical capillary according to the solution of [26] for $(a/\lambda_D) = 1,10,30$; here a is the capillary radius and λ_D is the Debye length. In the figure $\kappa = 1/\lambda_D$ (not the dielectric constant)

2.6 Axially Inhomogeneous Channels

Electrophoretic separation of macromolecules such as DNA, RNA and proteins is an indispensable tool in modern biology, health care and forensics. Due to the dissociation of molecular groups in the presence of water, a macro-molecule in aqueous solution often acquires a charge. If an external electric field is applied, the molecule migrates along the electric field lines with a velocity (v) that is proportional to the field strength (E). The constant of proportionality (μ_{ep}), which can have either sign, is called the electrophoretic mobility of the species, thus $v = \mu_{ep}E$. The separation of a mixture of chemical species into its components by taking advantage of the differences in their electrophoretic mobilities is known as electrophoresis. The modern implementation of electrophoresis is usually done through a platform known as 'Capillary Electrophoresis' a schematic diagram of which is shown in Fig. 2.3. A micro-capillary, usually made of fused silica is stretched between two reservoirs of relatively large capacity. Usually capillaries of 25–75 μm internal diameters and 10–100 cm length are used. The interior of the capillary as well as the two reservoirs are filled with an electrolyte (the buffer). The buffer serves to provide a conducting path for the electric current and also serves to preserve pH stability which is important for many biomolecules. The sample to be analyzed (the analyte) is introduced at one end of the capillary in the form of a plug, which travels down the capillary due to a combination of the electroosmotic flow generated in the capillary as well as due to its own electrophoretic mobility. The differences in electrophoretic mobility of the sample constituents cause them to separate and travel to a detector placed near the outlet in the form of distinct bands. Most often, the detector is of the UV

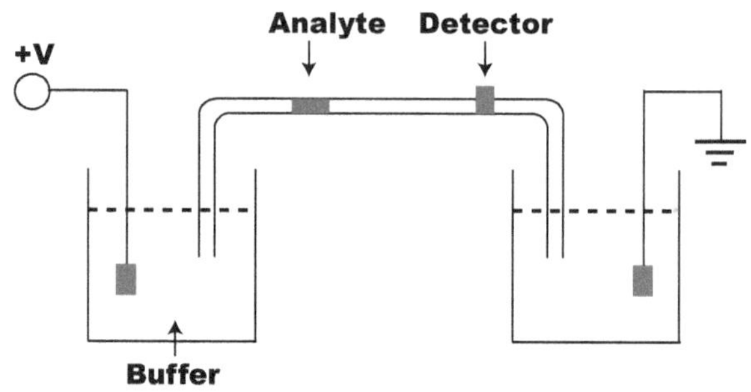

Fig. 2.3 Schematic diagram of a CE experiment

absorbance type: the capillary passes between a small UV light source and a photodetector, the passage of a band is signaled by a drop in the UV light intensity. For specialized applications, various other detection techniques such as LIF (Laser Induced Fluorescence) may be employed. Further, the CE output may be interfaced with other analytical devices, such as mass spectrometers, in order to learn about the chemical identity of sample components. The entire set up: capillary, reservoirs and sample injector may all be etched together on a single glass or silicon substrate to make a microfluidic chip. CE has many advantages over the traditional Slab-Gel-Electrophoresis techniques that it evolved from. An important advantage is that it can be easily integrated into a 'Lab on a Chip' platform.

We have seen so far, that except for a very thin sheath around the channel walls, EOF in a microfluidic channel has a uniform flow profile. This is a great advantage in CE applications since it implies that the presence of EOF does not lead to significant added dispersion due to shear (also known as Taylor-Aris dispersion). This conclusion however is valid as long as all of the parameters involved, namely the electric field $\mathbf{E}$, dielectric constant ε, the zeta-potential ζ and viscosity μ are constants. Variability in any of these parameters could induce axial pressure gradients which perturb the flow and distort the uniformity of the flow profile. Calculating the perturbations in the flow due to such causes and the resultant axial dispersion is a fundamental fluid mechanics problem of considerable interest in CE. Some of the common causes of axial variability in CE are [15]:

1. *Non-uniform zeta-potential:* A common cause of axial variation of the ζ-potential is adsorption of contaminants at the capillary wall [31, 30, 13, 12]. The problem is particularly severe in the case of proteins which are often cationic at physiological pH and tend to be deposited on the negatively charged walls effectively reducing the ζ-potential.
2. *Thermal variations:* The flow of current in the capillary produces a significant amount of Joule heat which is then lost through the walls. Any non-uniformity in heat generation or loss could lead to temperature variations along the capillary. Even if no such inhomogeneity is present, the temperature varies across the capillary from high values at the center to low values near the wall. Since the

viscosity coefficient, μ is a function of temperature, variations of viscosity in the radial as well as axial directions are possible.

3. *Variations in the Electric field:* In the steady state, the average current density over any cross-section must be constant $\bar{J} = \sigma \bar{E} =$ constant where σ is the electrical conductivity. Except for very dilute samples, the electrical conductivity of the buffer in the sample zone could be altered, due to the presence of the analyte, thus $\sigma = \sigma(x, t)$. Thus the imposed electric field, $E = \bar{J}/\sigma(x, t)$ could vary in the axial direction as well as in time. In sample stacking an interface between high and low conductivity buffer is used to build up the concentration or 'stack' the sample. The electric field then undergoes a jump across the interface.
4. *Variable geometry and curved channels:* Microfluidic channels could contain various features such as nozzles that involve cross-sectional shapes and sizes that vary in the axial direction. The requirement to fit a long separation length into a compact chip requires sinuous 'serpentine' channels.

2.6.1 Exactly Solvable Models

In the remainder of this section, we consider two problems involving axially inhomogeneous channel properties that admit analytical solutions. In both cases the limit of infinitely thin Debye Layers is adopted via the HS slip boundary conditions.

2.6.1.1 Cylindrical Capillary with Axial Variations in Zeta-Potential

Anderson and Idol [5] considered the problem of EOF through a uniform, infinite, straight cylindrical capillary with a ζ-potential that varies solely in the axial direction, $\zeta = \zeta(x)$. A uniform external electric field and zero imposed pressure gradient was assumed. An exact solution to the Stokes flow problem was derived under the assumption of thin EDL. It was shown that the velocity field $\mathbf{u} = u\hat{\mathbf{x}} + v\hat{\mathbf{r}}$ may be expressed in terms of the stream function ψ,

$$u = -\frac{1}{r}\frac{\partial \psi}{\partial r} \tag{2.35}$$

$$v = \frac{1}{r}\frac{\partial \psi}{\partial x} \tag{2.36}$$

where ψ is given by a series expansion

$$\psi = -\frac{\varepsilon_0 \kappa E}{\mu}\left[\frac{r^2}{2}\langle\zeta\rangle - 2\sum_{m=1}^{\infty} a_m^{c}(r)\cos\left(\frac{2m\pi x}{L}\right) - 2\sum_{m=1}^{\infty} a_m^{s}(r)\sin\left(\frac{2m\pi x}{L}\right)\right] \tag{2.37}$$

where

$$\frac{a_m^c}{\widehat{\zeta_m^c}} = \frac{a_m^s}{\widehat{\zeta_m^s}} = \frac{rI_0(\alpha_m)I_1(\alpha_m r) - r^2 I_0(\alpha_m r)I_1(\alpha_m)}{\alpha_m I_1{}^2(\alpha_m) + 2I_0(\alpha_m)I_1(\alpha_m) - \alpha_m I_0{}^2(\alpha_m)} \tag{2.38}$$

and

$$\widehat{\zeta_m^c} = \frac{1}{L}\int_0^L \zeta(x)\cos\left(\frac{2m\pi x}{L}\right) dx \tag{2.39}$$

$$\widehat{\zeta_m^s} = \frac{1}{L}\int_0^L \zeta(x)\sin\left(\frac{2m\pi x}{L}\right) dx \tag{2.40}$$

are the cosine and sine transform of the ζ function $\zeta(x)$, $\alpha_m = 2m\pi/L$, and $\langle \quad \rangle$ indicates the average over the length of the capillary:

$$\langle f \rangle = \frac{1}{L}\int_0^L f\, dx. \tag{2.41}$$

In Eq. (2.38), I_n denotes the modified Bessel function of integer order n.

The above solution implies a remarkably simple formula for the cross-sectional average of the axial velocity, $\bar{u}$ (or equivalently, the volume flux per unit cross-sectional area),

$$\bar{u} = -\frac{\varepsilon_0 \kappa \langle \zeta \rangle E}{\mu}, \tag{2.42}$$

which follows on integrating (2.35) over the cross-section of the capillary. Thus, the flux per unit area at any instant over any cross-section is the same and equal to that of the flow through a uniform capillary with $\zeta = \langle \zeta(x) \rangle$.

2.6.1.2 Flow Between Parallel Plates

Ajdari [2] considered the problem of electroosmotic flow between a pair of parallel plates at $z = \pm h$ under the application of a uniform external electric field, E and arbitrary position dependent variations of the zeta-potential on the surface of the plates $\zeta = \zeta_{\pm}(x, y)$. Though the parallel plate geometry is not directly relevant to CE applications, it may serve as a reasonable approximation to flow in shallow rectangular channels etched on chips. The only assumptions in the analysis are those of low zeta-potentials ($\zeta \ll k_B T$, the Debye-Huckel approximation) and the assumption of low Reynolds numbers (Stokes Flow). On account of the linearity of the fluid flow equations in the Stokes flow limit, and, since the problem for the potential in the EDL is decoupled from the fluid problem, an exact solution could be obtained by Fourier transforming the equations along the planes parallel to the plates. In particular, it is shown that if the limit of thin EDLs and the 'lubrication limit' (which we consider in the next section as a powerful and general approach for handling EOF problems in arbitrary geometries), are assumed, the flow reduces to a superposition of a flat profile (a 'pure' EOF) and a parabolic component driven by the induced pressure gradients. In a subsequent paper [3] Ajdari generalized the solution to include the case of small (compared to plate separation) amplitude irregularities on the surface of the plates.

Ajdari's solutions are highly instructive in the sense that they not only provide a clear intuitive picture of the nature of the various limiting cases of importance in EOF theory, but also unravels several non-intuitive, perhaps counter-intuitive aspects of EOF. For example, for certain distributions of zeta-potentials recirculating regions of convective rolls are obtained. These flow patterns could be useful in the design of microfluidic mixers. In fact, Stroock et al. [29] constructed such mixers using electroosmotic flow in a long channel of rectangular cross-section (260 μm $\times$ 130 μm) with a patterned surface charge of alternating sign that was fabricated using soft lithographic techniques [6]. It was further shown by Ajdari that surface irregularities and variations in the zeta-potential in combination could generate net forces on the plates which could even be perpendicular to the applied electric field and need not vanish even if the net charge on either plate vanishes! The theory of Ajdari was applied by Long et al. [21] to obtain analytical solutions in the neighborhood of localized 'defects' in the zeta-potential for both the parallel plate as well as cylindrical geometries. These solutions are useful in providing an intuitive understanding of the perturbations in EOF that may result from various local surface inhomogenieties in the zeta-potential.

2.7 The Lubrication Approximation

When the Helmholtz-Smoluchowski slip velocity is variable over the capillary surface an analytical solution for the flow field is difficult except for the special geometries discussed in the last section. Generally one may need to resort to the more expensive process of full numerical simulation. However, if the variations are 'slow' in the axial direction; a term that will be made more precise later, classical fluid mechanics allows certain approximations often leading to explicit analytical solutions for the flow problem. The method is known as 'lubrication theory'; the name derives from early applications to the study of the flow of lubricant in the narrow gap between moving machine parts [7]. The method has since been applied to varied realms of fluid mechanics, from analysis of micro-pumps for mechanical pumping of fluids in microfluidic channels [10] to blood flow in capillaries [20, 27]. One advantage of the lubrication theory approach is that the channel geometry and distribution of the slip velocity can be quite arbitrary as long as it satisfies the requirement of 'slow' axial variations [14].

Let us consider an infinitely long straight channel, or one whose length is very much larger than a characteristic width, which we will denote by a_0. The channel could have any cross-sectional shape, which, could possibly vary in the axial (x) direction. The geometries possible could include situations such as a converging nozzle of square cross-section that gradually tapers to a round cross-section at the tip. Further, the distribution of ζ (or more generally of the slip velocity given by Eq. (2.34)) over the channel walls could be arbitrary. We make the following assumptions:

1. The characteristic length scale for the variation of the cross-sectional shape and area in the x direction is much larger than a_0 .
2. The characteristic length scale for the variation of the ζ-potential (or more generally, the slip velocity) in the x direction is much larger than a_0 .
3. The characteristic time-scale for any temporal variations are of order T, where $T \gg t_d \equiv a_0{}^2/\upsilon$ (υ being the kinematic viscosity of the fluid).
4. The thin EDL limit is appropriate.

Changes of cross-sectional geometry of channels (such as nozzles) are characterized by a typical length scale $\sim L \gg a_0$ so that the assumptions of lubrication theory are valid. In time varying situations, flow properties often vary in time on a scale $t_e \sim L/u_e$; the flow through time (u_e is a characteristic electroosmotic speed). In this case the assumption $t_e \gg t_d$ is equivalent to $Re(a_0/L) \ll 1$ where $Re = (a_0 u_e)/\nu$ is the Reynolds number. Since in microfluidic applications $Re \sim 0.01 - 1.0$, the above requirement is easily satisfied in most cases. In the analyte adsorption problem, the characteristic time for wall interactions (which is also the time scale on which the ζ-potential varies) is $t_a \sim t_e$. Indeed if $t_a \gg t_e$ adsorption is not significant at all and if $t_a \ll t_e$ the sample will be wholly adsorbed a few diameters from the inlet. In the latter case, of course, the issue of 'band broadening' is irrelevant as there is hardly any recovery. The ratio $t_a/t_d \sim t_e/t_d = \mathrm{Re}^{-1}(\mathrm{L}/\mathrm{a}_0)$. Using the estimates $L \sim$ 10–100 cm, $a_0 \sim$ 10–100 μm and Re $\sim 0.01 - 1.0$ we get $t_a/t_d \sim 10^3 - 10^7$. Thus, the assumption of slow variations is valid in such cases as well. However the theory is not suitable for all applications, for example the fine scale irregularities in a channel wall or the ζ-potential have a characteristic length scale very much *smaller* than a_0. Lubrication theory is inappropriate for such applications.

If the assumption of 'slow' variations, as defined above, are satisfied, a formal asymptotic solution to the problem of electroosmotic flow in terms of the ratio of characteristic radial distance to characteristic axial scale (a small parameter, ϵ) may be carried out [12]. The solution may be summarized as follows: $\mathbf{u} \sim \hat{\mathbf{i}}u(x, y, z) + O(\epsilon)$, $\mathbf{E} \sim \hat{\mathbf{i}}E(x) + O(\epsilon)$,

$$u = -\frac{u_p}{\mu}\frac{dp}{dx} + \frac{\varepsilon_0 \kappa F}{\mu}\frac{\psi}{A(x)}, \tag{2.43}$$

$$Q = -\frac{\bar{u}_p}{\mu}A(x)\frac{dp}{dx} + \frac{\varepsilon_0 \kappa F}{\mu}\bar{\psi}, \tag{2.44}$$

$$E(x) = F/A(x). \tag{2.45}$$

Here F is a constant representing the electric flux through any cross-section, $A(x)$ is the cross-sectional area and the overbar indicates average over the cross-section, $\bar{f} = A^{-1}\int f\, dydz$. The constant Q represents the volume flux of fluid through any cross-section. The functions u_p and ψ are defined by:

$$\frac{\partial^2 u_p}{\partial y^2} + \frac{\partial^2 u_p}{\partial z^2} = -1, \tag{2.46}$$

$$u_p\big|_{\partial D(x)} = 0, \tag{2.47}$$

and

$$\frac{\partial^2 \psi}{\partial y^2} + \frac{\partial^2 \psi}{\partial z^2} = 0, \tag{2.48}$$

$$\psi\,|_{\partial D(x)} = -\zeta, \tag{2.49}$$

where $D(x)$ is the domain representing the cross-section of the channel and $\partial D(x)$ is the contour of its boundary, at axial location 'x'. Both of these functions u_p and ψ may be evaluated by quadrature from a knowledge of the Green's function, G, of the Laplace operator with zero boundary condition corresponding to the domain $D(x)$

$$u_p = \frac{1}{4\pi}\int_{D(x)} G(x; y, z, y_*, z_*) dy_*\, dz_*, \tag{2.50}$$

$$\psi = \frac{1}{4\pi}\oint_{\partial D(x)} \zeta(x, y_*, z_*)\left(m\frac{\partial G}{\partial y_*} + n\frac{\partial G}{\partial z_*}\right) ds_*, \tag{2.51}$$

where (m, n) are the direction cosines of the unit normal on $\partial D(x)$.

The physical content of the solution (2.43), (2.44) and (2.45) is clear. If the properties of the channel vary slowly in the axial direction, then, according to (2.43), the flow velocity, to a first approximation is purely axial, and, it may be expressed as a linear superposition of a purely pressure driven flow, and a purely electroosmotic flow. Further, in calculating the local electroosmotic flow component one must use an 'effective' ζ-potential 'ψ' which is a certain weighted average, (2.51), of the actual ζ-potential around the contour of the cross-section. If ζ does not vary along such a contour, then clearly $\psi = -\zeta$. The pressure driven component and the electoosmotic component are proportional to the local pressure gradient and the local electric field strength respectively. The local pressure gradient is calculated by using the condition, (2.44) for volume conservation of fluid, and the local electric field is calculated by using the condition, (2.45) for electric flux conservation. The solution is completely specified by two independent physical constants, the volume flux of fluid, Q, and, the flux of electric field, F. These constants may be expressed, if desired, in terms of the total pressure drop, and, the total voltage drop, respectively, between the inlet and outlet sections.

2.7.1 Applications

For flow in straight channels where the lubrication approximation is justified, Eqs. (2.43), (2.44) and (2.45) provide a greatly simplified approach to computing the flow compared to solving the full three-dimensional partial differential equations describing the problem. In the lubrication formalism, one only needs to solve the sequence of two dimensional problems defined by Eqs. (2.46), (2.47), (2.48) and (2.49). Further, if the cross-sectional shapes and boundary distribution of ζ-potential at different axial locations can be made congruent (e.g. flow through a conical nozzle) through a similarity transformation, the two dimensional boundary value problems (2.46), (2.47), (2.48) and (2.49) need only be solved once rather than for each x location. Furthermore, for certain cross-sectional shapes the Green's function G or equivalently the solutions ψ and u_p may be available in analytical form. In such cases the flow field may be obtained without the need of numerically integrating partial differential equations [12].

2.7.1.1 Fluidic Resistance

Effective impedances between two given points is a concept that is widely used in analyzing electronic circuits. A similar concept may be introduced for microfluidic circuits if we recognize that in addition to the 'electric current' a 'fluidic current' due to electroosmotic flow is present along each branch of a circuit. This fluidic current between any two points 'a' and 'b' is driven by either an applied pressure $p_a - p_b$ or an applied Voltage $V_a - V_b$ or a combination of both. If we integrate Eq. (2.45) along the channel, we get

$$V_a - V_b = \int_{x_a}^{x_b} E(x)\,dx = FL\langle A^{-1}\rangle \tag{2.52}$$

where $x = x_a$ and $x = x_b$ are the locations of points 'a' and 'b' and $L = x_b - x_a$ is the channel length. If we solve (2.44) for dp/dx and integrate along the channel, we get

$$p_a - p_b = -\int_{x_a}^{x_b} \frac{dp}{dx} dx = \mu QL\langle \bar{u}_p^{-1} A^{-1}\rangle - \varepsilon_0 \kappa FL\langle A^{-1} \bar{u}_p^{-1} \bar{\psi}\rangle. \tag{2.53}$$

If we use (2.52) in (2.53) to eliminate 'F' and rewrite the resulting equation with Q on the left hand side, we get an expression for the volume flow rate in the channel in terms of the applied pressure and voltage difference:

$$Q = \frac{p_a - p_b}{8\mu L}\pi a_*^4 - \frac{\varepsilon_0 \kappa \zeta_*}{\mu}\pi a_*^2 \frac{V_a - V_b}{L}, \tag{2.54}$$

where for convenience

$$a_* = \left[\frac{8}{\pi\langle \bar{u}_p^{-1} A^{-1}\rangle}\right]^{1/4} \tag{2.55}$$

$$\zeta_* = -\frac{1}{\sqrt{8\pi}} \frac{\langle \bar{\psi} \bar{u}_p^{-1} A^{-1}\rangle}{\langle A^{-1}\rangle \langle \bar{u}_p^{-1} A^{-1}\rangle^{1/2}}. \tag{2.56}$$

The Eq. (2.54) can be interpreted to mean that the volume flux through any straight microfluidic channel is equal to that of the flux through an 'equivalent' cylindrical capillary of radius 'a_*' and ζ-potential 'ζ_*' subject to an identical pressure and voltage drop. The parameters 'a_*', and 'ζ_*' depend purely on the geometry of the channel and the charge distribution on its walls. This concept of an 'equivalent' capillary could be quite useful in the analysis of microfluidic circuit components, and could be considered analogous to the concept of 'effective impedance' in the analysis of electrical circuits.

2.7.1.2 Circular Cross-Sections

As an example consider a channel with circular cross-section, but we will let the radius, $a(x)$, be a function of the axial co-ordinate. In order not to violate the limits of applicability of lubrication theory, the length scale, Δx over which $a(x)$ varies significantly is assumed very much larger than the largest value of $a(x)$. The boundary value problems for u_p for all cross-sections may be made congruent through the similarity transformation

$$u_p = a^2(x)U_p(\rho), \tag{2.57}$$

$$\rho = r/a(x), \tag{2.58}$$

and the resulting equation for $U_p(\rho)$ may be integrated to give

$$U_p(\rho) = \frac{1}{4}\left(1 - \rho^2\right). \tag{2.59}$$

The equation for ψ may be solved in cylindrical co-ordinates (r, θ, x) as

$$\psi = -\bar{\zeta} - \sum_{m=1}^{\infty} \left[\tilde{\zeta}_m \exp(im\theta) + \tilde{\zeta}_m^* \exp(-im\theta)\right] \rho^m \tag{2.60}$$

where $\tilde{\zeta}_m$ is the complex Fourier Transform:

$$\tilde{\zeta}_m = \frac{1}{2\pi} \int_0^{2\pi} \zeta(x, \theta) \exp(-im\theta) d\theta \tag{2.61}$$

and $\bar{\zeta} = \tilde{\zeta}_0$ is the ζ-potential averaged over the perimeter.[1] From Eq. (2.60) we have

$$\bar{\psi} = -\bar{\zeta}(x) \tag{2.62}$$

[1] overbar will indicate cross-sectional average except where the variable is defined only on the boundary of the cross-section in which case it would indicate average over the perimeter

so that, Eqs. (2.55) and (2.56) may be evaluated to give the following expressions for the effective radius and ζ-potential

$$a_* = \frac{1}{\langle a^{-4}\rangle^{1/4}} \tag{2.63}$$

$$\zeta_* = \frac{\langle \bar{\zeta} a^{-4}\rangle}{\langle a^{-2}\rangle \langle a^{-4}\rangle^{1/2}} \tag{2.64}$$

In the absence of an applied voltage, Eqs. (2.54) and (2.63) give the well known [7] result for pressure driven flows in slowly varying channels

$$p_a - p_b = \frac{8\mu Q}{\pi} \int_{x_a}^{x_b} \frac{dx}{a^4(x)}. \tag{2.65}$$

When there is no external pressure difference, $p_a = p_b$, Eqs. (2.54), (2.63) and (2.64) imply

$$Q = -\frac{\varepsilon_0 \kappa}{\mu} \frac{\langle \bar{\zeta} a^{-4}\rangle}{\langle a^{-4}\rangle \langle a^{-2}\rangle} \frac{V_a - V_b}{L}. \tag{2.66}$$

When ζ is independent of θ and $a(x) = a_0$ is a constant, the above expression reduces to

$$Q = -\frac{\varepsilon_0 \kappa \langle \zeta \rangle}{\mu} \frac{V_a - V_b}{L} \pi a_0^2, \tag{2.67}$$

which is seen to be identical to Eq. (2.42) derived by Anderson and Idol.

The nature of the lubrication approximation becomes clear on comparing with a problem that admits a full analytical solution. Such a problem may be formulated by stipulating that the capillary considered above be of uniform radius, $a(x) = a_0$ and that the ζ-potential varies sinusoidally in the x-direction, $\zeta(x) = \zeta_0 + \Delta\zeta \sin(2\pi x/\lambda)$. In this case lubrication theory results in the following analytical formula for the axial velocity u:

$$\frac{u}{u_0} = 1 + \frac{\Delta\zeta}{\zeta_0} F_0(\rho) \sin(\alpha X) \tag{2.68}$$

where

$$F_0(\rho) = 2\rho^2 - 1. \tag{2.69}$$

Here $\rho = r/a_0$ and $X = x/a_0$ are dimensionless radial and axial co-ordinates, u_0 is the EOF speed in a uniform channel of the same radius with $\zeta = \zeta_0$ and $\alpha = 2\pi(a_0/\lambda)$ (the lubrication limit corresponds to $\alpha \ll 1$). The exact solution corresponding to the sinusoidal distribution of ζ in the Stokes flow limit may be written down as a special case of the series solution presented by Anderson and Idol:

$$\frac{u}{u_0} = 1 + \frac{\Delta\zeta}{\zeta_0} F(\rho) \sin(\alpha X) \tag{2.70}$$

where

$$F(\rho) = \frac{\alpha^{-1} I_0(\alpha\rho)\left[1 - \frac{\alpha I_0(\alpha)}{2I_1(\alpha)}\right] + \frac{\rho}{2} I_1(\alpha\rho)}{\alpha^{-1} I_0(\alpha) + \frac{1}{2} I_1(\alpha) - \frac{I_0^2(\alpha)}{2I_1(\alpha)}}. \tag{2.71}$$

Here I_0 and I_1 are the modified Bessel functions of order zero and one respectively. In the limit of small α, one may replace I_0 and I_1 by their asymptotic forms for small argument [1], so that, one may readily verify,

$$\lim_{\alpha \to 0} F(\rho) = 2\rho^2 - 1 = F_0(\rho) \tag{2.72}$$

which is consistent with (2.69) derived using lubrication theory.

In Fig. 2.4 we compare $F(\rho)$ and $F_0(\rho)$ for several values of the ratio a_0/λ. It is seen, that, for $a_0/\lambda \ll 1$ (in practice 0.1 or less), the prediction of the lubrication analysis is in excellent accord with the exact solution as expected. For $a_0/\lambda \sim 1$, the exact solution deviates significantly from the lubrication solution. In the opposite limit of $a_0/\lambda \gg 1$, the fluid essentially does not 'see' the rapidly fluctuating ζ-potential except very near to the wall, and the lubrication limit solution is qualitatively incorrect. The latter situation may describe random fine scale inhomogeneities in the wall charge.

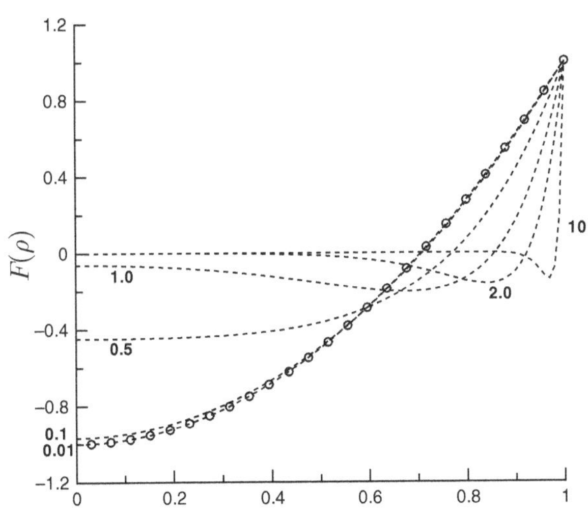

Fig. 2.4 Comparison of the exact solutions (*lines*) with the result of the lubrication approximation (*circles*) for an infinite cylindrical capillary (radius a_0) with sinusoidal variation (wavelength λ) of ζ along its length for $a_0/\lambda = 0.01, 0.1, 0.5, 1.0, 2.0$ and 10.0

2.8 Summary and Conclusions

Electroosmosis and other related electrokinetic effects are interfacial phenomena that are observed whenever charge separation occurs at the interface of the substrate and buffer. Generally EOF is present together with electrophoretic migration of individual species in a CE channel, even though it is possible to suppress it by applying special chemical coatings to the substrate. EOF presents both an advantage as well as a hindrance to efficient electrophoretic separation. The presence of EOF in the CE microchannel enables single point detection (species of either charge elutes at the same end), reduces analysis times and enables operation of the microdevice in a continuous mode. On the other hand, the disadvantage is that any effect that causes the EOF to deviate from its classical flat profile lead to Taylor dispersion and consequent band broadening.

Mathematically, electrokinetic flows are described by the incompressible Navier-Stokes equations with an electric body force term together with the equation of continuity. These equations are coupled to Poisson's equation relating the potential to the charge distribution and 'drift-diffusion' type conservation equations for each of the ionic species. This system needs to be supplemented by an advection-diffusion equation for the analyte. The resulting system of equations are quite complex and nonlinear. Fortunately, a series of simplifications can be made to these equations, at each step exploiting a certain disparity in scales inherent in the problem.

The first level of simplification comes about through the assumption of thin Debye layers. This is justified if characteristic channel radii are much larger than the Debye length, a condition that is most often satisfied quite well in microfluidics applications. This disparity in scales allows us to drop the term representing the electrical force in the Navier-Stokes equations and instead, to replace the classical 'no slip' boundary conditions at the solid fluid interface by the 'Helmholtz-Smoluchowski slip boundary conditions'. Thus, within the realm of this approximation, the electrical forces are described by a single parameter, 'the ζ-potential' that enter the fluid flow description solely through the new 'slip' boundary conditions and the Poisson-Boltzmann (or Gouy Chapman) equations for the EDL can be ignored.

The second level of approximation becomes possible due to the smallness of channel diameters compared to overall capillary length (10–100 cm). This allows us to invoke a well developed branch of fluid mechanics, namely 'lubrication theory' for the description of the fluid flow. The consequent reduction in complexity enables a rational description of an important class of problems involving EOF; namely the problem of EOF through channels that are not homogeneous in the axial direction. Axial inhomogeneity can arise due to a variety of reasons, in particular due to adsorption of charged sample components to the wall (which in turn changes the ζ-potential), variations in temperature due to non-uniform heating or cooling, alteration of the electrical conductivity of the buffer by the sample or axial variation in buffer pH (as in sample stacking or isoelectric focussing). In general, axial inhomogeneities 'threaten' the fluid by challenging the 'law of continuity' for fluid flow, and the fluid responds by creating appropriate pressure fluctuations in the axial direction. These pressure fluctuations give rise to a 'Poiseuille' type of flow with a parabolic profile which through the mechanism of Taylor-Aris dispersion leads to greatly enhanced effective axial dispersion. If the channel diameter is much smaller than the capillary length, it is shown that except very close to the inlet or outlet sections, the flow field admits an analytical description, namely as a superposition of a certain 'pure' EOF and a parabolic component.

Though the basic analytical machinery for treating EOF in a wide variety of situations of interest appear to be available, there are many areas where theoretical understanding is in a relatively primitive state. Important open areas of research include analysis of EOF and dispersion in curved microfluidic channels, the effect of conductivity and pH gradients in problems involving EOF, the effects of Joule heating on flow and band broadening and flow through channels of non-circular geometry. Advances in the study of this new area of fluid mechanics should facilitate the development of software and numerical tools that could be very useful in the quest to develop 'lab on chip' technologies and more efficient separation methods.

Acknowledgment Support from the NIH under grant R01EB007596 is gratefully acknowledged.

References

1. Abramowitz M and Stegun IA Eds. (1970) Handbook of Mathematical Functions. Dover Publications Inc., New York, USA.
2. Ajdari A (1995) Electro-osmosis on inhomogeneously charged surfaces. Phys. Rev. Lett. 75:755–758.
3. Ajdari A (1996) Generation of transverse fluid currents and forces by an electric field: Electro-osmosis on charge-modulated and undulated surfaces. Phys. Rev. E 53:4996–5005.
4. Anderson JL (1985) Effect of nonuniform zeta potential on particle movement in electric fields. J. Coll. Int. Sci. 105(1):45–54.
5. Anderson JL and Idol WK (1985) Electroosmosis through pores with nonuniformly charged walls. Chem. Eng. Commun. 38:93–106.
6. Anderson JR, McDonald JC, Stone HA and Whitesides GM (2001) Integrated components in microfluidic devices in pdms: A biomimetic check valve, pressure sensor & reciprocating pump. *Preprint*.
7. Batchelor G (2000) An Introduction to Fluid Dynamics. Cambridge University Press, Cambridge, UK 2000.
8. Burgreen D and Nakache FR (1964) Electrokinetic flow in ultrafine capillary slits. J. Phys. Chem. 68(5):1084–1091.
9. Bush JWM and Hu D (2005) Walking on water: Biolocomotion at the interface. Annu. Rev. Fluid Mech. 11:207–228.
10. Day RF and Stone HA (2000) Lubrication analysis and boundary integral simulations of a viscous micropump. J. Fluid Mech. 416:197–216.
11. Eisner T and Aneshansley DJ (1999) Spray aiming in the bombardier beetle: Photographic evidence. Proc. Natl. Acad. Sci. USA p. 97059709.
12. Ghosal S (2002) Band broadening in a microcapillary with a stepwise change in the -potential. Anal. Chem. 74(16):4198–4203.
13. Ghosal S (2002) Effect of analyte adsorption on the electroosmotic flow in microfluidic channels. Anal. Chem. 74:771–775.

14. Ghosal S (2002) Lubrication theory for electroosmotic flow in a microfluidic channel of slowly varying cross-section and wall charge. J. Fluid Mech. 459:103–128.
15. Ghosal S (2006) Electrokinetic flow and dispersion in capillary electrophoresis. Annu. Rev. Fluid Mech. 38:309–338.
16. Helmholtz HV (1879) Stüdien über electrische grenscations. Ann. der Physik und Chemie 7:337–387.
17. Kirby BJ and Hasselbrink EF (2004) Zeta potential of microfluidic substrates: 1. theory, experimental techniques, and effects on separations. Electrophoresis 25:187–202.
18. Kirby BJ and Hasselbrink EF (2004) Zeta potential of microfluidic substrates: 2. data for polymers. Electrophoresis 25:203–213.
19. Landau LD and Lifshitz EM (2002) Course of theoretical physics Vol 10. Physical Kinetics. Butterworth-Heinenann, Oxford, UK.
20. Lighthill MJ (1968) Pressure-forcing of tightly fitting pellets along fluid-filled elastic tubes. J. Fluid Mech. 34:113–143.
21. Long D, Stone HA and Ajdari A (1999) Electroosmotic flows created by surface defects in capillary electrophoresis. J. Coll. Int. Sci. 212:338–349.
22. Probstein R (1994) Physicochemical Hydrodynamics. John Wiley and Sons, Inc., New York, USA.
23. Probstein RF (1972) Desalination: Some fluid mechanical problems. Trans. ASME J. Basic Eng., 94:286–313.
24. Probstein RF and Hicks RE (1993) Removal of contaminants from soils by electric fields. Science, 260:498–503.
25. Reuss FF (1809) Sur un nouvel effet de le électricité glavanique. Mémoires de la Societé Impériale des Naturalistes de Moscou, 2:327–337.
26. Rice CL and Whitehead R (1965) Electrokinetic flow in a narrow cylindrical capillary. J. Phys. Chem. 69:4017–4024.
27. Secomb TW, Skalak R, Özkaya N and Gross JF (1986) Flow of axisymmetric red blood cells in narrow capillaries. J. Fluid Mech. 163:405–423.
28. Smoluchowski MV (1903) Contribution à la théorie de l'endosmose électrique et de quelques phenoménes corrélatifs. Bulletin International de l'Academie des Sciences de Cracovie, 8:182–200.
29. Stroock AD, Weck M, Chiu DT, Huck WTS, Kenis PJA, Ismagilov RF and Whitesides GM (2000) Patterning electro-osmotic flow with patterned surface charge. Phys. Rev. Lett. 8415:3314–3317.
30. Towns JK and Regnier FE (1992) Capillary electrophoretic separations of proteins using nonionic surfactant coatings. Anal. Chem. 91:1126–1132.
31. Towns JK and Regnier FE (1992) Impact of polycation adsorption on efficiency and electroosmotically driven transport in capillary electrophoresis. Anal. Chem. 64:2473–2478.

Chapter 3
Microscale Transport Processes and Interfacial Force Field Characterization in Micro-cooling Devices

Sunando DasGupta

Abstract The enhancement of transport rates due to the smaller length scales, transfer areas and high surface to volume ratio in miniature devices are explored along with the evaluation of additional shape-dependent forces at small length scales. The thickness of the adsorbed film, apparent contact angles and the curvature profiles of the evaporating film and the contact line velocities of the oscillating thin liquid film on silicon and high refractive index glass surfaces are accurately measured using image analyzing interferometry. The measurements are consistent with previous models describing fluid flow and heat transfer in thin liquid films based on interfacial concepts. The oscillating velocities of the thin film and the contact angle variations accurately capture the forward and backward motion of the film during oscillation.

Keywords Ultrathin evaporating film · Contact line · Dispersion constant · Image analyzing interferometry · oscillating meniscus

Miniaturization of devices where the characteristic lengths can be comparable to that of boundary layers leads to significant enhancement of transport rates. For example, microscale heat exchange is currently an active area of research due to its possible applications in several technologically important processes, e.g. in the electronic packaging industry, cooling of solar panels in microgravity environments, and spacecraft thermal control. The increases in the required power dissipation of higher level integrated devices will result in increased thermal gradients and higher mean operating temperatures of the device. Thus it is necessary to develop innovative thermal control schemes capable of handling higher heat fluxes. In cases where large amounts of heat must be removed, the use of change of phase heat transfer mechanism can prove to be a promising choice. This technique has been widely used in various cooling operations, through the utilization of micro-heat pipes for the

S. DasGupta (✉)
Department of Chemical Engineering, Indian Institute of Technology Kharagpur, Kharagpur-721302, India
e-mail: sunando@che.iitkgp.ernet.in

S. Chakraborty (ed.), *Microfluidics and Microfabrication*,
DOI 10.1007/978-1-4419-1543-6_3,

thermal control of electronic components. The smaller length scales, transfer areas and high surface to volume ratio in micro-devices can provide excellent opportunities for process intensification and innovative solutions for a variety of specialized situations.

However, additional forces start to become important in smaller length scales and influence the basic physics of the transport processes. The relevant governing equations cannot be deduced directly, e.g., from the force balance of fluid mechanics as in macroscopic physics. In order to design such devices, extensive study of the solid-liquid-vapor contact line region incorporating intermolecular interaction in the three phase contact line region has been undertaken by a number of researchers. In thin liquid films, the shape-dependent interfacial stress field controls fluid flow and heat transfer. The intermolecular interactions between a thin film of liquid, its vapor, and a solid surface are crucial to many equilibrium and nonequilibrium processes such as adsorption, spreading, evaporation and condensation, wetting, and stability of thin films. The applications can be as varied as coating, surface cleaning, and microscale transport processes. Furthermore, the ability to understand and control these interactions is growing in importance for the understanding and optimization of lab-on-a-chip processes.

There is a "pressure jump" at the liquid-vapor interface, due to the anisotropic stress tensor near interfaces. The classic Young-Laplace equation of capillarity has been successfully used to describe the pressure jump at a curved liquid-vapor interface [1–9]. Examples of its use to describe the fluid dynamics in a micro heat pipe can be found in [11–13] where the pressure jump is a function of the liquid-vapor surface tension and the interfacial radius of curvature. However, near the liquid-solid interface, additional changes in the stress field within the liquid occur because of changes in the intermolecular force field due to solid molecules replacing liquid molecules. This leads to the augmented Young-Laplace model for the pressure jump at the liquid–vapor interface. These long-range van der Waals forces have been found to be extremely important in that they lead to the concept of an extended evaporating meniscus [14]. In a completely wetting evaporating system, a thin adsorbed film extends for a long distance beyond the classic equilibrium meniscus. The thin film controls the important processes of spreading and wetting.

Evolution and instabilities of liquid–vapor interfaces were studied by various researchers. For example, de Gennes [15] presented a unified model of dry spreading by considering a precursor film around a spreading drop. The effects of hydrodynamics and spreading on contact line motion were also studied in the past. Zheng et al. [16] have presented data on an unstable oscillating, evaporating thin film of pentane where moving velocities of the oscillating film were obtained. A force balance for the oscillating meniscus based on intermolecular and shape-governed forces was used to describe the oscillating velocities. However, the details of the region below a film thickness of 0.1 mm were not adequately addressed. In a subsequent study [17] image-analyzing interferometry with an improved data analysis technique is used to study experimentally the liquid vapor interfacial profile, including the profile in the contact line region during condensation and evaporation. Herein, the use of the developed techniques for force field characterizations of evaporating and oscillating menisci on a silicon substrate and high refractive index glass are demonstrated with a brief description of the forces involved in such systems.

3.1 Basic Concepts

The principal objective of this section is to provide a very brief description of the concepts of intermolecular forces, their classifications and roles in the interfacial phenomena in a thin liquid film. The discussion will relate the idea of disjoining pressure to the more classical interfacial concepts.

Intermolecular forces play a major role in determining many of the bulk properties of the matter in all its phases. The surface tension, heat of vaporization and diffusion in liquids are examples of the physical properties which are a consequence of intermolecular forces. Intermolecular forces also determine the three dimensional structure of molecules. One way to classify intermolecular forces is by the nature of the interactions as was proposed by Israelachvilli, J. in his famous book [18] – "First there are those that are *purely electrostatic* in origin arising from the Coulomb forces between charges. The interactions between charges, permanent dipoles, quadrupoles, etc., fall into this category. Second, there are *polarization forces* that arise from the dipoles moments induced in atoms and molecules by the elastic fields of nearby charges and permanent dipoles. All interactions in a solvent medium involve polarization effects. Third, there are forces that are *quantum mechanical* in nature. Such forces give rise to *covalent bonding* (including charge-transfer interactions) and to the repulsive *exchange interactions* (due to Pauli exclusion principle) that balance the attractive forces at very short distances."

Intermolecular forces are also divided into two categories based on their operating ranges as short range and long range forces. Short range forces include chemical bonds, metallic bonds, hydrogen bonds etc, and although their range of interaction is small (one or two atomic distances), these forces are usually more powerful and sometimes dominate the long range forces. Long range forces operate in a distance range of 1–100 nm. They comprise mainly of London dispersion forces and in the case of polar liquids, the Keesom dipole-dipole forces and the Debye dipole-induced dipole forces. These three forces are collectively known as van der Waals forces. Long range forces can significantly affect the properties and the behavior of the thin liquid films on a solid substrate.

Dispersion forces make up the most important contribution to the total van der Waal force between atoms and molecules, and because they are always present they play a role in a host of important phenomenon such as adhesion, surface tension, physical adsorption, the properties of gases and liquids, the strength of solids, the flocculation or aggregation of particles in aqueous solutions, and the structures of condensed macromolecules such as proteins and polymers. Dispersion forces are long-range forces and may be repulsive or attractive. Dispersion forces are quantum mechanical in origin and arise due to the presence of instantaneous dipole moment and induced dipole even in non-polar atoms.

There are two methods for the calculation of van der Waals forces between condensed media. They are the microscopic method derived by Hamaker in 1937 and the macroscopic method by Dzyaloshinskii et al. [19]. By assuming that the microscopic properties are additive, Hamaker was able to integrate the pair-wise interactions between molecules to obtain their overall dispersion force. In effect, he has neglected the many – body interactions which propagate over an appreciable intermolecular distance. Furthermore, this theory does not take into account the relativistic retardation effect, the keesom and Debye forces (for polar liquids) and the effect of an intervening layer in between the two condensed media.

On the other hand DLP theory is more rigorous with fewer assumptions, although it is a continuum model with their inherent assumptions. The DLP theory has no adjustable parameters and the only required inputs into the DLP theory are the temperature of the system, and frequency dependent dielectric susceptibilities which characterize the electromagnetic fluctuations within the materials involved. Therefore, to calculate the van der waals forces from the DLP theory, one only needs to know the macroscopic properties of bulk materials. The modern general theory for the attractive force per unit area for two similar bodies of large extend, separated by a small gap, was given by Lifshitz in 1955, using the methods of quantum electrodynamics. The extension to thin liquid films between two dissimilar bodies was given by Dzyaloshinskii et al. [19]. The general formula is quite complex, involving an infinite sum of integrals over all frequencies of the complex dielectric susceptibilities of all three media evaluated on the imaginary frequency axis. However, the limiting cases, where the gap thickness is small or large compared to the principal absorption wavelengths, leads to simpler expressions in which the force per unit area is proportional to h^{-3} and h^{-4} respectively, where h is the separation. This shift in exponent is related to electromagnetic retardation, owing to the finite speed of light. As shown by Troung and Wayner [20], the van der Waals forces can be obtained directly from the DLP theory, if the zero frequency dielectric constants, the major absorption peaks in the frequency spectrum, and the indices of refraction, which can be obtained from spectroscopic optical data, are known. In actuality, the effective exponent of the film thickness varies smoothly from -3 to -4 as the film thickness increases. The agreement of the theories in the limiting cases is good when the only interactions are the London dispersion forces, but not when the Debye and Keesom contributions' of polar materials have to be taken into account. However, works of Beaglehole et al. [21] and Beaglehole and Christenson [22] demonstrate the limitations of the DLP theory in predicting the interactions in very thin liquid films (below 2 nm) and their results suggest that structural effects are present in some adsorbed films at room temperature, provided the substrate are smooth and homogeneous.

Another major phenomena affecting the spreading of a liquid film, which in turn affects the cooling potential of a thin evaporating film is surface tension which arises at the interface. The molecule at the interface experiences unbalanced forces since it is no longer surrounded symmetrically by other similar molecules. The direction of these forces which act on the molecules forming the interfacial layer posses an excess free energy since work has been done to bring them to the surface and to maintain them there. This energy is defined as surface tension for a liquid-vapor interface and it has a unit of force per unit length (N/m).

The molecules in the liquid–vapor interface will experience a weaker force than they would have experienced if the vapor region has been replaced by a fluid. This is due to the fact that the density of vapor is much smaller than that of a bulk fluid. Consequently, these molecules will experience a net force pulling them back into the bulk fluid. This resultant force will have the effect of reducing, or rather, minimizing the surface free energy and the surface area of the fluid. For this reason, liquid droplets and gas bubbles tend to form spherical shapes.

3.2 Disjoining Pressure

The disjoining pressure (π) concept is very useful and informative concerning the general wetting characteristics of a system it allows fluid flow concepts in an ultra-thin film to be introduced and evaluated because it can be viewed as an effective pressure resulting from a body force acting between the substrate and the mobile liquid film. An excellent review on this subject was performed by Israelachvili [18]. Briefly, the disjoining pressure is (minus) the potential energy per unit volume due to intermolecular forces, F, and is a function of the film thickness, δ:

$$\Pi(\delta) = -F(\delta) \tag{3.1}$$

Disjoining describes the physical process whereby a completely spreading liquid naturally tends to disjoin a solid from the vapor by spreading an equivalent change in energy per unit area, E_S, with film thickness can be given as

$$\Pi = \frac{dE_S}{d\delta} \tag{3.2}$$

Therefore, for a completely wetting system, the "surface energy" decreases with an increase in film thickness. This leads to a positive disjoining pressure, π, and a negative potential energy, F.

For a completely spreading system, the case of a solid flat plate partially immersed at an angle θ in a pool of wetting fluid as shown in Fig. 3.1 will be discussed first. The presence of interfacial forces in the thin wetting film and capillary forces in the thicker film stabilizes it against the hydrostatic forces. Therefore, the pressure, F, in the liquid, decreases with an increase in height. For a completely wetting system the adsorbed thin film can extended for a very long distance. The interfacial forces, $F(\delta)$, in the thin film start to become important at a thickness of approximately 10^{-7} m. the continuous film extends until it is a monolayer. At equilibrium, the chemical potential energy per unit volume in the thin flat film differs from the bulk liquid by an amount often returned to as the excess potential. Derjaguin has studied this potential energy experimentally and theoretically, and defined it in units of pressure as the disjoining pressure [1]. The Dzyaloshinskii-

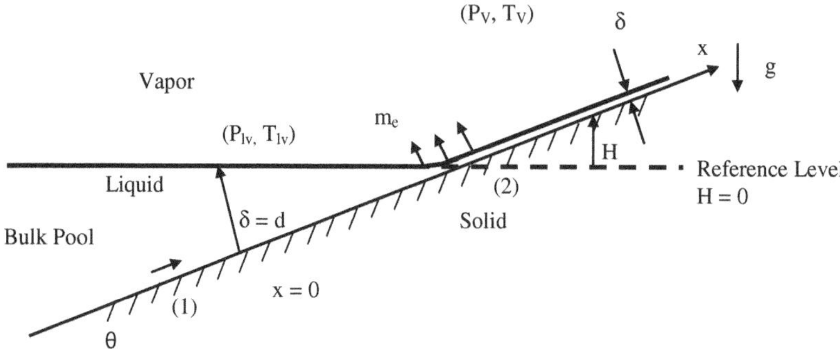

Fig. 3.1 An inclined flat plate immersed in a liquid at an angle θ to the horizontal

Lifshitz-Pitaevskii (DLP) theory enables one to calculate the force (per unit area), $F(\delta)$, from the film thickness, δ, and the optical and thermo physical properties of the solid, vapor, and the thin liquid film. Although the DLP theory is a powerful tool, extensive data on the optical properties as a function of frequency are needed. The works of Troung and Wayner [20] provided expanded discussions on this material.

In the limit of a thin film (non retarded regime, δ<10 nm) one can represent the dispersion force as

$$F(\delta) = \frac{A}{6\Pi\delta^3} = \frac{\bar{A}}{\delta^3} \tag{3.3}$$

While for the thicker (retarded) regime (δ > 20 nm),

$$F(\delta) = \frac{B}{\delta^4} \tag{3.4}$$

In Eq. (3.3), A is the well known Hamaker constant while B is defined as the retarded dispersion force constant. For stable wetting films, both A and B must be negative. This implies that the potential energy of the system increases when the wetting film increases in thickness. In a horizontal film or under zero gravity conditions, fluid naturally flows from the thicker to the thinner region and of film of uniform thickness forms. Calculation of these forces indicate that the above two equations are simple approximation s but as such are highly useful. The constants A and B are weak functions of thickness in their respective regimes. The equilibrium condition for the system presented in Fig. 3.1 requires that the following form of the chemical potential (per unit volume) throughout the liquid is zero:

$$-F(\delta) + \sigma K - \rho_1 g H = 0 \tag{3.5}$$

Where the three terms represent the contributions from the van der waals force, the capillary force, and the hydrostatic force. The curvature of the film is represented by K, the density by ρ, and the gravitational force per unit, mass by g. when only the liquid-vapor interfacial free energy, σ_{iv}, is used the subscripts are suppressed. The above equation was derived from a consideration of interfacial hydrodynamics. The theoretical film profile obtained using Eq. (3.5) can be compared to the measured equilibrium profile to determine the important interactive constant A and B. Therefore, a simple hydrostatic experiment can be done to determine the force field in a thin film. The hydrostatic term gives the pressure as a function of height and, therefore, the equation gives the profile as a function of pressure. Both A and σ are important here. However, the use of the theoretically difficult liquid–solid and solid–vapor interfacial free energies has been replaced by the use of the Hamaker constant. In the adsorbed thin film above the curvature controlled region, the pressure can vary by a large amount because H can be very large. In an evaporating system H and δ decrease dramatically but the pressure is still a function of the thickness and shape. Therefore, large pressure gradients are possible in evaporating thin film with a change in shape which usually occurs over a very short distance.

Under zero gravity conditions, Eq. (3.5) becomes

$$-F(\delta) + \sigma K = \text{constant} \tag{3.6}$$

The film thickness can then be obtained from the following equation, if A or B is known.

$$\frac{A}{6\Pi\delta^3} \text{ or } \frac{B}{\delta^4} = -\sigma K \tag{3.7}$$

3.3 Evaluation of Hamaker Constant

In the previous sections, the central importance of the Hamaker constant to the analysis of wetting , meniscus shape and the heat sink characteristic of evaporating thin films have been discussed. Fortunately, the Hamaker "constant" is a well known function of the film thickness and the optical properties of the liquid and the substrate. However, the necessary raw data on the optical properties are only well known for simple and ideal systems. For example, the properties are well known for non-polar simple fluids on quartz. In addition there has been an extensive discussion concerning water because of its importance even though it is an extremely complicated fluid. Some data for other polar liquids like the alcohols exist.

In general, the total disjoining pressure, $\pi(\delta)$ is the sum of a van der waals component (Π_{vdw}), an electrostatic component based on charge (Π_{el}) and a structural component (Π_s). The van der waals component is made up of keesom energy (due to dipole-dipole interactions based on orientation), Debye energy (due to dipole-nonpolar interactions based on polarizability), and the London dispersion energy due to electrodynamic interactions between the electrons in all atoms and molecules). The structural component is presumed negligible herein.

The substrate of primary importance herein is a single crystal of silicon. A surface oxide forms on the silicon that protects it from the environment and the ideal silicon substrate is really silicon with a native surface oxide approximately 3 nm thick. Theoretically, it should be possible to start with intermolecular forces (as described by the optical property of materials) and calculate the Hamaker constant, the ideal heat transfer coefficient and the heat sink of an evaporating ultra-thin film from first principles. However, the exact composition of the real experimental interface is unknown because it is a strong function of its history which is only approximately known because of possible interaction with water and other contaminants. Therefore, although all the theoretical concepts are available for overall direction, experimental characterization of an "engineering" surface is still required. The experimental section demonstrates this characterization.

3.4 Experimental

The details of the experimental system for the high refractive index glass surface (Fig. 3.2a) can be found elsewhere [17]. The experimental set-up for silicon surface (Fig. 3.2b) is a circular aluminum chamber with top glass view-port to enclose a silicon wafer. Through the glass plate the evaporating meniscus can be viewed at all times. An insulated strip heater is attached at the back of the silicon substrate to provide heat input to the silicon substrate. Monochromatic light of wavelength 546 nm is used and interference fringes can be viewed immediately near the contact line region through the objectives of a Leica DM-LM microscope and the images are captured at regular intervals of time.

The cleaning techniques are extremely important and are carried out in a Class 100 clean hood. The cell parts except the silicon wafer, is dipped in ethanol for 3–4 h to get rid of any adsorbed impurities and is dried in an oven at 110°C for 1 h. The silicon wafer is cleaned using piranha solution – a freshly prepared solution of 30% hydrogesn peroxide and concentrated sulphuric Acid (1:3) and repeatedly rinsed with demonized water to remove residual acid. The interferometric fringes can be viewed clearly and their response to heat input is monitored by the motion of the fringes. As the heat input to the strip heater is gradually increased the fringes come close together signifying an increase in the film curvature. Beyond a certain heat input the film starts to move periodically backwards and forwards and the frequency and amplitude of these oscillation increase with increase in heat input. The oscillations are continuously captured for subsequent analysis. For experiments with high refractive index glass, the system is perturbed slightly from equilibrium situation to study the interplay between the suction potential and the capillary forces.

3.5 Measurement Techniques

A digitized picture of the interferometric fringes and the gray value profile is shown in Fig. 3.3

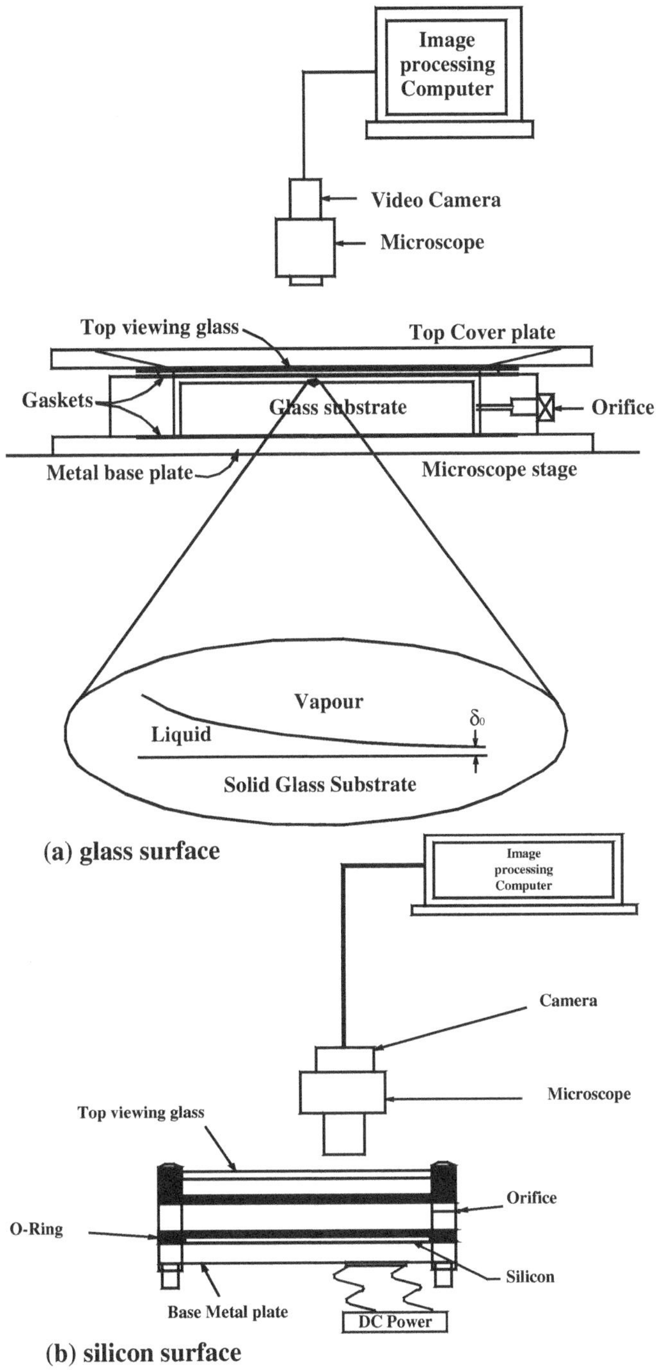

Fig. 3.2 Schematic of the experimental setup

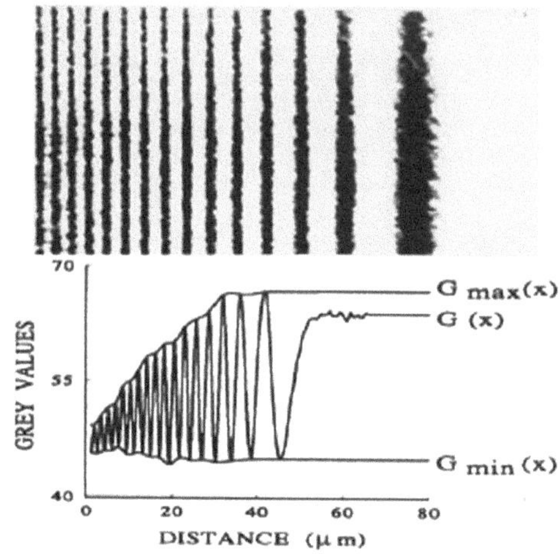

Fig. 3.3 Digitized picture of the interferometric fringes

To automate data acquisition, improve data resolution and enhance data analysis, the interferometric images are analyzed using image analyzing interferometry. Interference fringes are analyzed to determine the profile of the capillary meniscus in the thickness range, $\delta > 0.1$ μm. The images are digitized into 1024 (horizontal) × 768 (vertical) pixels and assigned one of 256 possible gray values representing intensity from 0 (black) to 255 (white). The grey value at each pixel is a measure of the reflectivity. Thus each microscopic pixel acts as an individual, simultaneous light sensor. For the 20X and 50X magnification used in this study, each pixel represents an area of 0.33 μm and 0.144 μm in diameter respectively. From each image a plot of the pixel gray value vs. distance was extracted. As is evident from Fig. 3.2, the reflectivity underwent a cyclic change with increase in film thickness. The computer program scanned the peaks and valleys and filtered the noise from the real peaks/valleys. It then interpolated peak/valley envelopes and by analyzing the relative reflectivity of any pixel with respect to these (dark and light pixel envelopes) determined a film thickness at *every* pixel. The fact that the extended capillary meniscus merged smoothly to an adsorbed flat film was utilized to estimate the adsorbed film thickness from the gray value data and the peak/valley envelopes.

The relevant equations and procedure for image analysis are presented extensively in [23] and are presented very briefly herein. The experimentally obtained gray value plot is scanned for sharp local variations which may be present due to dust particles and smoothing is applied, if required, without affecting the positions of the maxima and minima. A MATLAB code written for this finds the maxima and minima and generates the interpolatory envelopes for the maxima and minima. A relative gray value ($\overline{G}$), is defined as

$$\overline{G}(x) = \frac{G(x) - G\min(x)}{G\max(x) - G\min(x)} \tag{3.8}$$

Where $G\min(x)$ and $G\max(x)$ are the interpolatory envelopes to the various order minima and maxima (constructive and destructive fringes. Since reflectivity is a function of the adsorbed thin film thickness, the change in reflectivity in the flat adsorbed film region, given by the difference between G_0 (the reflectivity of a bare silicon surface) and $G(x)$, provides an estimate of the thickness of the adsorbed thin film. The reflectivity of the liquid film can be calculated at each pixel position using the following equation

$$RL(x) = \overline{G}(x)[RL_{\max} - RL_{\min}] + RL_{\min} \tag{3.9}$$

The reflectivity of a thin liquid film of refractive index n^1, on a solid surface of refractive index n_s, is related to the film thickness by the following relation

$$RL = \frac{\alpha + \beta\cos 2\theta_l}{\kappa + \beta\cos 2\theta_l} \tag{3.10}$$

Where

$$\theta_l = \frac{2\pi n_l \delta}{\lambda}; \quad r_1 = \frac{n_l - n_v}{n_l + n_v}; \quad r_2 = \frac{n_s - n_l}{n_s + n_l}; \tag{3.11}$$

$$\alpha = r_1^2 + r_2^2; \quad \beta = 2r_1 r_2; \quad \kappa = 1 + r_1^2 r_2^2 \tag{3.12}$$

Here n_v is the refractive index of the vapor and λ is the wavelength of light. The equation predicts that reflectivity undergoes a cyclic change with the minima and maxima given by the following equations:

$$RL_{\max} = \left(\frac{r_1 + r_2}{1 + r_1 r_2}\right)^2 = \frac{\alpha + \beta}{\kappa + \beta}, \quad \theta_l = 0, L\pi; \tag{3.13}$$

$$RL_{\min} = \left(\frac{r_1 - r_2}{1 - r_1 r_2}\right)^2 = \frac{\alpha - \beta}{\kappa - \beta}, \quad \theta_l = \frac{\pi}{2}, \left(L + \frac{1}{2}\right)\pi \tag{3.14}$$

Combining Eqs. (3.2) and (3.3), the film thickness at each pixel location is related to the gray value at that pixel location as

$$2\theta_l = \cos^{-1}\left(\frac{\beta + \kappa(1 - 2\overline{G}(x))}{\beta(2\overline{G}(x) - 1) - \kappa}\right) \tag{3.15}$$

Thus the film thickness at each pixel location is obtained based entirely on the experimental data. Once the film thickness is obtained at every pixel position, the slope ($d\delta/dx$) of the film thickness profile (local tangent angle) and the curvature are obtained at every pixel. Thus the technique successfully evaluates the slope and curvature at every pixel position. The oscillation of the film (backward and forward movement) is captured as a video and frame by frame analysis is performed to calculate the contact angle, curvature, wetted length and the advancing or receding velocities.

The liquid wall wetted length, c_i, at any moment t_i is given as

$$c_i = \frac{\sin\left(\frac{\phi}{2}\right)}{K_i \sin\left(\frac{\alpha}{2}\right)} \tag{3.16}$$

where $\Phi = 2\theta c$ angle covered by the arc of the meniscus on the wafer surface, α is the vertex angle of the surface equal to 180°, θc is the apparent contact angle between liquid and substrate. The estimated wetted wall length (c_i) from the above relation is used to find the moving velocity of the oscillating film. It can be calculated from the change of wetted length between instants of time t_{i-1} and t_i. The moving contact line velocity (during oscillation) (Ui) at any time instant "i" can be calculated from the relation [23]

$$U_i = \frac{c_i - c_{i-1}}{t_i - t_{i-1}} \tag{3.17}$$

This velocity may be positive or negative. If the contact line velocity is positive, it indicates that the contact line is moving towards the adsorbed flat film. If negative, the contact line is moving backwards.

3.6 Theory

The Augmented Young Laplace equation can be written for a point in the thicker portion of the meniscus, where the disjoining pressure effects are negligible. For the isothermal non-evaporating (no additional Heat input, $Q = 0$) case considered here, the liquid pressure will remain same, irrespective of the position at a constant gravitational level (different inclined positions). At equilibrium situation, where no evaporation or condensation is taking place, the Augmented Young–Laplace equation can be written as [17]

$$P_l - P_V = -\sigma K - \Pi \tag{3.18}$$

Where,

$$K(\text{Curvature}) = \frac{d^2\delta}{dx^2}\left[1 + \left(\frac{d\delta}{dx}\right)^2\right]^{-3/2} \tag{3.19}$$

$$\Pi(\text{DisjoiningPressure}) = \frac{-B}{\delta^n} \tag{3.20}$$

In these equations, δ represents the film thickness, σ represents the surface tension and B, a modified Hamaker constant ($B < 0$ for completely wetting systems). In the limit of very thin film of pure simple fluid (non retarded region) $n = 3$ and $B = A/6\pi = \bar{A}$, in which A is classical Hamaker constant; while in the thick film region (retarded region) $n = 4$ and B is a dispersion constant. Equation (3.6) can be used to construct the slope profiles from the known values of $\bar{A}$, δ_0, the curvature at the thicker end of the meniscus (K_∞) and the physical properties of the fluid. The following equation can be written for two points, one in the adsorbed region and the other in the transition region.

$$\sigma K - \frac{B}{\delta^4} = \sigma K_\infty Q = 0 \tag{3.21}$$

The curvature at the thicker portion of the meniscus (K_∞) is nearly constant in the region. Using the simplified form of curvature Eq. (3.21) can be modified as

$$\sigma \frac{\mathrm{d}^2\delta}{\mathrm{d}x^2} - \frac{B}{\delta^4} = \sigma K_\infty \tag{3.22}$$

The following non-dimensional variables are defined and introduced in Eq. (3.22) to obtain Eq. (3.24).

$$\eta = \frac{\delta}{\delta_0} Z = x\left(\frac{K_\infty}{\delta_0}\right)^{1/2} \tag{3.23}$$

$$\frac{\mathrm{d}^2\eta}{\mathrm{d}Z^2} + \left(\frac{-B}{\sigma K_\infty \delta_0^4}\right)\frac{1}{\eta^4} = 1. \tag{3.24}$$

A dimensionless parameter, α, is defined next as,

$$\alpha^4 = \frac{-B}{\sigma K_\infty \delta_0^4} \tag{3.25}$$

The factor α is a measure of the deviation of a specific meniscus from the equilibrium situation. For the equilibrium case, $Q = 0$ and $\alpha = 1$. Equation (3.24) can be integrated to obtain the following expression for the slope of the meniscus.

$$\left(\frac{\mathrm{d}\eta}{\mathrm{d}Z}\right)^2 = 2\eta + \frac{2}{3}\frac{\alpha^4}{\eta^3} + C_1 \tag{3.26}$$

Where C_1 is the constant of integration and using the boundary condition that at $\eta = \alpha$, $\mathrm{d}\eta/\mathrm{d}Z = 0$, the slope of the meniscus can be expressed as

$$\frac{d\delta}{dx} = -(K_\infty \delta_0)^{1/2}\sqrt{2\eta + \frac{2}{3}\frac{\alpha^4}{\eta^3} - \frac{8}{3}\alpha} \tag{3.27}$$

Hence if the curvature at the thicker end of the meniscus, K_∞, along with B, δ_0, and σ are known, the slope of the meniscus can be directly calculated as a function of the film thickness, using only the augmented Young–Laplace Equation. The minus sign in Eq. (3.27) is indicative of the fact that for the reference frame selected, the meniscus slope should always be negative (film thickness decreases as distance increases).

3.7 Experimental Results

The film thickness profiles on the silicon substrate are measured for four inclination angles (5, 7, 8 and 9°) of the set-up (against gravity) at isothermal conditions. Figure 3.4 illustrates the change in the film thickness profile with changes in angle of inclination from 5 to 9°. From Fig. 3.4, it is clear that as angle of inclination increases the adsorbed film thickness decreases appreciably and the shape of the film become steeper due to the film adjusting itself to a higher opposing body force field (gravity) acting in the opposite direction that drains the film and results in a thinner film. The variations in adsorbed film thickness and curvature of the capillary meniscus with angle of inclination are calculated and are shown in Table 3.1.

When the heat input to the system is increased to 1.8 W, the film starts to oscillate in forward and backward manner. The oscillation is captured in video and frame

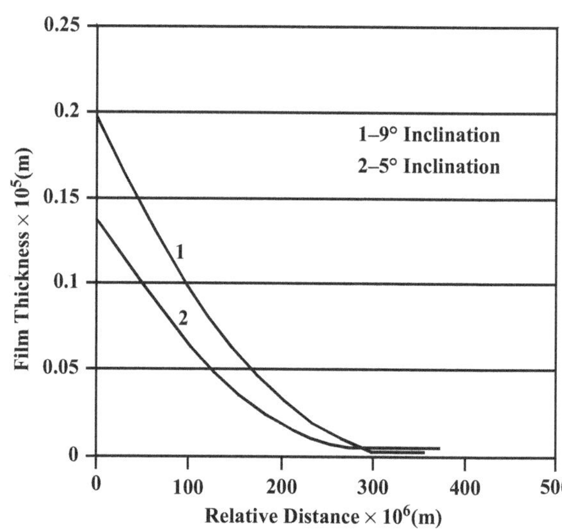

Fig. 3.4 Film thickness profile at isothermal condition for two inclinations on a silicon substrate

Table 3.1 Variation of adsorbed film thickness (δ_0) and curvature in the capillary region (K_∞) with angle of inclination for zero heat input cases on a silicon substrate

Angle of Inclination (°)	Adsorbed Film Thickness (A)	Curvature (m^{-1})
5	347	60.96
7	193	73.20
8	127	77.47
9	107	95.87

by frame analysis is used to evaluate the film thickness, slope and curvature profiles during different stages of oscillation. The adsorbed film thickness increases during the forward movement and decreases during the backward movement. The variations in adsorbed film thickness and contact angles with time for oscillatory movements of the film are presented in Figs. 3.5 and 3.6 respectively.

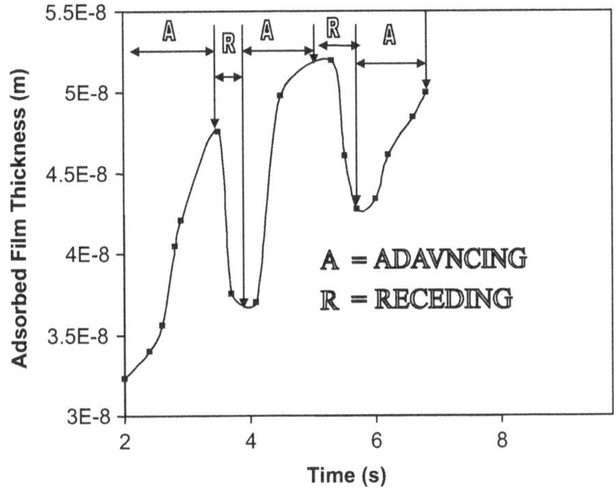

Fig. 3.5 Variation of adsorbed film thickness during oscillation on a silicon substrate

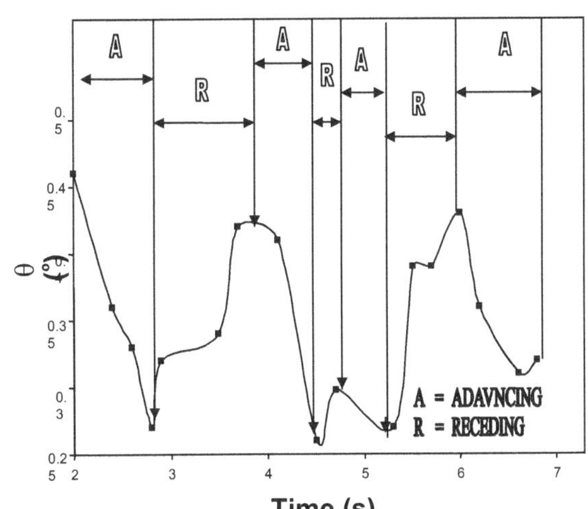

Fig. 3.6 Variation of contact angle during oscillation on a silicon substrate

The contact angle decreases during the forward movement as is expected during spreading of a meniscus. The curvature of the film increases when the film moves backward and decreases during the forward movement (spreading). The curvature varies from 130 to 250 m^{-1}.The oscillating film velocity is calculated from the estimated wetted wall lengths. The positive value of the moving velocity of the film signifies that the film is in the advancing state i.e., the contact line moves towards the flat film. The negative value of the velocity signifies that that film is receding. A representative velocity variation during the oscillation is presented in Fig. 3.7

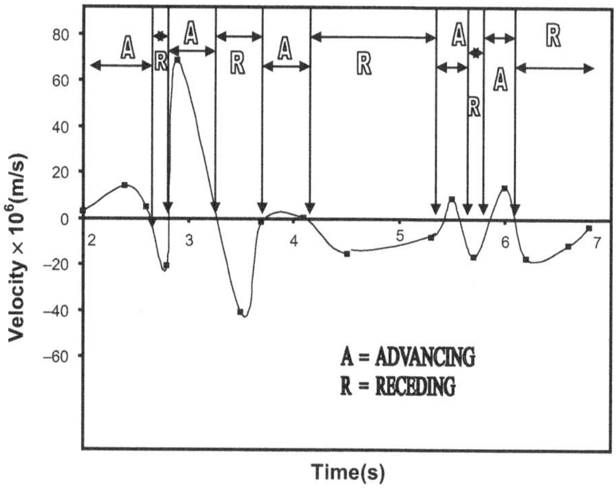

Fig. 3.7 Variation of velocity during oscillation on a silicon substrate

The method described in the theoretical section along with the experimental results can be used to evaluate the intermolecular force parameter, in situ, for a micro-device where evaporation from an extended thin liquid film is taking place. The unknown values of Hamaker constant (or dispersion constant, depending on the adsorbed film thickness) for a system can be obtained by comparing the slope of the meniscus through numerical analysis of the experimental data and the slope predicted by the augmented Young-Laplace equation (Eq. 3.27). The slope predicted by Eq. (3.27) is a function of α and the α corresponding to the closest match between these two slopes is selected to determine the value of the dispersion constant. The results for a heptane meniscus on a high refractive index glass from a related study [13] are presented in Table 3.2. It is to be noted that the required wavelength dependent optical properties for theoretical calculation of Hamaker constants are not

Table 3.2 Characteristics of the Heptane meniscus (zero heat input) (From [17])

Angle of Inclination(◦)	δ_0(nm)	$K_\infty(m^{-1})$	α	B_{expt} (Jm)
0	20.1	213	0.832	-3.21×10^{-31}
5.21	18.3	403	0.881	-5.24×10^{-31}
11.32	15.1	430	0.941	-3.38×10^{-31}
12.64	10.5	408	0.821	-0.434×10^{-31}
13.94	10.93	486	0.861	-0.738×10^{-31}

available for the high refractive index glass. However, approximate calculations suggest the obtained values of Hamaker constant are smaller than those predicted by the DLP theory. This trend has been reported by a number of researchers [17, 20] and is a direct result of possible contamination of the surface that drastically reduces the surface energy and underscores the need for in-situ evaluation of Hamaker constants for interfacial force field characterization.

3.8 Conclusions

Significant enhancement of transport rates can be achieved by device miniaturization due to the smaller length scales, transfer areas and high surface to volume ratio. However, additional forces start to become important in smaller length scales and influence the basic physics of the transport processes. For example, the shape-dependent interfacial stress field controls fluid flow and heat transfer in thin liquid films and controls the important processes of spreading and wetting. Even though the theoretical concepts are available, the extremely sensitive nature of these forces and their dependence on surface cleanliness necessitate experimental characterization of a surface.

The developed techniques describe a method for in-situ force field characterizations of evaporating and oscillating menisci on a silicon substrate and high refractive index glass. Image analyzing interferometry technique is used for accurate measurement of the liquid film thickness profile, including an estimate of the adsorbed film thickness. Adsorbed film thicknesses, curvature and contact line velocities of the oscillating film are obtained by an improved data analysis procedure. A model based on the augmented Young-Laplace equation is used to study the change in the effective pressure at the liquid-vapor interface. The model accurately captures the physics of the process and can be used to characterize the interfacial force field by evaluating the dispersion constant in-situ for the vapor-liquid-solid system. The instabilities associated with an evaporating, extended meniscus on the silicon surface are studied. As the heater power input increases, both the apparent contact angle and curvature increase and the meniscus start to oscillate at higher power inputs. The film thickness profiles demonstrated the spreading of the meniscus during advancing situation as well as the presence of a curvature gradient near the contact line region.

References

1. Derjaguin, BV and Zorin AM (1957) *Proc.* of 2nd Int. Congr. Surface Activity (London) 2, Butterworths Scientic Publications Ltd., London, England, 145.
2. Derjaguin BV and Churaev NV (1976) Colloid J. USSR 38:438.
3. Derjaguin BV, Nerpin SV and Churaev NV (1965) Bull Rilem 29:93.
4. Wayner PC Jr. (1991) Colloids Surf. 52:71.
5. López PG, Miksis MJ and Bankoff SG (1997) Phys. Fluids 9:2177.
6. Sharma A (1998) Langmuir 14:4915.

7. Poulard C, Benichou O and Cazabat AM (2003) Langmuir 19:8828–8834.
8. Wee S, Kihm KD, Pratt DM and Allen JS (2006) J. Thermophys. Heat Transfer 20:320.
9. Wang H, Garimella SV and Murthy JY (2007) Int. J. Heat Mass Transfer 50:163.
10. Panchamgam S, Chatterjee A, Plawsky JL, and Wayner PC Jr. (2008) Intl. J. Heat Mass Tran. 51:5368.
11. Babin BR, Peterson GP and Wu D (1990) ASME J. Heat Tran. 112(3):595.
12. Li J and Peterson GP (2007) Intl. J. Heat Mass Tran. 50:2895.
13. Park K, Noh K and Lee K (2003) Intl. J. Heat Mass Transfer 46:2381.
14. Wayner, PC Jr., Kao YK and LaCroix LV (1976) Int. J. Heat Mass Tran. 19:487.
15. de Gennes PG (1985) Rev. Mod. Phys. 57:827.
16. Zheng L, Plawsky JL, Wayner PC Jr. and DasGupta S (2004) ASME J. Heat Tran. 126:169.
17. Argade R, Ghosh S, De S and DasGupta S (2007) Langmuir 23(3):1234.
18. Israelachvili JN (1992) Intermolecular and Surface Forces, 2nd ed., Academic Press, New York.
19. Dzyaloshinskii IE, Lifshitz, EM and Pitaevskii LP (1961) Adv. Phys. 10:165.
20. Troung G and Wayner PC Jr., (1987) J. Chem. Phys. 87:4180.
21. Beaglehole D, Radlinsh EZ, Ninham BW and Christenson HK (1991) Langmuir 7:1843.
22. Beaglehole D and Christenson HK (1992) J. Phys. Chem. 96:3395.
23. Panchamgam S, Gokhale SJ, Plawsky JL, Wayner PC Jr. and DasGupta S (2005) ASME J. Heat Transfer 127:232.

Chapter 4
Bio-Microfluidics: Overview

Coupling Biology and Fluid Physics at the Scale of Microconfinement

Tamal Das and Suman Chakraborty

Abstract With a view to establish unique interfacial synergistic interactions between two seemingly distant fields of microfluidics and biology, Bio-microfluidics has become a progressive arena of research in recent times. Bio-microfluidic tools in the format of lab-on-a-chip devices have been extensively utilized to uncouth hitherto un-illuminated regions of cellular-molecular biology, biotechnology and biomedical engineering. This chapter elaborately delineates the linking between the fundamental microscale physics and biologically relevant physico-chemical events and how, in practice, these relations are exploited in microfluidic devices. Finally, potential directions of future biomicrofluidic research are also discussed.

Keywords Lab-on-a-chip · Micro Total Analysis System · Microfabrication · Microchannel · Combinatorial Chemistry · Micromixing · Microarray · Cellomics · Genomics · Proteomics · Diffusion · Dispersion · Drug Delivery · Microneedle · Polymerase Chain Reaction · Field-Flow Fractionation · Reaction Kinetics · Enzyme Assay · Electroosmosis · Electrowetting · Dielectrophoresis · Electrophoresis · Ferrofluid · Fluorescene · Microscopy · Confocal Microscopy · Optofluidics · Flow Visualization · Biosenor · Structural Biology · Capillary · Electrophoresis

4.1 Introduction

Microfluidics deals with the transport of minute volumes of fluid (typically, sub-nanoliter) through channels having at least one of three dimensions of the order of micrometer [1]. Though, initially microfluidics stemmed out of two distant fields, namely, analytical chemistry [2] and microfabrication [3], soon potentials of the subject were to be unleashed in the field of biology [4]. During the past few years, its

S. Chakraborty (✉)
Department of Mechanical Engineering, Indian Institute of Technology, Kharagpur, Kharagpur-721302, India
e-mail: suman@mech.iitkgp.ernet.in

S. Chakraborty (ed.), *Microfluidics and Microfabrication*,
DOI 10.1007/978-1-4419-1543-6_4, © Springer Science+Business Media, LLC 2010

scope has stretched beyond exploring exotic transport phenomena and low Reynolds number fluid physics into the domains of biochemical analysis [5, 6]. The urgency of invoking microfluidic devices in solving relevant chemical and biochemical problems has been obviated because of two major beneficiaries that they have in promise. First, within microfluidic confinements, the assay volume requirement of liquid analytes has been discovered to be unprecedentedly low. Second, due to its intrinsic augmented surface area to volume ratio over reduced length scales, microfluidic devices can offer enhanced throughput, reduced reaction time and higher sensitivity [7, 8]. Subsequently, adopting microfluidics in biology has revolutionized the paradigms of molecular biology, biochemistry and bioengineering in such a magnanimous extent that relevant fundamental science and applications are classified by the researchers under the tenet of a separate subject hailed Biomicrofluidics [4]. The initial flourish of Biomicrofluidics has been facilitated by the expanding the necessity of achieving faster and high throughput Genomics during the past decades [9]. However, in post genomic era, Biomicrofluidics based applications seem to sprout everywhere in the vistas. Its spectrum encompassed vast bio-domains ranging from cell biology [10] to protein crystallization [11], from nucleic acid isolation [12–14] to lethal virus detection [15]. Now, the subject has become progressive enough to miniaturize bulk of the analytical experiments performed in laboratory scale within few square centimeter space of a monolithic platform and specific jargons such as Lab-on-a-Chip (LoC) and micro Total Analysis Systems (μTAS) have become cliché in the scientific world [16–18]. Microfluidic systems have been demonstrated to possess potential in diverse spectra of biological applications, encompassing molecular separations, enzymatic assays [19], polymerase chain reactions [20], and immunohybridization reactions [21]. These are outstanding individual instances of down-scaled methods of laboratory techniques, but there also exist stand-alone functionalities, analogous to a single component integrated circuit. Given that the present day industrial approaches to address pertinent large-scale biological integration have emerged in the form of gigantic robotic based fluidic platforms which consume substantial space and expenses, Biomicrofluidics has a straightway objective of replacing them with powerful miniaturization [6]. This way, its projected functionalities are quite similar to the silicon based intergrated circuit that replaced spacious valve-devices during VLSI (Very Large Scale Integration) chip revolutions of computation industry.

The very purpose of microfluidics devices is to deal with samples often dissolved in an aqueous phase and then, manipulate the system through characteristic analytical procedures such as heating, mixing and separation. Subsequently, processed solutions may be transported to some form of a detector or sensor and the data is acquired. Microfluidic channel networks [22], generally and economically fabricated in a monolithic platform [23] made of a moldable silicon based polymer such as Polydimethylsiloxane (PDMS) or glass, include features such as separators, mixers, valves and injectors which are essentially microscale counterparts of existing macroscale analytical and bioreactor process components. However, in an inexpensive mode, microchannel networks may be fabricated in compact disc like platform and therein, fluid flows may be actuated by centrifugal and coriolis forces

[24, 25]. Once the device has been designed and manufactured, what becomes indispensable here is to achieve apposite micro-macro interfacial connections to user accessible macroscale input-output components. The microfluidic technology incomplemented with macroscale interface, permits the consistent maneuvering of small sample volume with ensured reproducibility and accuracy.

Two of the most early Biomicrofluidics appliances were microfluidics based genomic microarrays, i.e., gene chips and microchannel enhanced capillary electrophoresis. In gene chips, disease specific marker complementary oligonucleotides are arrayed on the microchannel system and while, Deoxyribonucleic acid (DNA) samples isolated from a patient are passed through the system, any presence of disease specific DNA fragment can be promptly detected through chemical, fluorescence or electrical sensing means [12, 26, 27]. What microfluidics imparts here is the enhanced effective reaction rate, augmented mass transfer and low sample volume requirement (Fig. 4.1). Arraying multifarious disease specific markers together in a microfluidic based genomic microarray system, excellent parallel processing capability has been attained yielding simultaneous prognosis of several ailments. If gene chips have been able to transform the arena of DNA detection and quantitation, capillary electrophoresis [28, 29] has undoubtedly influenced separation and isolation of targeted gene specific DNA sequences. Juxtaposing them in a single platform, researchers have amazingly reduced the assay time from hours to seconds. With time, ever widening applications of microfluidic technology have also

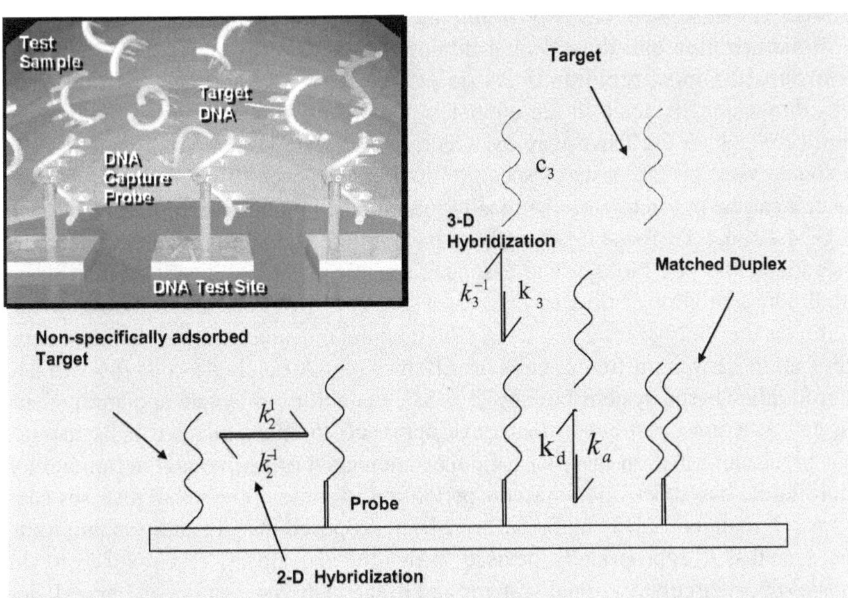

Fig. 4.1 Schematic delineation of DNA hybridization in a microfluidic platform. Surface phase (2D) hybridization in conjunction with bulk phase (3D) reaction augments the effective hybridization rate [27]

included isolation and detection of gene specific RNAs [30]. However, though capillary electrophoresis technique has been instrumental in unleashing high throughput genomics, its applications have been most prevalent in detection and separation of proteins which are physiologically important and expressed in trace level. Because of the fact that proteins of disparate functionalities may not only differ in sequence or size but also in structural conformation; separation, purification and identification of protein samples (collectively known as proteomics) are inarguably more complicated than genomic samples [31–33]. Given that a human sample contains over twenty thousand different proteins and each of them has its own characteristics isoeletric point, solubility, polarity and identifying reagents, designing a generic macro-scale proteomic assay system remains insurmountably elusive. On this very drawback, high-throughput complex biomicrofluidic systems for proteomics are emerging rapidly [34]. Microfluidic systems either performing or coupled with isotachophoresis [35], isoelectric focusing [36, 37] and mass spectrometry systems [38, 39] have become the solutions to new generation proteome biology. In another corner of proteomics, microfluidic devices have provided the most proficient solutions for protein crystallography. Three dimensional structures being uniquely deciphered from Nuclear Magnetic Resonance (NMR) and X-Ray Crystallography analysis, synthesis of protein crystal is the most indispensable stage in structural biology [40, 41]. Protocols to generate protein samples are seldom unambiguous and robust, requiring delicate control over process temperature, ionic concentration, pH and evaporation rate. Such enormity of preciseness in temporal variation of the parameters can be achieved only within microfluidic confinement.

Miniaturization and dimensional diminution in case of microchannel network have imparted most prolific effects on cell biology. Considering that biological cells dimensionally scale in the order few tens of micrometers, microfluidic platform provides an exclusive way by which they can be handled individually [10]. In some cases, on the basis of specific requirement, even different parts of a single cell can be physically or chemically manipulated though microfluidics [42–44] (Figs. 4.2 and 4.3). These functionalities have implicated utterly unfathomed niches in fundamental cell biology and associated medical diagnosis [45]. While differential addressability of diverse parts of a single cell or several neighboring cells facilitates the study of intra and intercellular signal transduction (i.e. chemical communication between different cells or different parts of a single cell) [46–50] and microfluidics based system biology [51, 52], the ability to isolate and study a single cell at a time bestows a distinctive approach to study infected cells in vitro [53–56]. Relevantly, on the basis of either attenuated or augmented deformability, microfluidic based detection systems pertaining to some of the lethal diseases such as cancer, malaria, AIDS and SARS have been proposed [57]. In addition, microfluidic systems, if appropriately devised, providing the closest resemblance to the physiological circulatory-renal systems and tissue matrices, confers an unparalleled platform for in vitro simulation of in vivo cellular behavior [48]. Not only that; the undeniable similitude between microchannel systems and blood vessels has encouraged researchers to adopt the technology of microfluidics in the domain of artificial tissue engineering [58]. Having explicated the need and the promising applications

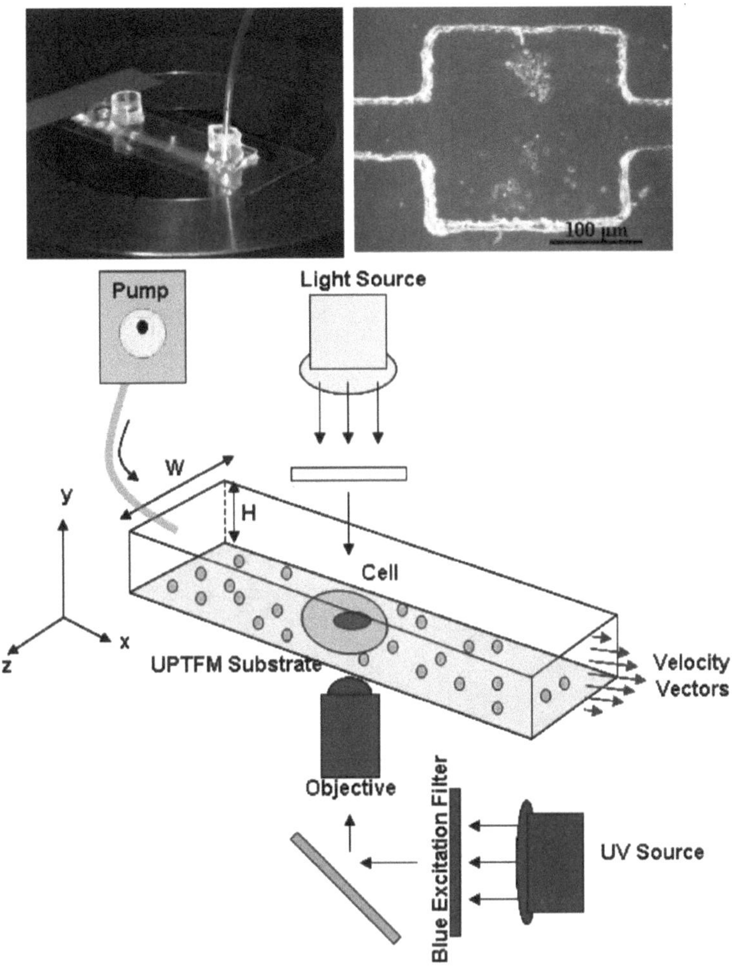

Fig. 4.2 An integrated microfluidic cell culture platform coupled with traction force microscopy [44] system. (*Top Right*) Square shaped microwells with adhered HeLa cells

of microfluidics technology in biological paradigm, following we discuss different physical methods and factors predominant in microfluidic species transport, with properly emphasizing relevant biological utilizations.

4.2 Diffusive Transport of Biochemical Species

Diffusive transport is intrinsic. Diffusion is the process by which a concentrated group of particles in a volume will, by Brownian motion, spread out over time so that the average concentration of particles throughout the volume is constant [59]. Under a finite temperature, diffusion arises because of the collisions between solute particles or molecules and solvent molecules. Between two successive collisions, solute

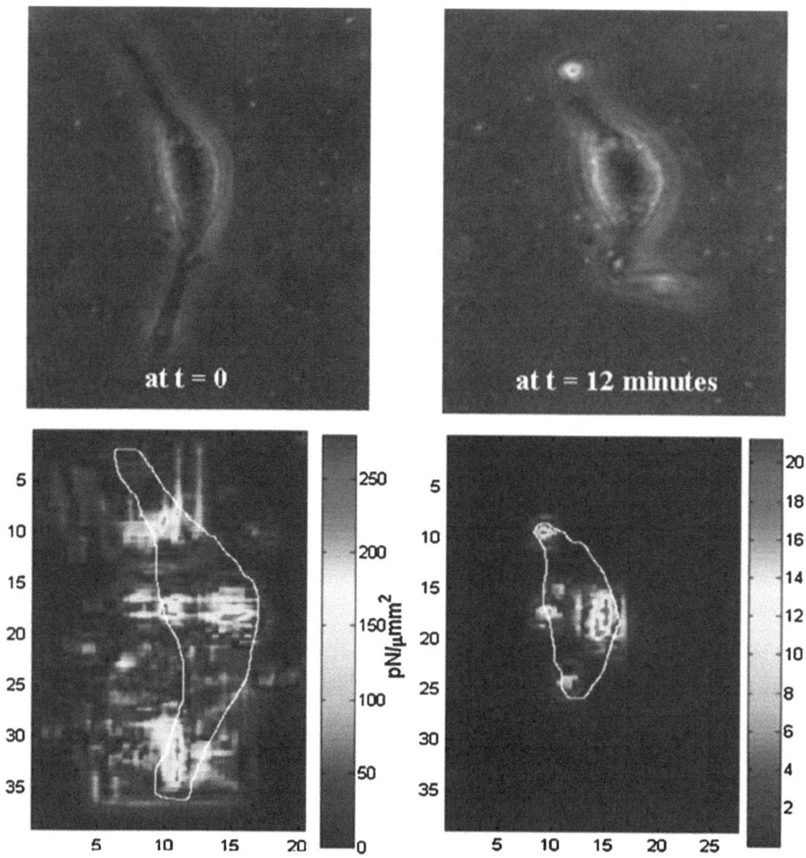

Fig. 4.3 Change in the traction force landscape as L929 mouse fibroblast cell is detached by enzymatic treatment

molecules, in this case biological macromolecules such as proteins, nucleic acids, hormones and other organic molecules, move in a straight line, ensemble average of which is called mean free path. In a collision, molecules change their velocity and direction stochastically, spreading over all around the solvent volume and homogenizing solution concentration. Mathematically, diffusive flux (J) i.e. concentration or number of molecules crossing unit surface during unit time is proportional to the gradient of concentration (∇c), with proportionality coefficient called as diffusion coefficient (D). This law is known as Fick's law and is given as follows:

$$J = -D\nabla c \tag{4.1}$$

In this equation minus sign arises out of the fact that diffusion proceeds opposite to the concentration gradient of solute molecules. The argument goes like this. If we imagine a hypothetical barrier between two volume segments of the solvent

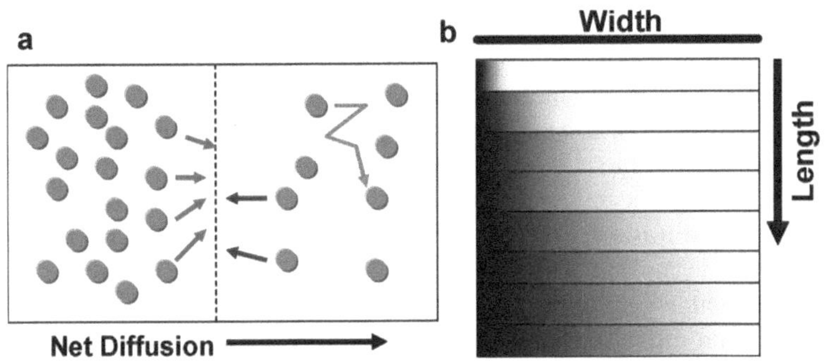

Fig. 4.4 (**a**) Net diffusion takes place opposite to the concentration gradient. (**b**) Diffusive mixing along the length of a microchannel

(Fig. 4.4), molecules from both sides should cross the barrier. However number of crossing molecules is higher from high concentration to low concentration direction than the reverse one. Thus net molecular transport occurs against the concentration gradient till the solution becomes homogeneous. The dimensionality and unit of diffusion coefficient is given as [Length]^2/[Time] i.e. m^2/s respectively. Then, a generalized convection-diffusion equation is obtained by putting Fick's law into the mass balance equation, yielding

$$\frac{\partial c_i}{\partial t} + v.\nabla c_i = \nabla.(D\nabla c_i) + R_i \tag{4.2}$$

And in expanded form

$$\frac{\partial c_i}{\partial t} + \left(v_x \frac{\partial c_i}{\partial x} + v_y \frac{\partial c_i}{\partial y} + v_z \frac{\partial c_i}{\partial z} \right) = D \left(\frac{\partial^2 c_i}{\partial x^2} + \frac{\partial^2 c_i}{\partial y^2} + \frac{\partial^2 c_i}{\partial z^2} \right) + R_i \tag{4.3}$$

Here, c_i is the concentration of a ith biomolecular species, v_x, v_y and v_z are x, y and z components of velocity$v$. R_i is the reactive term of ith biomolecular species. In general, for standard biomolecular species, typical magnitudes of D range in order of 10^{-10}–10^{-9} m^2/s. In general, approximating dilute solution, Diffusion coefficient can be evaluated as:

$$D = \frac{k_B T}{6\pi \mu R_H} \tag{4.4}$$

Where, k_B, T, μ and R_H are Boltzmann Constant, absolute temperature, solvent viscosity and hydrodynamic radius of solute molecules. For Brownian motion in dilute solution, root mean displacement is directly proportional to the time interval (Δt) during which the displacement has been studied, yielding the following relationship:

$$\langle x^2 \rangle = 6D\Delta t \tag{4.5}$$

However, in pertinence to intracellular diffusion or diffusion of molecules over biomembrane, this does not hold true for most of the studied macromolecules. There are three specific reasons namely molecular crowding, facilitated transport and diffusive inhomogeneity of solvent. Firstly, cytosol can not be approximated as dilute solution. In fact it is molecularly "crowded" to such an extraordinary extent that the diffusion is significantly damped and the molecules are consequently thought to undergo "subdiffusion" [60, 61]. In contrasting situation, intracellular propagation of specific molecules can be facilitated by molecular motors or secondary messenger. Though this is intrinsically a reactive mechanism, on the ground of coupled reaction-diffusion mass transport theory, reactive terms can be absorbed into the diffusion part, yielding a resultant augmented diffusive transport, known as "superdiffusion" [62, 63]. There is another way which is commonly believed to be a prime source for anomalous diffusion. Cytosol or biomembranes are highly heterogeneous and the heterogeneity varies in both space and time. Though, in principle, diffusion in heterogeneous medium can be modeled in the following way:

$$\begin{Bmatrix} j_x \\ j_y \\ j_z \end{Bmatrix} = - \begin{bmatrix} D_{xx} & D_{xy} & D_{xz} \\ D_{yx} & D_{yy} & D_{yz} \\ D_{zx} & D_{zy} & D_{zz} \end{bmatrix} \nabla c \tag{4.6}$$

These equations are significantly cumbersome and computationally expensive for obtaining even a numerical solution let alone any possibility of achieving an analytical expression. In the wake of such imposed problem, the diffusion equation (4.5) is empirically modified as

$$\left\langle x^2 \right\rangle = 6D_{eff}\Delta t^{\alpha} \tag{4.7}$$

Where D_{eff} is the effective diffusion coefficient and α is scaling exponent parameter. Sub and super diffusion is respectively defined for two regimes of α i.e. $0 \leq \alpha < 1$ and $\alpha > 1$.

While most of the problems in complete form of Eq. (4.3) are only numerically solvable, there are few idealized cases for which the relatively simple analytical solutions are attainable. One of cases we consider here is the diffusion of biomacromolecules in one-dimension (x) from a point source. The point source at $x = 0$ is mathematically modeled as

$$\begin{aligned} c(x, t = 0) &= c_0 \text{ at } x = 0 \\ &= 0 \text{ elsewhere} \end{aligned} \tag{4.8}$$

Where c_0 is initial concentration of the point source. Subsequently, $c(x,t)$ is solved as following

$$c(x, t) = \frac{c_0}{\sqrt{4\pi Dt}} e^{-\frac{x^2}{4Dt}} \tag{4.9}$$

The above equation suggests that with time, the solute font will take the shape of a progressively spread Gaussian curve with its maxima at $x = 0$. This spreading is known as band broadening and is one of the most common phenomena encountered while investigating biological species transport through microcapillary. It is important to realize that with an imposed velocity as in the case of capillary electrophoresis (CE), high performance liquid chromatography (HPLC), the shape evolution of solute band remain grossly unperturbed; only it maxima gets displaced along x-direction with time.

The effect of diffusive transport is most appreciably featured in biomicrofluidic version of Polymerase Chain Reaction (PCR) [20, 64]. This is a technique that has changed the direction and momentum of molecular biology which since 1950s has been predominantly dominated by reductionist ideology. PCR has enabled the researchers to fathom into the hitherto unexploited terrain of individual gene and their function. Note that the PCR and its variants have opened up new vistas of systems biology where interaction between many gene products (i.e. proteins) and the underlying regulation dynamics are being elucidated with enormous rabidity. PCR basically amplifies the chosen sequence portion of a nucleic acid chain (it may be chromosome or any DNA fragment). The working principle is based upon DNA polymerase's (the enzyme that copies a DNA sequence) dependency on the presence of a short nucleic fragment, called primer bound to its complement sequence segment of the genome. Granted an appropriate physicochemical environment, DNA Polymerase binds to the primer and commences it copying mechanism at that very location. In PCR, these primer sequences are strategically chosen such that only the desired stretch of nucleic acid sequence (i.e. template) is copied. Once produced, these newborn copies start acting as templates, amplifying the sequence in subsequent cycles of reaction. It is estimated that within 20–30 PCR reaction cycles, the desired sequence can be amplified as much as 10^6 times in number. Though it is quite fiddling to handle one single nucleic acid chain, one can, sure enough, work with one million identical copies of it. This is the manipulation of life's digital information at its best. Technically, single PCR cycle comprises of three major reactive substeps – denaturation, annealing and extension which mainly differs by its operating temperature. In typical PCR reaction cycle, these temperature values are 95°, 55–60° and 72°C respectively. Hence, for successful completion of single PCR reaction, a device should be able to ramp up and down the reaction zome temperature at frequent basis with uncompromisable precision and the advent of microfluidics based PCR devices has banked upon this very constraint. It is understandable that within the purview of augmented surface area per unit volume of microfluidic confinement, heat transfer in and out of the reaction mixture must occur at faster rate than what is experienced in conventional macroscale PCR devices, reducing the whole operation time from hours to seconds. Microfluidics based PCR systems have been initially implemented as chamber stationary component [64, 65] and therefore, lacked the flexibility of controlling reaction rate. They have been eventually replaced by flow-through and thermal convection-driven microfluidic devices where reactive fluid is transported back and forth between three distinct temperature zones maintained within a monolithic platform (Fig. 4.5).

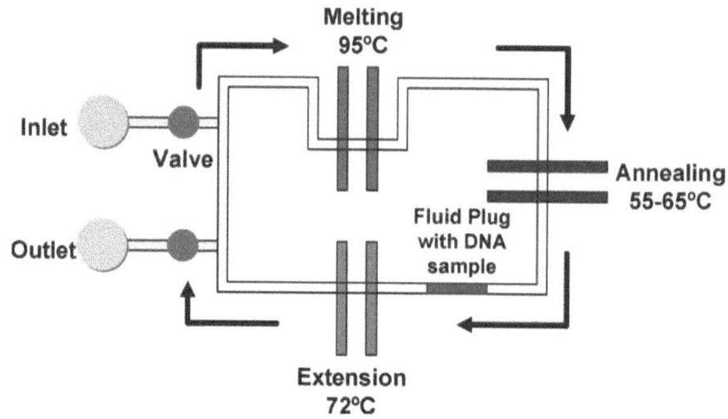

Fig. 4.5 Schematic representation of micro PCR

This clan of PCR systems precedes several variants such as Single straight capillary based flow-through PCR microfluidics, On-chip serpentine rectangular channel based flow-through PCR microfluidics and Circular arrangement of three temperature zones for flow-through PCR microfluidics. If it appreciated that the cycle time of PCR depend on the synthesis rate of the polymerase and subsequent diffusion time, diffusion plays a pivotal role in microfluidic PCR systems. Not only this, diffusion possesses its importance in mixing of several ingredients required for a successful PCR. Essentially these components are transported as separate entities dissolved in either similar or different buffers. Within microfluidic framework they coalesced or merged and in consequence, mixing occurs by diffusion. In absence of turbulence which is often serves as facilitator of mixing in larger scale devices, Brownian dynamics governed diffusive mixing emerges as the key player. One must note that rapid formation of a homogeneous solution is the most important step to achieve a uniform and maximum efficiency reaction. For microchannels diffusion accelerates the process.

While in macro-systems, random Brownian displacement scaling $\sqrt{6Dt}$, diffusive mixing can take intolerably long time to serve as the sole method of mixing, in microscale system the situation is quite reversed because of mainly two reasons. First, due to the reduced dimensions, characteristic time scale for diffusive propagation is comparable to the time scale of convective transport and second, diffusion requires no external power input. Being an intrinsic process which solely stems out of path of increasing entropy, it is energetically inexpensive. Hence, for natural Microsystems which has been evolving for billion years to maximizing it input versus output, diffusive mixing is an inexorable solution to mass transport problem. Most of the intracellular transport processes ferrying small molecules such as potassium, calcium ions or comparatively bigger macromolecules such as proteins, RNAs, takes the advantage of molecular diffusive transport (also know as passive transport). Only when the transport in the direction of increasing chemical concentration appears inevitable, diffusive mixing is replaced by the active processes involving consumption of intracellular energy in form of adenosine tri-phosphate (ATP) molecules. In case of extracellular transport, active processes being inaccessible, the role of diffusion becomes even more vital. One medically relevant case is the diffusion of drug molecules through biological cell and tissues. There are three sets of governing parameters that influence drug adsorption-namely its physicochemical properties, chemical formulation and route of administration. There exist several dosage forms such as tablets, capsules and solutions which additionally include other ingredients. A specific dosage form is preferred for a particular drug which is then appropriately administered by various routes encompassing oral, sublingual, parenteral, buccal, rectal, topical and inhalational. Irrespective of the administration route, drugs should be in solution to be absorbed into the specific cell or tissue. Relevantly, drug molecules are required to cross numerous semi-permeable cell membrane barriers prior to reaching the systemic circulation. The process is executed through passive diffusion, facilitated passive diffusion, active transport, receptor mediated endocytosis or pinocytosis. Among these mechanisms, the passive diffusion is the most common and energy inexpensive manner by which the intracellular inclusion of drug molecules takes place. Drug-diffusion predominantly occurring between high to low concentration, diffusion rate is expected to be solely proportional to the concentration gradient. However, in physiological systems, effective diffusive coefficient of the drug molecules depends upon lipid solubility of drug, molecular density of extracellular space, molecular ionization and the area of absorption surface. Cell membrane being composed of lipid bilayer, diffusion of small un-ionized lipophilic drugs is favored to the highest extent. Given that the ionized molecules are weakly lipophilic, the administration route of a particular drug depends on pKa of the drug and pH of the relevant physiological fluid onto which the drug is predominantly absorbed. Hence, while for weak acids, administration through low pH medium such gastric fluid (pH 1.5) is preferred, weak bases are principally injected directly into the blood stream (pH 7.4).

The ability to perform assays in miniaturized scale and high-throughput screening of large number of chemicals simultaneously have opened a new vistas in drug discovery and screening with microfluidics based analytical devices [66, 67]. There are several novel and upcoming spectra in this realm of application including single cell based protein isolation, analysis and ligand screening [68, 69], rapid compound generation by microfluidic combinatorial chemistry [2, 70, 71] and high-speed selection of active and pharmaceutically relevant compound. The science of microfluidics has also contributed extensively in field of drug delivery. In recent years, microfluidics-based strategies have been utilized in drug delivery devices to improve the operation time and accuracy of the process [72]. Sophisticated photolithographic procedures have been exploited to fabricate futuristic drug delivery instruments. Devices with an array of microneedles competently preserve the chemical activity of a drug compound and enable administration with characteristic cellular scale precision and most importantly, with minimal pain. Researches related to the application of microneedles (Fig. 4.6) for gene and drug delivery are essentially categorized into three broad classes namely local delivery, systemic delivery and cellular delivery [73, 74]. Microfluidics based highly localized drug delivery systems have been able to reduce the applicable doses and side effects due to non-specific drug adsorption. In addition to low molecular weight organic compounds, there are several classes of macromolecules such proteins,

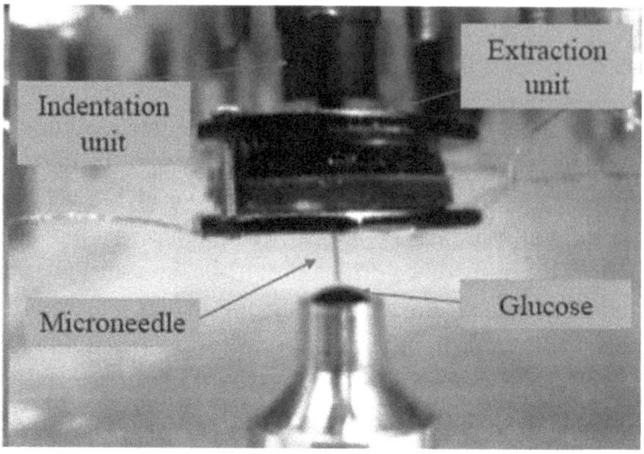

Fig. 4.6 Working principle of a microneedle which mimics the blood sampling technique of female mosquito [74]

peptides and oligonucleotides which can be successfully injected using microneedles. Performance of modern microfluidic drug delivery systems and the aptitude of sustained drug release have been significantly augmented by invoking complex fluidic circuits including a cluster of micropumps, microvalves and feedback sensors into a single microfabricated platform. Moreover, self-regulated drug delivery microdevices whose operating mechanism relies upon the biomolecular detection by "smart" polymeric compounds with interconnected feedback loops have unleashed a unique array of novel possibilities towards future advancements.

4.3 Particle Transport, Dispersion and Mixing in Biomicrofluidics

Comprehensive understanding of the local fluid dynamics in terms of the specific functional objectives of a microdevice is the key to implement a novel and effective design. In the regime of micrometer length scale where surface forces override volume forces, from the perspective of design and fabrication, the projected transport, mixing, separation and manipulation of particles hold the positions of utmost importance. One useful tool of the systemic approach is definitely the dimensional analysis. Here, one may ignore the terms where Reynolds number appear as a multiplying pre-factor after normalization. Instead, flow physics in the confined system are thought to evolve around other dimensionless parameters, representing the mutual ratio of surface, electrical, magnetic and thermal forces. In addition, in microscale flow physics, channel geometry has been noted to play an important role [75–77].

4.3.1 Dispersion

The supremacy competition between convection and diffusion is fatefully significant for many Biomicrofluidics application involving mass transport and chemical reactions. In a pressure-driven flow, dispersion of the solute generally occurs much more elaborately than the theoretical prediction considering the sole effect of diffusion. In this case, every streamline of parabolic flow profile is browsed by solute particles due to their inherent Brownian motion [78]. The aforementioned events are analytically accounted in Taylor-Aris approach and hence, this type of dispersion is recognized as Taylor-Aris dispersion [79]. In Taylor-Aris approach it is not presumed that in advection-diffusion field, net solute transport can be explained simply by superimposing advection and diffusion. Instead, effective diffusion is appreciated to be dependent on the imposed velocity field. This occurs due to the finite transverse gradient of velocity field which becomes predominant in cased pressure-driven flow situation. For fluid carrying microparticles of concentration c through a microcapillary of radius R, the axisymmetric (x,r) advection-diffusion equation can be described as

$$\frac{\partial c}{\partial t} + u(r)\frac{\partial c}{\partial x} = D\left[\frac{\partial^2 c}{\partial x^2} + \frac{1}{r}\frac{\partial c}{\partial r} + \frac{\partial^2 c}{\partial r^2}\right] \tag{4.10}$$

where the pressure-driven velocity field is assumed to vary only in radial direction and therefore, can be approximated as

$$u(r) = 2\bar{u}\left(1 - \frac{r^2}{R^2}\right) \tag{4.11}$$

where $\bar{u}$ is the mean velocity. Further, a no flux boundary condition should provide $\partial c/\partial r\,|_{r=R} = 0$ at the capillary surface. In Lagrangian coordinate system moving with average velocity of the fluid, the equation (4.10) can be rewritten as

$$\left[\frac{\partial c}{\partial t} + \bar{u}\frac{\partial c}{\partial x}\right] + \bar{u}\left(1 - \frac{2r^2}{R^2}\right)\frac{\partial c}{\partial x} = D\left[\frac{\partial^2 c}{\partial x^2} + \frac{1}{r}\frac{\partial c}{\partial r} + \frac{\partial^2 c}{\partial r^2}\right] \tag{4.12}$$

Now, in Taylor-Aris approach it is assumed that the concentration gradient along x-axis constant i.e.

$$\frac{\partial c}{\partial x} = \frac{\partial \bar{c}}{\partial x} = \text{constant} \tag{4.13}$$

where $\bar{c}$ is average concentration over the capillary cross-section and formulated as

$$\bar{c} = \frac{1}{\pi R^2}\int_{r=0}^{r=R} c2\pi r dr \tag{4.14}$$

It further implies that the term $D\frac{\partial^2 c}{\partial x^2}$ can be dropped from the equation (4.12). In fact for bulk of situation axial change in particle concentration has been revealed to be negligibly small in comparison to the radial change, superbly vindicating the aforementioned approximation. Subsequently, utilizing the relation given in equation (4.13) the solution for the reduced version (i.e. $D\frac{\partial^2 c}{\partial x^2} = 0$) of Eq. (4.12) should be obtained as

$$c = \bar{c} + \frac{\bar{u}R^2}{4D}\frac{\partial \bar{c}}{\partial x}\left[\frac{r^2}{R^2} - \frac{r^4}{2R^4} - \frac{1}{3}\right] \tag{4.15}$$

after little mathematical manipulation and rearrangement. From this relation, the net flux describing the average mass flow (Q) through unit capillary cross-section is evaluated as

$$j = \frac{Q}{\pi R^2} = \frac{1}{\pi R^2}\int_{r=0}^{r=R} uc2\pi r dr = -\left(\frac{R^2\bar{u}^2}{48D}\right)\frac{\partial \bar{c}}{\partial x} \tag{4.16}$$

comparing it with one dimension version of Fick's equation (4.1) obtained as $j = -D\frac{\partial c}{\partial x}$, one can easily notice that the effective diffusion coefficient (D_{eff}) can be analogously determined as

$$D_{eff} = \frac{R^2\bar{u}^2}{48D} \tag{4.17}$$

It is pertinent to note that Taylor-Aris approximation is valid only when $\bar{u}$ is sufficiently large to impose a strong velocity gradient in radial direction. When this is applicable, one may automatically expect that $D_{eff} \gg D$, implicating

$$\frac{R^2\bar{u}^2}{48D} \gg D \,\text{i.e.}\left(\frac{\bar{u}R}{D}\right) \gg 7 \,\text{i.e.}\, Pe \gg 14 \tag{4.18}$$

Here, Pe is the Peclet number $\left(\frac{2\bar{u}R}{D}\right)$ which represents the relative strength of convective and diffusive transport. Thus, in the regime of convective-diffusive transport through microchannel Peclet number is major non-dimensional number.

For the applications where dispersion is undesired as in solution purification and isolation, electrokinetic strategy performs superior than pressure driven mechanism owing to its innate uniform velocity profile across the channel width. However, a caveat must be sounded that in nanochannel or in very dilute ionic medium where electric double layers formed due to the surface charges overlap, even electrokinetic flow becomes parabolic yielding significant dispersion. Also the residence time of solute molecules inside the intended zone becomes a governing factor and in reduced time scale, suspended solute particles may not diffusively encompass the entire channel. It leaves Taylor-Aris dispersion analysis invalid in this case and the effectual chemical reactions are controlled by the local velocity profiles within microchannel. This generally generates an increased band broadening towards the microchannel center.

4.3.2 Mixing

It is important to note that characteristics biomolecular diffusion coefficients are of the order of 10^{-10} m^2/s. Hence, for a microchannel with 10 μm width and 1 mm/s average flow velocity, approximately 100 channel widths i.e. 1 mm length is required for the completion of mixing solely by diffusion. In order to circumvent such low-strength diffusive mixing, different inventive strategies targeting mixing by incipient transverse flow have been designed. Further, forced mixing becomes inevitable at small scales because of the macroscale-specific turbulence dependent mixing is no more accessible. In biomicrofluidic systems, mixing is achieved either by passive mechanisms [80] such as Hydrodynamic focusing, flow separation, flow split-recombination and chaotic advection. It appears spontaneously rational that augmenting the mixing of a tracer in a fluid or two fluids will be facilitated considerably by chaotic advection mechanism [81]. The essential principle of chaotic mixing is the generation of secondary flow or transverse flow patterns exploiting the action of either centrifugal forces in curved channels or purposefully patterned microchannel surfaces. In order to produce chaoatic flow profiles, stream lines should cross each other at different times. Again, according to the theory of dynamic systems, in two dimension (2D), chaotic particle motion arises only in cases with time-variant velocity field. In three dimensional flow, however, chaotic motion may be observed for even time-invariant velocity field. The incidence of chaotic advection intrinsically facilities fast distortion and elongation of fluid elements, thereby increasing the two-fluid interfacial area across which diffusion takes place. The most prominent advantage with passive mixing such as chaotic advection is the ability to be operational without requiring external energy input. However, the passive mixing process being ingrained within the architecture of the microchannel network, it is not controllable in spatio-temporal scale. Here, one may have to invoke active mixing strategies. Active micromixing depends on applications of external energy which induces disturbance in the flow fields. In active mixing, transverse flow is generated by using either hydrodynamic or electrokinetic approaches. Approaches that have been previously employed to enhance mixing performance in active format include piezoelectrics, pneumatics, acoustics, electroosmosis, dielectrophoresis, magnatohydrodynamics, electrowetting of droplets and time dependent generation of transverse flow. Among the aforementioned strategies, variations of zeta potential at the channel wall have been the most promising protocol [82]. Alternative spatial patterns of induced zeta potential create a microcirculation within the flow field, which in turn catalyses the mixing process (Fig. 4.7).

4.3.3 Separation Processes

Separation of different kinds of biomolecules of varying in size molecular weight, chemical composition or even optical isometry stays at the epicenter of analytical biochemistry. In microfluidics, for effective and fast separation of

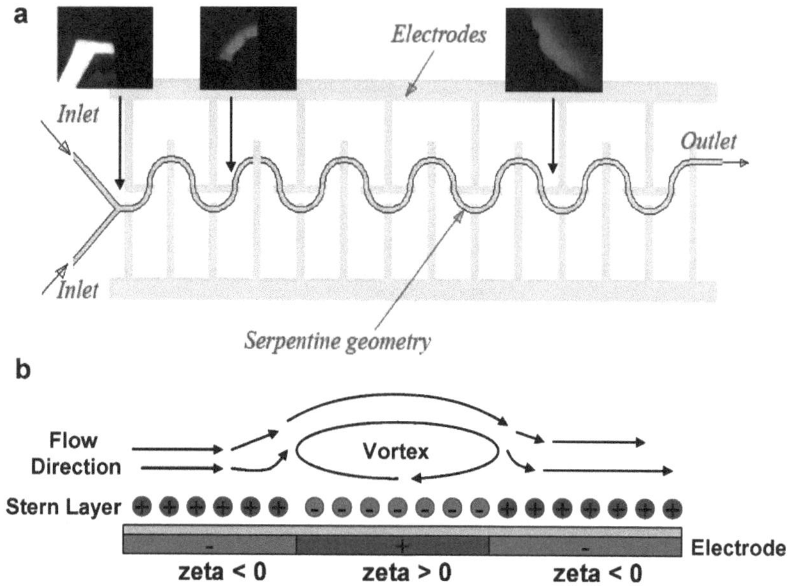

Fig. 4.7 Combined active-passive micromixing in a serpentine microchannel with alternatively charged embedded electrodes

biomolecules, electric and thermal cross fields are generally employed to achieve augmented spreading of an injected solute in the flow direction. Other methods for separation enhancement include surface modification, hydrodynamic interaction and micro-nano sized post array for preferentially hindered motion of solute particles.

Field-Flow Fractionation (FFF) is defined as the broad clan of separation methodologies where solute zones are primarily layered at the side of a microchannel by the appliance of external field [83]. Interaction between field and solute governs the layer thickness (Fig. 4.8). Subsequently, longitudinal flow mediated displacement of solute layer takes place. Since in a typical pressure driven flow, flow velocity decreases away from the channel center, the displacement solute layers is differentially retarded depending upon their proximity to channel wall. Hitherto used perpendicular fields include Thermal gradients (Thermal FFF), electric field (Electrical FFF), magnetic field (Magnetic FFF), electrothermal gradient (Electrothermal FFF) [84] and centrifugal forces (Sedimentation FFF). The applicability of each of the aforementioned FFF technologies is guided by the potency and specificity of field-solute interaction. Owing to field-solute interaction, solute is transported in transverse direction with a velocity u_T. In steady state condition, diffusion acts in the reverse direction, along the negative u_T axis and the formed layer possesses characteristic thickness l given by the relation $l = D/\left|u_T\right|$.

In functional applications, FFF has startling similarity with chromatography yet it encompasses a broader range of solute purification with enhanced resolution,

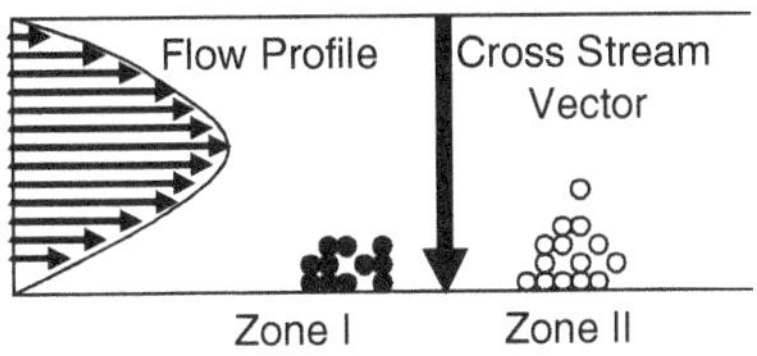

Fig. 4.8 Schematic representation of field flow fractionation. Cross stream vectors are created by electro, thermal, electro-thermal, magnetic mechanisms

augmented process rapidity, easy operation and interface with other sensor-detecting device. With relevance to Biomicrofluidics [85], Hyperlayer mode FFF (HyFFF) is the most important technique [86] belonging to FFF class with its capability separate a wide range of particles with variable diameters (0.5–50 μm). In this mode of methodology, particles are transported by the shear force or hydrodynamic lift force yielding faster dilution of larger particles than the smaller ones.

The Electric field mediated FFF (EFFF) relies upon application of an electric field in the transverse direction with respect to flow [87]. In tune with the underlining working principle of FFF technology, EFFF works by forcing the particles towards different locations between microchannel center and the solid wall. The localization of particles is governed by their intrinsic electrophoretic mobility or surface charge with highly mobile or highly charged particles moving in proximity of the solid wall while relatively uncharged particles form a diffused cloud close microchannel center. The displacement of particles toward the channel being counteracted by diffusive effects, equilibrium average thickness of the particle layers is obtained from a magnitude balance between diffusive and electric forces. Owing to the parabolic profile of the longitudinal velocity field, particles staying close channel center are moved further than the particles layering close to the channel wall. Evidently, in this method particles are separated on the basis of their electrophoretic mobility or zeta-potential which are pivotal parameters deciding a particle's transport across cell membrane, hormonal control and antigen-antibody interaction. EFFF has been applied to separations cells, large molecules, colloids, emulsions and delicate macromolecular composutes such as liposomes for which electrophoretic separation had been infeasible method.

In Thermal gradient mediated FFF (TFFF) system [88], the solute particles are differentially layered with application of spatially varying temperature field across the channel width. Here, particles are segregated depending upon their relative thermal diffusophoretic mobility or thermal diffusion coefficient (D_T). As in EFFF, the particles having higher D_T layers close to channel wall and in consequence, their longitudinal motion is highly impeded. The disparity in average velocity influences the spatio-temporal separation of the sample solute components at the output side of the TFFF channel. Thermal field-flow fractionation (TFFF) has generally been utilized in polymer separation, purification, and analysis.

4.4 Biochemical Reactions in Bio-Chips

4.4.1 General Reaction Scheme

For a generalized reaction of the form:

$$aA + bB \rightarrow cC + dD \tag{4.19}$$

The rate of the reaction can be given as

$$\begin{aligned} r &= \frac{1}{d} r_D \left(= \frac{1}{d} \frac{d[D]}{dt} \right) = \frac{1}{c} r_C \left(= \frac{1}{c} \frac{d[C]}{dt} \right) \\ &= \frac{1}{a} r_A \left(= -\frac{1}{a} \frac{d[A]}{dt} \right) = \frac{1}{b} r_B \left(= -\frac{1}{b} \frac{d[B]}{dt} \right) \end{aligned} \tag{4.20}$$

Further, given that for most of the cases, it is indispensable to determine the kinetics of a reaction from known parameters such change the is concentration of a particular component, the rate of the reaction can be expressed as

$$r = k[A]^{a_1}[B]^{b_1} \tag{4.21}$$

k is the rate constant and a_1 and b_1 are the coefficients which may or may not be equal to (a,b). Subsequently, the order of reaction is defined as

$$n = a_1 + b_1 \tag{4.22}$$

Evidently the unit of rate constant is dependent on n and is expressed as $[mol/m^3]^{1-n}/\mathrm{s}$ i.e. for zero order reaction it is mole/m^3/s; for first order reaction s^{-1}.

The rate constant strongly depends upon the Temperature of the reaction and this dependence is analytically expressed by Arrhenius equation

$$k = A_r e^{-E_a/RT} \tag{4.23}$$

Here E_a is the activation energy which is generally obtained from the difference between energy states of the reactants and the products. A_r is the Arrhenius frequency factor and represents the rate of collisions in the reactive mixture. Though for bulk of the reactions, obtaining the temporal variation in reactant concentration (as determined by analytical integration of rate equation) is impossible, it can be solved for unimolecular reaction of following type

$$A \rightarrow B \tag{4.24}$$

In this case, the generalized expression for a first order reaction is evaluated as

$$[A] = [A]_0 e^{-kt} \tag{4.25}$$

4.4.2 Michaelis-Menten Kinetics

Michaelis-Menten Kinetics represents a wide variety of biochemical reactions including enzyme-substrate interaction, antigen-antibody binding, DNA hybridiziation, protein-protein interaction and many more. Following, for sake of simplicity we describe the kinetics for enzyme-substrate interaction while other should follow similar methodology of formulation. In the beginning, it is formalized that an enzyme E which is a biochemical catalyst, acts upon a chemical species S (i.e. Substrate) to get the later converted into the product P. In the process, E is transients associated with S forming a complex ES which is consequently converts into the product and the unchanged enzyme E again. The scheme is represented as

$$E + S \underset{k_{-1}}{\overset{k_1}{\rightleftharpoons}} ES \xrightarrow{k_2} E + P \tag{4.26}$$

Assuming, rate constant are given as k_1, k_{-1} and k_2, the change in the concentration of complex ES can be summarized as

$$\frac{d[ES]}{dt} = k_1[E][S] - k_{-1}[ES] - k_2[ES] \tag{4.27}$$

Michealis-Menten scheme presumes that ES achieves a steady state under its rapid conversion to P. Using this approximation, Eq. (4.27) can rewritten as

$$\frac{d[ES]}{dt} = k_1[E][S] - k_{-1}[ES] - k_2[ES] = 0 \tag{4.28}$$

Which can be manipulated further to

$$[E][S] = \frac{k_2 + k_{-1}}{k_1}[ES] \tag{4.29}$$

$$\text{Or } [ES] = \frac{k_1[E][S]}{k_{-1} + k_2} = \frac{[E][S]}{K_m} \tag{4.29a}$$

K_m is the Michaelis-Menten constant and represents the enzyme-substrate affinity in an inverse sense. Further, expressing total enzyme concentration $[E_0]$ as

$$[E_0] = [E] + [ES] \tag{4.30}$$

We obtain

$$[ES] = \frac{([E_0] - [ES])[S]}{K_m} \tag{4.31}$$

Or identically

$$[ES] = \frac{[E_0][S]}{K_m + [S]} \tag{4.31a}$$

Given that the velocity of the reaction is expressed as

$$v = \frac{d[P]}{dt} = k_2[ES] \tag{4.32}$$

The final Michealis-Menten kinetic equation is given as

$$v = k_2[ES] = \frac{k_2[E_0][S]}{K_m + [S]} = \frac{v_{\max}[S]}{K_m + [S]} \tag{4.33}$$

Here, $v_{\max}$ is the maximum velocity of the reaction and is assumed as $k_2[E_0]$. Michealis-Menten relation can be expressed in another alternative form, famously recognized Lineweaver-Burk expression

$$\frac{1}{v} = \frac{1}{v_{\max}} + \frac{K_m}{v_{\max}}\frac{1}{[S]} \tag{4.34}$$

It should be noticed from Eq. (4.34), that if $1/v$ is plotted against $1/[S]$, the intercept and the slope are given as $1/v_{\max}$ and $K_m/v_{\max}$ respectively.

One of the recognized utilities of Lineweaver-Burk expression is its ability to delineate the nature of inhibition. It is known that while the reactions allowing competitive inhibition alter the x-intercept value, maintaining the magnitude of y-intercept, noncompetitive inhibitory reactions do the reverse.

4.4.3 Lagmuir Adsorption Model

Similar to Michealis-Menten model which characterizes a representative biochemical reaction, the adsorption of a biomolecular species is, in general, is simulated by Langmuir model. In this model, the sample case is considered as the adsorption of a biomolecule from a solution onto a solid surface and the reverse desorption porcess. If we denote the molecules in solution and surface-adsorbed phase by S and Ω, then the process is given as:

$$S \underset{k_d}{\overset{k_a}{\rightleftarrows}} \Omega \tag{4.35}$$

If the maximum concentration of the surface-adsorbed molecule is assumed as Ω_0, the net rate of adsorption can be written as

$$\frac{d\Omega}{dt} = k_a[S]_w(\Omega_0 - \Omega) - k_d\Omega \tag{4.36}$$

$[S]_w$ is the concentration of S at the solid surface or wall as in case of a microfluidic system. The Eq. (4.36) can be integrated in time to delineate the evolution of Ω

$$\Omega(t) = \frac{k_a[S]_w\Omega_0}{k_a[S]_w + k_d}\left(1 - e^{-(k_a[S]_w+k_d)t}\right) \tag{4.37}$$

which asymptotically reaches to a value of $\Omega = \frac{k_a[S]_w\Omega_0}{k_a[S]_w+k_d}$ as $t \to \infty$. Whereas for small time, Ω takes a much simpler form of

$$\Omega = k_a[S]_w\Omega_0 t \tag{4.38}$$

Though Michaelis-Menten kinetics and langmuir models are suitable approximations, for general biochemical reaction occurring in microfluidic system, the appropriate contribution of fluid flow should be appreciated. The system then becomes an advection-diffusion-reaction system which modeled in similitude to Eq. (4.2) with the reactive term R_i obtained from the underlining rate law. For example if we consider a simple reaction of type $mA + nB \to pC$ with reaction rate k, the mass transport equations for three chemical become

$$\frac{\partial c_A}{\partial t} + v.\nabla c_A = \nabla.(D_A\nabla c_A) - mkc_Ac_B \tag{4.39a}$$

$$\frac{\partial c_B}{\partial t} + v.\nabla c_B = \nabla.(D_B\nabla c_B) - nkc_Ac_B \tag{4.39b}$$

$$\frac{\partial c_C}{\partial t} + v.\nabla c_C = \nabla.(D_C\nabla c_C) + pkc_Ac_B \tag{4.39c}$$

which under simplified and reduced condition may behave like a Lotka-Volterra system.

Microfluidics systems have revolutionized the field of enzyme kinetics, particularly inhibition kinetics, because they require low volume of analytes and allow unprecedentedly rapid output generation [89, 90]. Microfluidic assay systems are grouped into two main broad categories namely offline and online inhibition studies. Online inhibition studies can further be divided into homogeneous and heterogeneous reactive systems. In offline inhibition studies, reaction is performed separately in macroscale and after a scheduled time, a sample volume is withdrawn, which is then separated and analyzed using microfluidics based capillary electrophoresis method. For example, offline enzyme inhibition assays have executed to investigate inhibition kinetics of several biologically relevant enzymes including β-glucuronidase [91, 92], β-galactosidase, protein kinase A [93] and src kinase [90]. These systems seemingly offer two major advantages. Firstly, microfluidic based CE increases the efficiency and the resolution of component separation-identification and secondly, it eases a very low sample volume requirement. However, being predominantly operated in macroscale, the offline assay systems fail to savor the complete package of microfluidic advantages and in order to eliminate this drawback; online enzyme assay devices have been implemented, which critically juxtapose both reaction and separation processes in a monolithic microchannel network system. In an online assay system, one of its components may be immobilized into the solid microchannel wall or all of them may be dissolved into the fluid phase. They are known as heterogeneous and homogeneous inhibition assay systems respectively [90]. While the heterogeneous type offers ultra-low component (enzyme, substrate or inhibtor) volume constraint and higher controllability of spatiotemporal reaction dynamics, the process is often restricted by the spontaneous partial or complete inactivation of a chemical component undergoing immobilization. The system requirement varying case to case, apparently, both of them have found their application niches. The list of enzymes studied under the heterogeneous system includes uridine diphosphate glucuronosyltransferase, angiotensin converting enzyme, HIV protease and β-galactosidase. Concurrently, inhibition kinetics of adenosine deaminase, acetylcholinesterase, liver rhodanase have been scrutinized using homogeneous biochip based inhibition assay methods [90].

4.5 Bio-Micromanipulation Using Electrical Fields

The electroosmtic flow arises when bulk fluid motion is powered by the stresses concentrated in charged layers near charged wall interface [94, 95]. The resulting velocity profile becomes uniform between channel walls except for very thin region near the interface. This kind of situation occurs when a charged surface is exposed to an ionic fluid. Owing to the strong attractive interaction of coulombic nature, counter ions migrate towards the charged wall and thereby, form an approximately immobilized ion layer called Stern layer. Subsequently, over this immobilized layer, a diffused layer of solubilized ions is created due to the counter-active effect of coulombic interaction and forces derived out of the entropic contribution. This segment is called the Gouy-Chapman layer (Fig. 4.9). Once an electric field is externally applied on the system, charged double layer acquires a velocity towards a definite electrode depending upon the net polarity of the wall charges. For electroosmotic flow, the velocity scales linearly with the electric field. Similar scaling relationship is also obtained for electrophoresis where charged particles moves under the direct interaction with imposed electric field and in turn, they drag the hydration layer formed around them. In contrast, for dielectrophoretic phenomena where movement originates essentially because an electrical dipole interacts with the gradient of the electric field, the response is deduced to be proportional to the square of the electric field.

From the very origin of microfluidics, electrokinetic methods have widely appreciated as the most suitable flow driving, actuation and component separation mechanism for several biologically relevant devices such as capillary gel electrophoresis, microchannel based liquid chromatography system and particle concentrator. In general, electrokinetic mechanisms are preferable because of several reasons enlisted below. First, for a microchannel of height h and width w, for

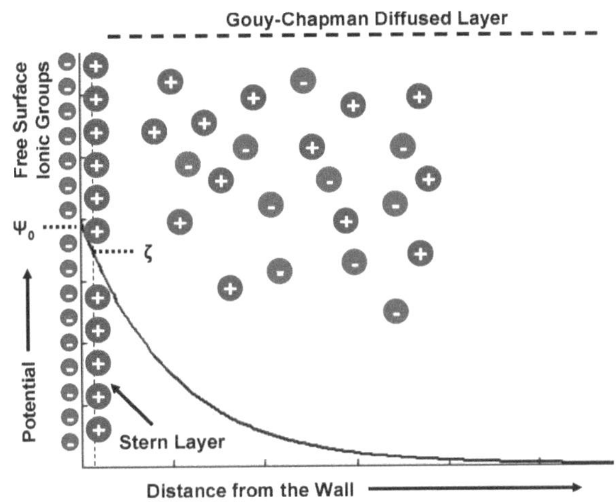

Fig. 4.9 Schematic representation of electric double layer formation at charged microchannel surface

a fixed potential difference, electrosmotic flow rate appears to be proportional to h×w which stands in contrast to pressure driven flow where volumetric flow rate is proportional to $h^3 \times w$. Second, diminished cross sectional area of microfluidic architecture enforces enhanced electrical resistance to ionic currents, which guarantees high electrical fields (>100 V/cm) to be persisted with low currents. Third, in small systems, thermal convection opposing the electrokinetic motion is appreciably prevented due to viscous damping. Lastly, for channels without significant design and compositional heterogeneity, the electrosmotic plug flow facilitates uniform transport of biological samples effectively eliminating any band broadening due to hydrodynamic dispersion.

In order to achieve efficient momentum and energy transfer for controlling the motion of fluids and molecules, it is important to have operational length-scale matching in an approximate scale. Pertinently, bulk of the biological entities of interest, such as DNA, proteins, and cells, have a characteristic length from nanometer to micrometer. Electrokinetics transport and flow actuation mechanisms are especially effective in this micron and sub-micron regime as they advantageously utilize the inherent small length scale. In addition, with the progress of micro-electromechanical systems (MEMS) fabrication technology, integration of micro or nano scale electrodes to polymer based fluidic device has become a trivial procedure. These effects, in combination, make electrokinetic forces ideal for manipulation and control of biological objects and performing desired fluidic operations. By utilizing electrokinetic effects, fluid flow can be manipulated, in general, using the electro-osmosis, AC electro-osmosis, electrophoresis, dielectrophoresis, electrowetting and electrothermal phenomena.

Before we proceed to describe each of aforementioned events in detail, evidently, as it becomes essential to comprehend the physical mechanism undergoing beneath the formation of Debye layer, before going to sub-categories of electric field based methods, we present an abridged formulation illustrating the counter ion distribution around a charged colloidal particle. From the Poisson equation describing the relation between the charge density ρ and potential ψ, it is known that

$$\nabla^2\psi = -\frac{\rho}{\varepsilon_0\varepsilon_r} \tag{4.40}$$

Where ε_0 and ε_r are vacuum and relative permitivity respectively. From this equation, in order to determine ψ, one must possess the knowledge of ρ at hand, which, in turn, should be available from the number densities of relevant ionic specie (n_i), as given by Boltzmann distribution

$$n_i = n_{i,0} \exp\left[-z_i e\psi/k_B T\right] \tag{4.41}$$

Here, z_i, $n_{i,0}$, e and k_B are ionic valency of ith species, ion concentration at bulk, elementary electronic charge and Boltzmann constant respectively. For symmetric electrolytes (e.g. NaCl, $CaSO_4$ etc.) net charge density is given as

$$\rho = n_0[ze \exp(ze\psi/k_B T) - ze \exp(-ze\psi/k_B T)] \tag{4.42}$$

i.e.

$$\rho = n_0 ze[\exp(ze\psi/k_B T) - ze \exp(-ze\psi/k_B T)] \tag{4.42a}$$

Theoretically putting the expression (4.42a) in (4.40) the variation in ψ can be solved. However, one may easily notice that, in this process, analytical solution is impossible to deduce. Hence it is further approximated that thermal energy is much higher than electric energy implying $k_B T \gg e\psi$. Under this assumption which is known as Debye-Hückel approximation, the Eq. (4.42a) can be linearized yielding an one-dimensional expression

$$\psi = \psi_0 \exp(-\kappa x) \tag{4.43}$$

Where ψ_0 is the potential at surface and κ^{-1} is the Debye length deduced as

$$\kappa^2 = 2\frac{z^2 e^2 n_0}{\varepsilon_0\varepsilon_r k_B T} \tag{4.44}$$

Evidently from Eq. (4.44), Debye length inversely depends upon the salt concentration. Typical value of Debye length lies in range of few nanometers for moderate concentrated salt solutions (100 mM) which are commonly used in biochemical experimentations.

4.5.1 Electroosmosis

Electroosmosis arises due to the development of electrical double layer at charged surfaces. When an ionic liquid is brought in contact with a charged solid surface, the surface charge is neutralized by counter ions in the ionic medium. The immediately layer of counter-ions which is approximated to immobilized is called Stern Layer. Beyond this immobilized and exclusive region of counter ions, there exists an outer region, where ions are in rapid thermal motion, and the layer is known as the diffuse electrical double layer (EDL) that spans a distance on the order of the Debye length (Gouy-Chapman Layer). An illustrative depiction of the aforementioned phenomenon has been given Fig. 4.9. Now, if an electric potential is externally applied along the channel, the diffused electrical double layer starts moving owing to the net electrostatic force. As the ions in the EDL move, they drag water molecules along themselves due to the cohesive nature of the hydrogen bonding of water molecules. The entire event then yields to a net movement of buffer solution. For an electroosmotic flow field, the reduced Navier-Stokes equation (neglecting convective and pressure terms) can be written as

$$\rho E + \mu \frac{\partial^2 v}{\partial z^2} = 0 \tag{4.45}$$

Where E is the applied electric field. The charge density can be obtained from Poisson Eq. (4.40) as

$$\rho = -\varepsilon_0 \varepsilon_r \nabla^2 \psi \tag{4.46}$$

Combining (4.45) and (4.46) we obtain

$$E\varepsilon_0 \varepsilon_r \frac{\partial^2 \psi}{\partial z^2} = \mu \frac{\partial^2 v}{\partial z^2} \tag{4.47}$$

And with the following boundary conditions

$$\begin{aligned} v\,|_{wall} &= 0 \\ \psi\,|_{wall} &= \zeta \\ \partial \psi / \partial z\,|_{bulk} &= \partial v / \partial z\,|_{bulk} = 0 \end{aligned} \tag{4.48}$$

The final expression for v yields

$$v(z) = \frac{\varepsilon_0 \varepsilon_r E}{\mu} [\psi(z) - \zeta] \tag{4.49}$$

As the potential exponentially decreases away from the wall surface, the electroosmotic velocity at the bulk can be deduced as

$$v_{eo} = -\frac{\varepsilon_0 \varepsilon_r E \zeta}{\mu} \tag{4.50}$$

From the deduction furnished above, it is evident than for electroosmotic flow, velocity remains constant across the channel width reducing the magnitude of sample dispersion. This stands in contrast to highly dispersive pressure driven flow where velocity profile assumes a parabolic shape with a maximum at channel center line. In order capitulate this benefit in microfluidic regime, electroosmotic flow has been recurrently used for sample injection in electrophoresis-based separation devices. In a generic design of such device, silica particles are packed in a fused silica capillary and the porous glass structure enables a high surface-to-volume ratio, augmenting the electroosmotic effect. Further, if the separation between bounding wall surfaces becomes commensurate to the Debye length, the effect of the electrical double layer overlap in governing the flow profile should be accounted [96]. Typically, glass substrates are utilized due to their well-investigated surface properties and surface modification techniques by chemical means. However, in tune with advancements of polymer-based microchannel systems, recently, the procedure has also been demonstrated in some appropriate polymeric materials such Polydimethylsiolxane (PDMS), Polymethylmethacrylate (PMMA) and Polystyrene. One plus point of these materials is that they can be easily molded into complex network architecture, required in many microfluidic high-throughput assay systems. With application of polymeric materials possessing chemically or physically tunable surface potential properties, electroosmotic pumping has been demonstrated in complex networks of intersecting capillaries where the fluid flow can be controlled quantitatively by simultaneously applying potentials at several judiciously chosen locations [97]. However, the flow actuation through complex fluidic surface frequently imposes challenging situations from analytical point-of-view. Fortunately, the electroosmotic flow being directly connected to the electric current, it can be trivially estimated by considering the equivalent resistive circuit and electroosmotic mobility in analogy to the well-established current analysis in electric circuits. Here it is to be appreciated, electroosmotic flow gifts us with another control in the form of surface potential which can be dynamically manipulated either by chemical treatment or by transverse electric field. A perpendicular electric field has been illustrated to alter the zeta potential. In conjunction, spatial modifications of electroosmotic mobility have been reported with protein adsorption and viscous polymer channel sidewall coatings. The situation becomes non-trvivially interesting when one used patterned surface charges. In fact, bi-directional electroosmotic flow and out-of plane vortices have been demonstrated with different surface charge patterns [98]. The capability of generating flow patterns offers the potential for achieving microscale mixing with enhanced efficiency.

4.5.2 AC Electroosmosis

AC electroosmosis is a recently identified electrokinetic phenomenon observed at frequency ranges below 1 MHz [99–101]. This observation has been reported for aggregation of yeast cells in interdigitated castellated electrodes. Though different in the nature of applied electric field, AC and DC electroosmosis possess an impending original similitude in mechanistic principle that they both exert a tangential force on electric double layer. Electric potential within the electrode forces the charges to build up concentration on the electrode surface. Subsequently the interfacial charge density is altered and formation of the electrical double layer takes place. The process is known as electrode polarization. The electrical double layer, now, experiences a net force due to the tangential component of the electric field and result in fluid movement. In alternating electric field, the sign of charges in the electrical double layer periodically oscillates with the oscillation of applied electric field and therefore, its tangential component. Therefore, the direction of the driving force for the fluid remains unaltered.

4.5.3 Electrophoresis

Electrophoresis describes the movement of charged particles in a liquid medium under an external electric field. When a particle with charge q is under a steady electric field E, the particle experiences an electrostatic force qE. The electrical force is balanced by a friction force, which can be estimated by Stoke's law, $6\pi\mu Rv$ for a spherical object. Velocity of the particle can be related to the applied field by the relation

$$V = \mu_{eph} E \tag{4.51}$$

Where μ_{eph} is the electrophoretic mobility. For particle with radius a and charge q_p, μ_{eph} is obtained as

$$\mu_{eph} = \frac{q_p}{4\pi\mu\kappa a^2} \tag{4.52}$$

Which is reduced to $\mu_{eph} = q_p/6\pi\mu a$ for cases with $\kappa a \ll 1$.

To cite the most relevant biomicrofluidic applications of electrophoresis, it is pertinent to mention that charged biomolecules such DNA, RNA, proteins migrate under the control of an externallly applied electric field. In free solution electrophoresis, a DNA molecule adopts a structure of free draining coil, which implies that the friction coefficient remain proportional to the length of the molecule. Simultaneously, it is important to note that the net charge and therefore, the force experienced by a DNA molecule is also proportional to the length. As a result, the effective mobility of the molecules, which is basically the ratio between the externally applied force and the friction coefficient, remains independent of molecular length at a given medium condition for large DNA molecules. However, the net charge of a DNA molecule is accessible only in reduced magnitude as a fraction of the molecular charge is counter-acted and neutralized by the counterions (as determined by counterion condensation theory). For small DNA molecules, typically containing 20–100 base pairs, it is envisioned that the length dependence of small DNA molecules arises due to the imposing consequence of counterion relaxation phenomenon. For protein or peptide molecules, the situation is not as simple as in electrophoresis of DNA. These molecules may be positively or negatively charged depending on the zwitter-ionic nature of amino acid side chains, the pH of the medium and non-intuitively, the secondary or tertiary structure of the protein. Amine side chains are predominantly exist in protonated form to give a positive charge at low pH (<5). In contrast, at high pH, carboxylic acid side chains are de-protonated to result in a net negative charge. The net charge of the protein at any pH is obtained depending upon the relative abundance of protonated and de-protonated forms of side groups. At a particular pH, characteristics of a particular protein or peptide molecule, the net manifested charge of the molecule may be nullified. This is known as isoelectric point (pI). As a result, proteins can be separated according to their mobility as well as isoelectric point, implicating a two-dimensional and high resolution separation system. In modern biological analysis which includes molecular fragment separation, sequencing and strutural analysis, 2D gel electrophoresis and capillary electrophoresis with an array of separating channels, have become precious tools. Another method which takes advantage of signature isoelectric point for a specific protein or peptide is Isoelectric focusing. Here, in developed gradient of pH between two electrodes, protein molecules preferentially accumulate in their respective isoelectric point. It has been demonstrated that generation of a pH gradient by electrolysis–driven production of H^+ and OH^- ions in a microfluidic device can be utilized in separating different components from mixture of proteins and peptides [102].

4.5.4 Dielectrophoresis

Subjected to an electric field, a dipole is induced in a polarizable particle. In a spatially diverging electric field, depending upon the sign of the polarizability difference between the particle and the bulk medium, it moves towards either towards or away from the increasing electric field. This electrokinetic phenomenon is known as dielectrophoresis (DEP) [103–105]. The dielectrophoretic force on a particle with radius a and relative permittivity $\varepsilon_{r,p}$ suspended in liquid with relative permittivity $\varepsilon_{r,l}$ can be expressed as

$$F = 2\pi a^3 \varepsilon_0 \varepsilon_{r,l} \, \mathrm{Re}(f_{CM}) \, |\nabla E|^2 \tag{4.53}$$

Where f_{CM} is the Clausius-Mossoti factor and is given by

$$f_{CM} = \frac{\varepsilon_p^* - \varepsilon_l^*}{\varepsilon_p^* + \varepsilon_l^*} \tag{4.54}$$

ε_p^* and ε_l^* are complex permittivities of particle and solvent and are formulated as

$$\varepsilon_p^* = \varepsilon_0 \varepsilon_{r,p} - j\frac{\sigma}{\omega} \tag{4.55a}$$

$$\varepsilon_l^* = \varepsilon_0 \varepsilon_{r,l} - j\frac{\sigma}{\omega} \tag{4.55b}$$

σ and ω being conductivity and frequency of the electric field.

As evident from Eqs. (4.53), (4.54) and (4.55), positive (\negative) dielectrophoresis i.e. the movement towards (\away from) the high field strength region occurs if the polarizability of bulk medium is lower (\higher) than that of the particle. DEP is commonly applicable for enforcing translational motions of different biological micro-objects. Using dielectrophoresis, different types of cells with distinct polarizabilities can be conveniently segregated. For example, it has been confirmed that a mixture of major leukocyte subpopulations, isolated from human blood sample, can be competently separated by dielectrophoresis. DEP has also been illustrated for maneuvering micron-sized biological entities such as DNA, proteins, bacteria and viruses. For small molecules, the ratio of dieletrophoretic and thermal forces governs the effective particle dynamics.

4.5.5 Electrowetting

An alternative of driving bulk fluid motion through a microchannel network is to employ droplet based digital microfluidics. Such digital fluidic system confers the performance of biological assays with minimum possible volume requirement. The droplet translation mechanism in a digital microfluidic circuit takes advantage of several physical effects such as thermocapillaries, dielectrophoresis and voltage or light mediated surface wetting, of which electrowetting on dielectric (EWOD) has been applied most abundantly due to its convenience of application, compatibility towards on-chip integration, inherent process reversibility and low power consumption. In this process, a thin dielectric film (typically, polymer materials or silicon oxide) is coated over the microfabricated electrodes in order to nullify the direct electrochemical contact-interaction between the fluid and the electrode. However, this compels the use of high voltage. If the permittivity and layer thickness of the dielectric thin film are given as ε_d and d respectively, the contact angle of liquid droplet having liquid-vapor surface tension γ_{lg} changes from its initial value θ_0 with applied potential V, obeying the following mathematical expression

$$\cos\theta = \cos\theta_0 + \frac{1}{2}\frac{\varepsilon_0\varepsilon_d}{\gamma_{lg} d}V^2 \tag{4.56}$$

By the application of spatially patterned electric field, different contact angles may be invoked in different parts of a single droplet which in turn, results in net droplet motion. Since its introduction, several designs have been realized for efficient driving droplets on a two-dimensional space [106, 107].

4.5.6 Electrothermal Flow

Electrothermal flow arises because of the temperature gradient in the medium which in turn, may be generated by joule heating of the fluid [108–111]. When such temperature gradient exists within liquid medium; the conductivity, viscosity, density and permittivity of the fluid varies spatiotemporally, exerting a net mobilization force on the fluid. By judiciously designing the heating elements and electrodes, electrothermally actuated micropumps can be fabricated. These pumps eliminate the requirement for moving parts. Electrothermal effects due to the applied AC potential at high frequency averts the thermochemical dissociation of the fluid.

4.6 Bio-Micromanipulation Using Magnetic Fields

Biomicrofluidic application involving magnetic forces are abound particularly as trapping and transport mode of magnetically tagged single cells in microchannel system. One main reason for exploiting magnetic field as the manipulator is its comparatively non-lethal effect on biological entities. There have been numerous means by which researchers utilized magnetic field in driving and actuation of microflows [112]. While Micropumps have been fabricated by Magnetohydrodynamic effect, magnetically doped polymer or magnetic materials such as ferrofluids are typically used for valving action. Other important fluidic operations such as micromixing have been performed by the regulated oscillation of magnetic microparticles in a two-fluid stream. From the biomicrofluidics perspective, magnetic particles have been used as solid supports for assaying bioreaction which are essentially integrated with magnetic system detection on lab-on-a-chip scale. Following, we will discuss major microfluidic methods involving magnetic field based manipulation.

4.6.1 Magnetic Field Flow Fractionation (MFFF)

Pioneered by Giddings et al. [83, 113], MFFF has since become well-utilized method for separation which fundamentally exploits the differences in magnetic susceptibility among various components in a mixture. Multifarious applications of MFFF have been appreciated in the realm of separating colloids, particles, and biological macrocmolecules including proteins and even, cells. Mechanistically, the operation principle of MFFF is very similar to other field flow fractionation variants with magnetic force standing as the major driving impetus. In MFFF systems, the wall-directed transverse magnetic force is counter-balanced by the diffusion and time-invariant suspended particle distribution is achieved. Subsequently, subjected to longitudinal hydrodynamic force, these particles traverses longitudinally according to the velocity distribution across the channel cross-section. As particles with different magnetic susceptibilities, thereby magnetic forces, attain different height locations, they experience different magnitudes of flow velocity and are consequently separated. Smaller component thickness yields to more significantly retained fractions and therefore longer elution times. In MFFF, to represent the separation, a dimensionless retention parameter λ [114] is defined which is related to the properties of the particles by

$$\lambda = (4d/w)\,(k_B T/\mu \Delta H)^2 \tag{4.57}$$

Where d, w, μ and ΔH are the particle diameter, capillary diameter, magnetic moment and drop of magnetic field across the capillary respectively. k_B is the Boltzmann's constant.

The most generic chemical means of synthesizing monodisperse magnetic particle is the thermal reduction of organometallic precursor comprising of pure metals such as Iron, Nickel, Cobalt as well as composites such as $CoFe_2O_4$, $FePt_3$ and $MnFe_2O_4$. Acetylacetonate has been the most commonly used organic group. Subsequently, synthesized particles are stabilized by organic ligands such as organic acid groups, amine terminated alkanes or phosphine oxide. Further, these conventionally used magnetic nano-particles being water insoluble and in some cases, bio-incompatible, researchers have developed synthetic protocol for obtaining water soluble monodisperse magnetic particle by using MFFF method. This technique is very distinctive in comparsion to electromagnetophoresis, where a magnetic field applied at right angles to an electrical current, influences the migration of nonmagnetic and neutral particles perpendicular to the current.

4.6.2 Magnetic Biomaterials

Most of the biological entities such as DNA, proteins, cell and other biologically relevant polymers are diamagnetic in nature. All materials are categorized, according to their magnetic susceptibility χ, into three broad magnetic classes namely diamagnetic, paramagnetic and ferromagnetic. Diamagnetic or nonmagnetic materials generally move opposite to the direction of magnetic field. In contrast, paramagnetic and ferromagnetic materials align themselves to the direction of net magnetic field [115]. In biology, a subclass of paramagnetic materials called supermagnetic particles is used very frequently. Supermagnetic particles generally consist of iron oxide crystal core and coated with polymer materials derivatized to produce free amine or carboxyl groups which may be tuned towards the exploitation of a particular biochemistry. For example, biological macromolecules such as proteins, DNA, mRNAs (messenger ribonucleic acids) can be tagged with such supermagnetic particles [112]. Subjected to a directional magnetic field, these bio-entities can be moved to the desired location. For magnetic manipulation of biological cells, particular cell specific surface protein markers are tagged with supermagnetic particles by antigen-antibody bonding which are effectively irreversible coupling with equilibrium constant value of 10^{14}. Of all commercially available supermagnetic substances, Dynal beads are the most popular from lab-on-a-chip scale application point of view, attributed to their inherent mono-dispersity in shape and size distribution. Dynal beads have, for the first time, enabled magnetic separation technology and have immediately generated fathomless possibilities for astounding variety of applications within the life sciences, biotech and healthcare. Functionally, these beads are supermagnetic particles tagged with cell-surface-marker specific antibodies. Presently several variants of dynal beads are available all of which are designed towards some targeted cell populations of mammalian immune system. To cite few examples, Protein A, Protein G, Mouse T-Activator CD3/CD28, Human T-Activator CD3/CD28, Human CD4, Human CD8, Human CD3, Mouse PanT, Mouse CD43, Human B Cells, Human Natural Killer Cells, Human Monocyte specific dynal beads are of wide use. In a mixed population of cells, generally isolated from blood sample as in pathological laboratories, cell specific dynal beads will segregate targeted cell types very conveniently. While labeling with magnetic particles of micron or nanometer size has been a plethora in biomicrofluidic assays, it must be mentioned that there are two specific cell types namely Red Blood Corpuscles (RBC) and Magnetotactic Bacteria which are intrinsically magnetic owing their special internal constitution and therefore, have found pertinent microfluidic applications.

4.6.3 Ferrofluids

From microfluidic applications, supermagnetic fluids or ferrofluids are probably the most suitable magnetic materials due to their fluid nature. Essentially they are suspensions of magnetic nanoparticles in water or in organic solvent and are coated to surfactant molecules. Surfactant coating inhibits particle aggregation, maintaining a homogenized magnetic behavior over a large length scale. Ferrofluids being stable hydrdynamically as well as magnetically over a wide range of shear and magnetic field strength, they render exquisite means of exploring the magnetohydrodynamics in lab-on-a-chip scale. The force experienced by the particles suspended in solution system is a strong function of the difference between susceptibilities (χ) between the particle and the bulk medium (commonly water), the absolute strength of the applied magnetic field (B) as well as its spatial gradient. Moreover, the force is directly proportional to trhe volume of the particle and can thus formulated as,

$$F_p = \frac{(4/3)\pi r_p^3.\Delta\chi}{\mu_0}(B\cdot\nabla)B \tag{4.58}$$

From the Eq. (4.58), it is very much evident that in a spatially homogeneous magnetic field, zero force is exerted on ferrofluids even if they are strongly magnetized. Microfluidic system with spatially variable magnetic is therefore employed in ferrofluid based lab-on-a-chip devices.

4.6.4 Magnetohydrodynamic Micropumps

Magnetohydrodynamic (MHD) pumps works on the principle of Lorentz force in mutually perpendicular electro-magnetic system [116]. Generically, micropump components consist of conducting fluid as the working liquid, which subjected to electric and magnetic field along the width and the height of microchannel, moves longitudinally due to the Lorentz force. Compatible to any shape of microchnnel geometry, flow rates through MHD pumps are conveniently regulated by tuning the externally applied electric and magnetic fields. MHD pumping is appropriate for any conducting liquid and moreover, does not necessitate any moving component. However, for specific applications, integrated switching circuits which can alter the magnitude and direction of the electro-magnetic fields may be incorporated. There is of course another class of MHD pumps which are operated encasing the movement of a plus ferrofluid is a moving magnetic field (Fig. 4.10). In this system, the immiscibility between organic ferrofluid solvent and the aqueous solution medium is used to segregate between pumping and transported liquid components. In a pioneering development, researchers [117, 118] have manufactured a circular shaped ferrofluidic micropump essentially consisting of two Nd magnets and two plugs of ferrofluid. One plug is held in a fixed location between the inlet and outlet channel while another plug is circulated by a rotating external magnet. Complementary merge and separation of two plugs in single cycle of plus revolution around the circular pump circuit enables pumping in and out of liquid solutions [119]. Further, very recently, Atencia and Beebe [120] have proposed a MHD pump based on the biomimetics of vortices generated in narrow fluidic confinements by various animals for swimming or flying. Uniquely, this design exploits the biological systems where pumping action is achieved with minimal energy expenditure. Mechanistically, a magnetic bar integrated into a microfluidic structure has been fixed by photolithographically fabricated microposts and the bar is suitably oscillated during operation to create vortices of required strength. In an important biomicrofluidic advancement, West et al. [121] have juxtaposed circular MHD micropump components with on-chip polymerase chain reactions (PCR) using concentrated buffer solutions.

4.6.5 Magnetic MicroValves

Magnetically derivatized or doped PDMS are mostly used for microvalving action [122]. Iron powders with intense magnetic properties are suspended into the base solution of polydimethylsiloxane precursor and thin membrane are then fabricated from it. When these membraned are exposed to the integrated and controllable electromagnetic field, they are deformed. This phenomenon is judiciously engineered in microfluidic circuits to attain the required valving operation. In magnetic valving system, advantageously, no physically moving part is obligatory.

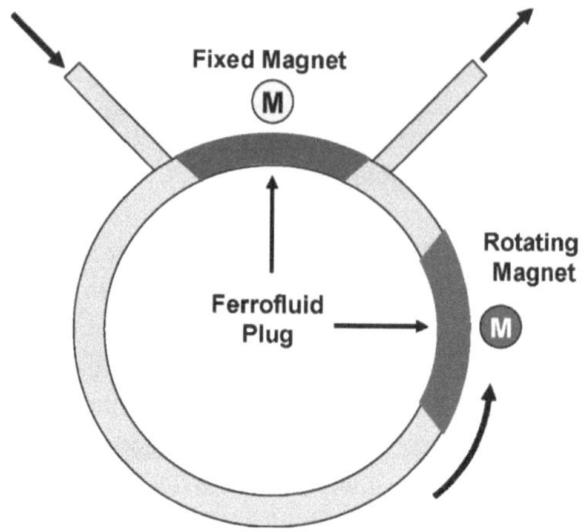

Fig. 4.10 Working mechanism of a representative ferrofluid micropump

4.6.6 Mixing Devices

Magnetic mixing devices work very much identical to the MHD pumps with oscillating magnetic bar system. Pioneeringly introduced to enhance the efficiency and transport kinetics in microchannel based DNA hybridization system by augmenting the mixing of DNA solution with hybridization buffer, Magnetic micromixing devices since then have attracted research attentions [123]. In this system, one end fixed bars commonly fabricated of magnetic permalloy are oscillated inside a microchannel system to enhance two-fluid mixing. Cycling magnetic field is created by four electromagnets tuned in sinusoidal fashion. Chain of magnetic microparticles, stably bonded together covalently or non-covalently with linker molecule, has also been used as the miniaturized stirring system [124]. It has been demonstrated that magnetic micromixing system comprising of pear-shaped chamber along with semicircular wall indentations performs with highest efficiency.

4.6.7 Magnetic Trapping and Sorting of Biomolecules

Biomolecules such as DNA, protein and entities like cells can be pulled along a magnetic field owing due their inherent characteristics [112]. While moved in spatially variable magnetic field, these objects can be uniquely confined into a particular domain. There have been numerous designs of the magnetic systems integrated with microfluidic network. To cite a few, Tapered electromagnet, sawtooth shaped tapered magnets and mech-type magnets have been extensively used. Here, the objective is to facilitate a rachet-like motion where magnetic particles are transported from field maximum to another. Several magnetic materials such as magnetic beads, magnetically labeled yeast cells and magnetotacticbacteria have been manipulated using aforementioned arrangements. Recently, functionally similar to optical tweezers, magnetic tweezers have been developed to trap magnetic particles or magnetically tagged biological entities within a very constricted spatial location, thereby increasing effective concentration. Alongside the magnetic transport and trapping, magnetic field based separation has also been investigated, where differential magnetic susceptibilities among different types of particles stand as the governing principle. Subsequently, in microfluidic systems, separation has been implemented by employing either H-type of filtering design or single channel based retarded system of separation.

4.6.8 Magnetic Particles for Bioassays

Magnetic particles are used for carriers in microfluidic based study of Ligand-Receptor interaction and screening of a cell specific type of Receptors. As a general procedure, magnetic nanoparticles which are dynamically trapped into a desired location, are tagged with Ligand of interest [125]. Subsequently, while receptor molecules or cells containing receptor molecules are transported through this region, they are captured by the magnetic particles. Hence, in principle, magnetic particles function as immobilization matrix; however with an advantage of spatiotemporal controllability of their manipulation. In addition, gifted with intrinsic high surface area to volume ratio, these micron or sub-micron sized particles non-trivially augment the binding rate and the sensitivity of the assay. One of most prolific use of these particles has been anticipated in microfluidics based immunoassays, as demonstrated by Hayes et al. [126]. They have essentially immobilized antibodies over NdFeB magnets and antigen containing sample has been assayed by studying the antigen-antibody binding phenomenon during sample transport through designed glass microcapillaries. Similar studies have been performed for various interleukins (e.g. IL-5), parathyroid hormone, DNA and RNA [112]. A plug of immobilized DNA onto a permanent external magnet embedded within microfluidic circuitry has been used to capture complementary secondary strands which are flushed through the microchannels. Taking advantage of a cluster magnetic particles in a microchannel network, disease specific mRNAs has been isolated from total RNA content with unprecedented capture efficiencies of 50% [127]. Relevantly, isolation of RNAs marking Dengue fever virus in a PDMS based microchip has been also illustrated [128]. With time, magnetically manipulated solid assays have covered the domain of marker specific cell isolation also. For example, an immortalized line of T lymphocytes called Jurkat cells have been separated from whole-blood sample by anti-CD3 tagged magnetic particles having 500–1000 μm radius [129]. The efficiency of isolation has been reported to be approximately 50%. Similar isolation method has also been used for capturing rare circulating T-lymphocytes [130].

4.7 Experimental Approaches

The objective of biomicrofluidics is to miniaturize the lab-scale analytical systems into a size of computer chip. Small device-length scale enables high surface to volume ratio, low sample volume requirement and augmented sensitivity which are critical requirements for detecting and analyzing low abundance samples. However, this objective has been overshadowed by the lack of dimensionally compatible optical, fluorescence and other detection systems. Till date, commonplace detection systems have been bulky with their size being 10^4–10^6 times that of the lab-on-a-chip devices. Hence, even if the process itself has been successfully miniaturized, the detection component still compels voluminous operation. In order to remove this drawback, in recent times, there has been a constant effort of fabricating on chip optical detection systems collectively known as "optofluidics". However, before we proceed to demonstrate the progresses in the field of optofluidics, let us briefly describe how in principle an optical or fluorescence microscope system works.

4.7.1 Optical and Fluorescence Microscopy

The type of microscope that utilizes visible light through a system of lenses in order to produce magnified images is called optical or light microscopes. The produced images are then viewed through eye-piece or captured by charge coupled device (CCD) camera and respectively, transferred into a computer screen. A typical optical microscope system consists of ocular lens or eyepiece, objective lenses, light illuminator, mirror, diaphragm and condenser (Fig. 4.11). An eyepiece is a cylindrical object with typically two or more lenses serially inserted inside to provide required observable focus of image. For convenience, they are placed in the top-portion of

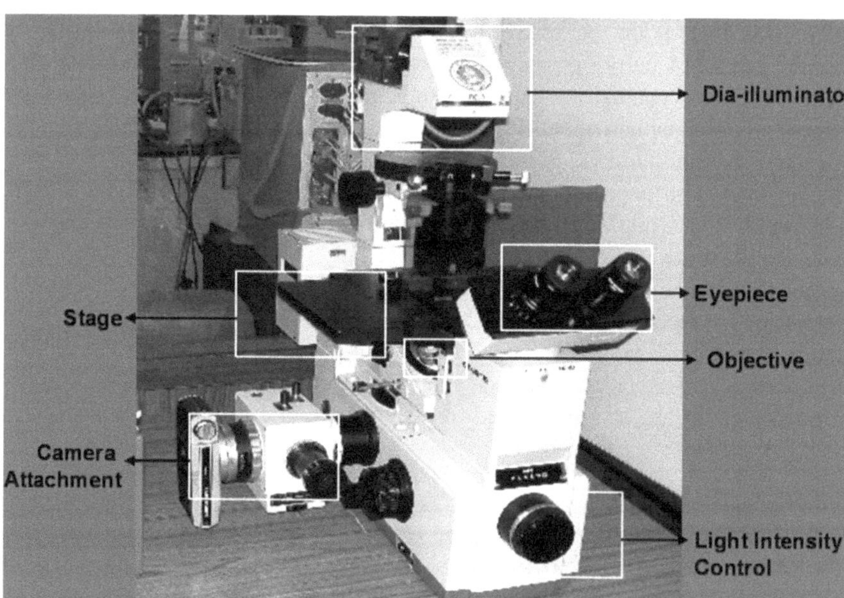

Fig. 4.11 Graphical dissection of an inverted phase contrast optical microscope

an optical microscope and come with 2–10x magnification in tune with the overall optical configuration of the light microscope system. The objective lenses are most important components of an optical microscopy system and are essentially cylinders containing embedded glass lenses to collect light from the sample. With a set of coarse and fine adjustment knob, objective lenses can be moved up and down to obtain the perfect focus of the object to be viewed. Typical objective lenses come with 4x, 5x, 10x, 20x, 40x, 50x and 100x amplification and with several different numerical aperture (NA) values. There is special version of lenses called oil immersion lenses with 50–100x magnification, which facilitate transmission of light from specimen to objective with minimal dissipative refraction. Next, the controllable illumination source which may be a mercury vapor lamp, xenon arc or tungsten halogen lamp, is placed either on the top (as in case of reflection microscope) or the bottom (as in case of inverted microscope) of microscope stage. Light coming from the source is then focused through an optical condensers, diaphragms and filters to produce variable quality and intensity. Optical microscopes have several variants including phase contrast, differential interference contrast (DIC) and stereo types. While phase contrast and DIC types are suitable to selective visualization of biological objects within different optical density than the bulk solution, stereo-microscopes are typically used to obtain a whole-device image of microfluidic systems. Compound optical microscopes are capable of constructing even 1000x magnified image of a specimen.

The resolving power or the resolution of an optical microscope is defined as the minimum distance between two neighboring airy disks due to diffraction, for which each of the disk can be effectively distinguished from it neighbor. This, in other way, represents the ability to resolve fine structural details. For an illuminating light with wavelength λ and objective with numeral aperture NA, the resolution is formulated as

$$d = \frac{\lambda}{2NA} \tag{4.59}$$

which for normal visible light falls in the range of 200–300 nanometers.

A normal brightfield microscope is limited by quite poor contrast as the refractive indices cell sample and bulk medium are approximately identically and only when the phase of the light passing through the sample is significantly altered, image is obtained with good contrast. With the unavoidable requirements of detecting a specific molecule or biological entity with a distinguished fluorescent emission, a process which assists sensitive detection with minimal background, fluorescent microscopes have been most essential visual analytic system for biomicrofluidics. In this class of microscopy, the specimen is illuminated with a light of narrow band of wavelengths belonging to a specific color (ultra-violet, blue, green, red) coming through an excitation filter to excite the desired type of fluophore which then emits the light of longer wavelength. For this purpose, either a particular component of the specimen is labeled with a fluorescent molecule such as fluorescein, rhodamine or a molecule e.g. green fluorescent protein (GFP) can be intrinsically fluorescent. Successively, an emission filter is used to selectively allow the emitted wavelengths. Generic components of a fluorescence microscope comprises of light source, the excitation filter, the dichroic mirror and the emission filter. Every dye comes with its specific range of spectral excitation and emission wavelengths and accordingly, the filters and the dichroic mirrors are selected. Multicolor images of several fluorophores can be constructed by super-imposing single fluorophore images. If the excitation and observation are made from the top the specimen, the microscope system is recognized as epi-fluorescence type. Although fluorescence microscope has enabled high resolution optical imaging, it is generally limited by gradual reduction in fluorescence intensity on exposure to excited light, a phenomenon called photobleaching. However, selective photobleaching and the intensity recovery thereafter are advantageously utilized in a technique called Fluorescence Recovery After Photobleaching (FRAP) [131] which measures the effective diffusion coefficient of a biological component. With time, several fluorescent molecules having reduced photobleaching and enhanced specificity towards a targeted biological or sub-cellular component are being chemically synthesized. In addition to epi-fluorescence microscopy, there are several advanced fluorescence imaging techniques such as confocal, multiphoton microscopy and the recent super-resolution techniques STED and 4Pi which enables visualization at single entity or at even single molecular level.

4.7.2 Confocal Microscopy

While most of the bright-field and fluorescence optical detection systems fulfill the need for biomolecular phenomena occurring within microconfinement, they are strongly impeded by their inability to resolve structures in z-dimension. This becomes specifically indispensable when the pattern flow or dynamics of biological entities is required to be monitored at three dimensional scales [132]. For example, there may be special microfluidic designs of fluidic components such as valves, pumps etc. whose operation principle takes advantages of every dimension. In this case, though planar two dimensional views may be elucidated by conventional optical detection system, the visualization of flow structure in third dimension remains unfortunately impaired. Also, complex high-throughput microfluidic system may be compiled of different component fabricated at multiple layers and thus, necessitates three dimensional visualization systems with capability of exclusively resolving the events occurring at each layer. Moreover, for microfluidic systems, in general, accurate characterization of device form holds the key for successful operation. Microfluidic devices characteristically contain an array of microfabricated components including channels, valves, pumps and mixers. The capacity to determine and control the dimensional design of these structures is a crucial obligation, not only to guarantee the functionality of operating devices, but also to offer a fundamental means for inventing and improving novel designs. From these very needs, depending upon the explicit rationale, laser scanning confocal microscopy has discovered many relevant applications in the realm of biomicrofluidics.

A confocal microscope system exploits a point illumination and a pinhole in an optically aligned plane in front of the electronic detector in order to diminish the out-of-focus or rather out-of-plane emission. Essentially, confocal microscopes possess a very low depth of focus. As they are integrated with highly-sensitive sensor coupled motorized system along each of x, y and z axis, image data from exclusively each focal plane can be conveniently accessed. In commercial confocal microscopy systems, highly collimated laser light sources of a specified wave-length and extremely narrow spectral spread, are transported along optical fibers to the specimen location. Operationally, only one point (or pixel in image notation) of the specimen being illuminated and imaged at a time, two or three dimensional imaging over a finite specimen section imposes sequential scanning over a regular raster-tracks in the specimen. This fundamental limitation impedes the rapidity of data acquisition which is a decisive factor governing the performance and efficiency of microfluidic systems. Relevantly, in recent times, fast scanning spinning or Nipkow disk confocal microscopes have been invented. The disk is composed of multiple pinhole-like apertures and it spins swiftly over the microscope stage in such as way each aperture traces a linear path through the image. A photosensitive detector collects data from each aperture, which are then assembled by computer program in form a image. Thus, advanced confocal imaging with spinning disk mechanism and multiple confocal apertures are capable of unleashing a high degree parallel data processing mechanism for image acquisition at an unprecedented speed.

The use of confocal microscopy as a detection system for biomicrolfuidics has opened up a vast range of molecular experimentations of which Fluorescence Recovery After Photobleaching (FRAP) and Förster Resonance Energy Transfer predomiantly worth definite emphasis. In FRAP [133], fluorescence intensity of a small region of the sample, referred as region of interest or ROI here in after, is bleached with high intensity (~1–20 mW) laser pulses. As the fluorescent molecules from neighboring region move into the ROI by the virtue of intrinsic diffusion, the ROI intensity get replenished to some extent (Fig. 4.12). Subsequently, analyzing the dynamics of fluorescence recovery in tune with trend predicted by transport equations, the effective diffusion coefficient (D_{eff}) can be determined. In biological systems where it is extremely unfeasible to decouple advective and diffusive parts of transport mechanism, D_{eff} empirically represents their compounded effect. The fluidity of a cellular component such as cell membrane or cytosol is then directly correlated with acquired D_{eff}.

FRET technology explores resonance energy transfer between two compatible fluophores by an inter-molecule dipole-dipole interaction mechanism and is used to assess the Ligand-Receptor binding and the coupling reaction kinetics at the molecular level. In a generic FRET assay [134], donor fluophore is excited, which then transfers its energy to its acceptor counterpart and emission characteristic of acceptor fluophore is recorded. As the dipole-dipole interactions are effective within only few nanometers of molecular separation, the energy transfer occurs only when molecules are in close-proximity typically encountered during binding or reaction.

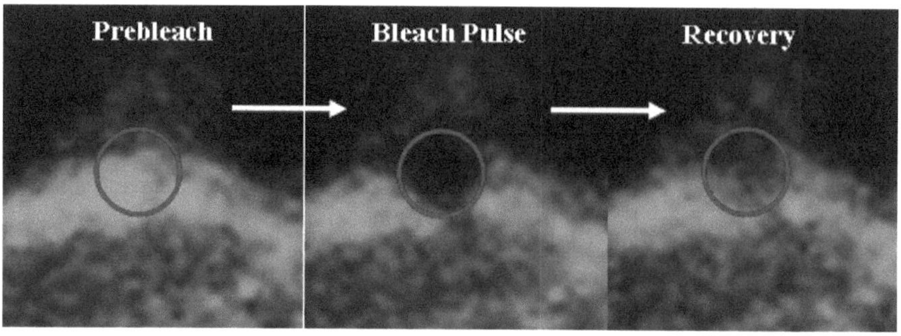

Fig. 4.12 Schematic representation of FRAP technology. Cells have been grown inside microchannel network for 36 hours and labeled with membrane binding carbocyanine dye DiO

In order to explore the fundamental biophysics at the single-molecule level, researchers have invented an enormously high resolution technology based on the low-background and narrow field-depth of evanescent waves. This observatory technology, known as Total Internal Reflection Fluorescence Microscopy (TIRFM), has enabled visualization down to single molecule length scale. Interestingly, owing to its intrinsic ability to resolve close-to-the-surface interfacial phenomena, the TIRFM system has been advantageously utilized in revelation of liquid slip length over a hydrophobic microchannel surface with nanometer resolution [135, 136].

4.7.3 Optofluidics

The classes of optical systems that are integrated with or synthesized of fluids are known as optofluidics [137, 138]. The technology has stemmed from the requirement of fabricating an optical detection system dimensionally commensurate with the micron-sized device itself. The high resolution imaging constraint in conventional microfluidic systems nullifies the cost-size benefits of micro-analytical systems. The most convenient solution in this respect is to integrate sensor-aperture arrays and possibly, illumination sources into the device itself. In principle, the detection parts are fabricated separately and then, coupled with the polymer based microfluidic device as a separate operational layer. With respect to the built-in materials, the optofluidics is broadly classified in three main categories – liquid-in-solid, liquid-in-liquid and systems. Liquid-solid systems work on the principle of highly contrasting refractive indices and total internal reflection in the interface which can be judiciously maneuvered to fabricate in situ diaphragm based lenses and optical-fiber like wave-guides. Moreover, within the domain of microfluidic technology, the focal length of adaptive integrated lenses can be dynamically manipulated by attenuating hydrostatic pressure on the diaphragm surface. A periodic array of voids where each void is selectively filled with absorbing dye can be utilized to manufacture any arbitrary shaped two-dimensional pattern. Another example of liquid-solid optofluidics technology, on-chip compact interferometer has been used as biosensors and chemical analyzers. Optofluidic devices that rely on evanescent-wave propagation can also perform as surface-plasmon sensors. High-Q resonators and the zero-mode waveguide sensors also belong to the class of liquid-solid optofluidic systems. In liquid-in-liquid systems, the total internal reflection between two immiscible liquid is utilized as the optical wave-guides with liquid–core and liquid-cladding. In low Reynolds number regime of microfluidics, flows are predominantly laminar and the liquids injected in parallel fashion continue to move side by side without any appreciable mixing. Now, if there exists a significant disparity of refractive indices between two liquids, transmitted light can effectively be confined within the domain encompassed by the liquid with lower refractive property, essentially yield a liquid form of optical cable which is then manipulated for in situ visualization and optical detection. The remaining class of optofluidic systems i.e. solid-in-liquid is mostly encountered when solid particles are dispersed within the bulk liquid and resulting localized gradients in optical properties serve as the underlining working ingredient. One popular example is the optical trapping of colloidal particle and biological cells where a highly focused beam of light is utilized to trap and move the suspended materials. In conjunction, researchers have also developed extremely sophisticated microfluidic devices including quantum dot based enhanced Raman scattering fluidic systems and fluidic memory devices.

4.7.4 Flow Visualization

Perhaps, for the subject of microfluidics which aims towards producing smaller and faster assay devices, the most essential and fundamental information is the complete knowledge of spatiotemporal pattern of fluid flow [139]. As a result of the radical declination in length-scales associated to miniaturization, microflows differ extensively from conventional macroscale flow theories with the incipient domination of extended interfacial phenomena. Thus, the prime objective of microscale flow visualization is to acquire the flow profile within designed fluidic circuits. The situation can be extremely interesting and non-intuitive if the flow is significantly perturbed by the presence of deformable biological objects such as cells. Theoretical investigations have demonstrated that in presence of dimensionally commensurate obstacles, surface stresses can be amplified in non-trivial manner. Till date, bulk of the microscale flow visualizations have been particle based of which microscale particle image velocimetry (μPIV) has been the most dominant type. In this system, the velocity vectors are computed by determining the displacement of dispersed particles between two successive time frame images [140]. The displacement calculation algorithm relies upon determining the peaks in averaged correlation landscape. Other particle based flow visualization techniques include particle streaking method where particle movement lines are delineated with finite time exposure during image acquisition [141] and Laser Doppler Velocimetry (LDV) [142]. In order to meet the criterion of using nanometer sized tracer particle compatible to cause least flow disturbance, one of the major challenges in particle based microflow visualization remains to eliminate the noise due to the enhanced relative contribution of stochastic Brownian motions in effective displacement [143]. In simultaneous advancements, scalar methods of microflow visualization which are primarily designed towards deciphering the flow patterns typically occurring during micromixing, flow instability and flow transformation, have been progressively developed. Here, the underlining principle is to either excite or bleach a strip-like fluid segment spanning across the microchannel cross-section and investigate the gradual deformation of the strip with time. For this purpose caged-fluorescence molecules and molecules with controllable bleach properties have been extensively used. While two-dimensional flow profiles are non-tedious to characterize, imposing challenges still persist in quantitative flow visualization in three dimensional forms and confocal based μ-PIV promises to solve pertinent hindrances [144, 145].

4.7.5 Non-Optical Detection

They say "seeing is believing" and while optical sensing techniques constitute the most common and form of detection module, these are predominantly limited by the inability to distinguish features below the wavelength of visible light spectrum. Optical signals because of their propagation in all directions, are generally collected in weak intensity and are needed to be amplified for satisfying all practical purposes. Moreover, from the optical measurement, it is very difficult to determine the number of targeted molecules actually present. With the background of these drawbacks in conjunction with the superfluous requirement of bulky optical microscopy components, the non-optical detection techniques for detecting very trace amount of desired specimens have gained increasing applications. In this clan of detection systems, there exists an array of integrated sensors which acquire and transmit some kinds of physical property changes in response to a chemical alteration in their vicinity. For the simplest of examples, if a molecule binds to an electronic sensing unit, it changes the conductivity, the resistance and the capacitance of the sensor, any of which can then be measured with auxiliary electronic circuits. In recent times, the performance and sensitivity of such element have been drastically augmented by implementing nanotubes or nanowires as the working segment. Specifically in applications related to on-chip DNA hybridization, nanowire based detection systems are of high demand.

Relevantly, macroscale high-end chemical detection methodologies such as Raman Spectroscopy have been miniaturized into a microfluidic confocal Raman spectroscopy device to impart the ability of performing local chemical detection and qualitative-quantitative analysis for various species present in infinitesimal amount. Other non-optical probing techniques such as atomic force microscopy (AFM) system have been quite functional in this respect [146]. Fundamentally, the deflection of the AFM probe with different forces has been used to examine the torsional force-velocity interrelation with reduced noise level and the non-intuitive consequences of the forces whose relative manifestation is appreciated exquisitely within micro-confinements. Label free methods of detection [147] such as those utilizes total internal reflection fluorescence microscopy (TIRFM) technology, Mach-Zehnder interferometer (MZI) and cantilever based biosensors [148] have also found their application in the realm of biomicrofluidics. However, challenges still remain pertaining to the limit of detection, precision, applicability and reproducibility of these biosensors.

4.8 Concluding Remarks

In the last 15 years or so, microfluidics has thrown its fathom in every corner of biology, in order to exploit the physico-chemical events at the level of single biological entities with unprecedented resolution, precision and elaboration. In the aforementioned sections, we have attempted to delineate the glimpse of technology, its elemental root, the governing dynamics and the future perspectives. It is to be summarized that the progresses in biomicrofluidic technology have been predominantly projected towards yielding lab-on-a-chip type appliances possessing enhanced device performance, increasing parallelization and augmented response. Simultaneously, there has been a parallel look out for fundamental sciences which are manifested only at the length-scale of microconfined domains. In this journey, there remain several challenges which biomicrofluidics must circumvent if its vistas are to be expanded in future. These include biocompatibility of microfluidic devices, sustenance and biochemical perseverance of biological entities within microconfinement and intrinsic limitation pertaining to molecular handling in trace concentration level. Original research endeavors on exploring microscale fluid physics, effect of physical forces on biological dynamics and instrumentation design for appreciating the aforementioned physico-biological correlations in practical dimensions need to be aptly emphasized in order to achieve the challenging feats of realizing biomicrofluidics as a sustainable technology in the future.

References

1. Whitesides GM (2006) The origins and the future of microfluidics. Nature. 442:368–373.
2. deMello AJ (2006) Control and detection of chemical reactions in microfluidic systems. Nature 442:394–402
3. Squires TM and Quake SR (2005) Microfluidics: Fluid physics at the nanoliter scale. Rev. Mod. Phys. 77:977–1026
4. Beebe DJ, Mensing GA, and Walker GM (2002) Physics and applications of microfluidics in biology. Annu. Rev. Biomed. Eng. 4:261–286
5. Hong JW and Quake SR (2003) Integrated nanoliter systems. Nat. Biotechnol. 21: 1179–1183
6. Melin J, and Quake SR (2007) Microfluidic large-scale integration: The evolution of design rules for biological automation. Annu. Rev. Biophys. Biomol. Struct. 36:213–231
7. McDonald JC, Duffy DC, Anderson JR, Chiu DT, Wu HK, Schueller OJA, and Whitesides GM (2000) Fabrication of microfluidic systems in poly(dimethylsiloxane). Electrophoresis. 21:27–40

8. Thorsen T, Maerkl SJ, and Quake SR (2002) Microfluidic large-scale integration. Science. 298:580–584
9. Wainright A, Nguyen UT, Bjornson T, and Boone TD (2003) Preconcentration and separation of double-stranded DNA fragments by electrophoresis in plastic microfluidic devices. Electrophoresis. 24:3784–3792
10. El-Ali J, Sorger PK, and Jensen KF (2006) Cells on chips. Nature. 442:403–411
11. Hansen CL, Skordalakes E, Berger JM, and Quake SR (2002) A robust and scalable microfluidic metering method that allows protein crystal growth by free interface diffusion. Proc. Natl. Acad. Sci. USA 99:16531–16536
12. Anderson RC, Su X, Bogdan GJ, and Fenton J (2002) A miniature integrated device for automated multistep genetic analysis. Nucleic Acid Res. 28:E60
13. Burns MA, Johnson BN, Brahmasandra SN et al. (1998) An Integrated nanoliter DNA analysis device. Science 282:484–487
14. Emrich CA, Tian H, Medintz IL, and Mathies RA (2002) Microfabricated 384-lane capillary array electrophoresis bioanalyzer for ultrahigh-throughput genetic analysis. Anal. Chem. 74:5076–83
15. Gu MB, Mitchell RJ, and Kim BC (2004) Whole-cell-based biosensors for environmental biomonitoring and application. Adv. Biochem. Eng. Biotechnol. 87:269–305
16. Reyes DR, Iossifidis D, Auroux P-A, and Manz, A (2002) Micro total analysis systems. 1. Introduction, theory, and technology. Anal. Chem. 74:2623–2636
17. Craighead H (2006) Future lab-on-a-chip technologies for interrogating individual molecules. Nature. 442:387–393
18. Janasek D, Franzke J, and Manz A (2006) Scaling and the design of miniaturized chemical-analysis systems. Nature. 442:374–380
19. Mao H, Yang T, and Cremer PS (2002) Design and characterization of immobilized enzymes in microfluidic systems. Anal. Chem. 74:379–385
20. Ottesen EA, Hong JW, Quake SR, and Leadbetter JR (2006) Microfluidic digital PCR enables multigene analysis of individual environmental bacteria. Science 314: 1464–1467
21. Verpoorte E (2002) Microfluidic chips for clinical and forensic analysis. Electrophoresis. 23:677–712
22. Mijatovic D, Eijkel JCT, and van den Berg A (2005) Technologies for nanofluidic systems: topdown vs. bottom-up – a review. Lab Chip. 5:492–500
23. Ng JMK, Gitlin I, Stroock AD, and Whitesides GM (2002) Components for integrated poly(dimethylsiloxane) microfluidic systems. Electrophoresis. 23:3461–3473
24. Madou M, Zoval J, Jia G, Kido H, Kim J, and Kim N (2006) Lab on a CD. Annu. Rev. Biomed. Eng. 8:601–628
25. Chakraborty D, Gorkin R, Madou M, Kulinsky L, and Chakraborty S (2009) Capillary filling in centrifugally actuated microfluidic with dynamically evolving contact line motion. J. Appl. Phys. 105:084904
26. Das S, Das T, and Chakraborty S (2006) Analytical solutions for the rate of DNA hybridization in a microchannel in the presence of pressure-driven and electroosmotic flows. Sens. Actuators B: Chem. 114:957–963
27. Das S, Das T, and Chakraborty S (2006) Modeling of coupled momentum, heat and solute transport during DNA hybridization in a microchannel in the presence of electro-osmotic effects and axial pressure gradients. Microfluid. Nanofluid. 2:37–49
28. Dolnik V, Liu S, and Jovanovich S (2000) Capillary electrophoresis on microchip. Electrophoresis 21:41–54
29. Heller C (2001) Principles of DNA separation with capillary electrophoresis. Electrophoresis. 22:629–643
30. Bhattacharyya A and Klapperich CM (2008) Microfluidics-based extraction of viral RNA from infected mammalian cells for disposable molecular diagnostics. Sens. Actuators B: Chem. 129:693–698

31. Lion N, Rohner TC, Dayon Loïc et al. (2003) Microfluidic systems in proteomics. Electrophoresis 24:3533–3562
32. Hu S, Michels DA, Fazal MA, Ratisoontorn C, Cunningham ML, and Dovichi NJ (2004) Capillary sieving electrophoresis/micellar electrokinetic capillary chromatography for two-dimensional protein fingerprinting of single mammalian cells. Anal. Chem. 76: 4044–4049
33. Freire SLS and Wheeler AR (2006) Proteome-on-a-chip: Mirage, or on the horizon?. Lab Chip. 6:1415–1423 (2006)
34. Wang YC, Stevens AL, and Han JY (2005) Million-fold preconcentration of proteins and peptides by nanofluidic filter. Anal. Chem. 77:4293–4299
35. Wainright A, Williams SJ, Ciambrone G, Xue Q, Wei J, and Harris D (2002) Sample Pre-Concentration by Isotachophoresis in Microfluidic Devices. J. Chromatogr., A. 979: 69–80
36. Tan A, Pashkova A, Zang L, Foret F, and Karger BL (2002) A miniaturized multi-chamber solution isoelectric focusing device for separation of protein digests. Electrophoresis 23:3599–3607
37. Li Y, Buch JS, Rosenberger F, DeVoe DL, and Lee CS (2004) Integration of isoelectric focusing with parallel sodium dodecyl sulfate gel electrophoresis for multidimensional protein separations in a plastic microfludic network. Anal. Chem. 76:742–748
38. Brivio M, Fokkens RH, Verboom W, Reinhoudt DN, Tas NR, Goedbloed M, and van den Berg A (2002) Integrated microfluidic system enabling (bio)chemical reactions with on-Line MALDI-TOF mass spectrometry. Anal. Chem. 74:3972–3976
39. Moon H, Wheeler AR, Garrell RL, Loo JA, and Kim C-J (2006) Integrated digital microfluidic chip for multiplexed proteomic sample preparation and analysis by MALDI-MS. Lab Chip. 6:1213–1219
40. Pollack L, Tate MW, Finnefrock AC, Kalidas C, Trotter S, Darnton NC, Lurio L, Austin RH, Batt CA, and Gruner SM (2001) Time resolved collapse of a folding protein observed with small angle x-ray scattering. Phys. Rev. Lett. 86:4962–4965
41. Hansen C and Quake SR (2003) Microfluidics in structural biology: Smaller, faster... better. Curr. Opin. Struct. Biol. 13:538–544
42. Andersson H and van den Berg A (2003) Microfluidic devices for cellomics: A review. Sens. Actuators B: Chem. 92:315–325
43. Gmez-Sjberg R, Leyrat AA, Pirone DM, Chen CS, and Quake, S.R (2007) Versatile, fully automated, microfluidic cell culture system. Anal. Chem. 79:8557–8563
44. Das T, Maiti TK, and Chakraborty S (2008) Traction force microscopy on-chip: Shear deformation of fibroblast cells. Lab Chip. 8:1308–1318
45. Helmke BP and Minerick AR (2006) Designing a nano-interface in a microfluidic chip to probe living cells: challenges and perspectives. Proc. Natl. Acad. Sci. USA 103:6419–6424
46. Abhyankar VV, Lokuta MA, Huttenlocher A, and Beebe DJ (2006) Characterization of a membrane-based gradient generator for use in cell-signaling studies. Lab Chip. 6:389–393
47. Lee PJ, Hung PJ, Shaw R, Jan L, and Lee LP (2005) Microfluidic application-specific integrated device for monitoring direct cell–cell communication via gap junctions between individual cell pairs. Appl. Phys. Lett. 86:223902
48. Lucchetta EM, Lee JH, Fu LA, Patel NH, and Ismagilov RF (2005) Dynamics of Drosophila embryonic patterning network perturbed in space and time using microfluidics. Nature 434:1134–1138
49. Takayama S, Ostuni E, LeDuc P, Naruse K, Ingber DE, and Whitesides GM (2001) Laminar flows – subcellular positioning of small molecules. Nature 411:1016
50. Wang SJ, Saadi W, Lin F, Minh-Canh Nguyen C, and Jeon NL (2004) Differential effects of EGF gradient profiles on MDAMB-231 breast cancer cell chemotaxis. Exp. Cell Res. 300:180–189
51. Breslauer DN, Lee PJ, and Lee LP (2006) Microfluidics-based systems biology. Mol. BioSyst. 2:97–112

52. Taylor RJ, Falconnet D, Niemisto A, Ramsey SA, Prinz S, Shmulevicha I, Galitski T, and Hansen CL (2009) Dynamic analysis of MAPK signaling using a high-throughput microfluidic single-cell imaging platform. Proc. Natl. Acad. Sci. USA 106:3758–3763
53. Becker FF, Wang XB, Huang Y, Pethig R, Vykoukal J, and Gascoyne PR (1995) Separation of human breast-cancer cells from blood by differential dielectric affinity. Proc. Natl Acad. Sci. USA 92:860–864
54. Yang MS, Li CW, and Yang J (2002) Cell docking and on-chip monitoring of cellular reactions with a controlled concentration gradient on a microfluidic device. Anal. Chem. 74:3991–4001
55. Brehm-Stecher BF and Johnson EA (2004) Single-cell microbiology: tools, technologies, and applications. Microbiol. Mol. Biol. Rev. 68:538–559
56. Maerkl SJ and Quake SR (2007) A systems approach to measuring the binding energy landscapes of transcription factors. Science 315:233–237
57. Yager P, Edwards T, Fu E, Helton K, Nelson K, Tam MR, and Weigl BH (2006) Microfluidic diagnostic technologies for global public health. Nature. 442:412–418
58. Andersson H and van den Berg A (2004) Microfabrication and microfluidics for tissue engineering: state of the art and future opportunities. Lab Chip. 4:98–103
59. Hatch A, Kamholz AE, Hawkins KR, Munson MS, Schilling EA, Weigl BH, and Yager P (2001) A rapid diffusion immunoassay in a T-sensor. Nat. Biotechnol. 19:461–465
60. Weiss M, Elsner M, Kartberg F, and Nilsson T (2004) Anomalous subdiffusion is a measure for cytoplasmic crowding in living cells. Biophys. J. 87:3518–3524
61. Banks D and Fradin C (2005) Anomalous diffusion of proteins due to molecular crowding. Biophys. J. 89:2960–2971
62. Caspi A, Granek R, and Elbaum M (2000) Enhanced diffusion in active intracellular transport. Phys. Rev. Lett. 85:5655–5658
63. Brangwynne CP, Koenderink GH, MacKintosh FC, and Weitz DA (2008) Cytoplasmic diffusion: molecular motors mix it up. J. Cell Biol. 183:583–587
64. Zhang C, Xu J, Ma W, and Zheng W (2006) PCR microfluidic devices for DNA amplification. Biotechnol. Adv. 24:243–284
65. Pal D and Venkataraman V (2002) A portable battery-operated chip thermocycler based on induction heating. Sens. Actuators A Phys. 102:151–156
66. Dittrich PS and Manz A (2006) Lab-on-a-chip: microfluidics in drug discovery. Nat. Rev. Drug Discov. 5:210–218
67. Kang L, Chung BG, Langer R, and Khademhosseini A (2008) Microfluidics for drug discovery and development: From target selection to product lifecycle management. Drug Discov. Today. 13:1–13
68. Gao J, Yin XF, and Fang ZL (2004) Integration of single cell injection, cell lysis, separation and detection of intracellular constituents on a microfluidic chip. Lab Chip. 4:47–52
69. Huang B, Wu H, Bhaya D, Grossman A, Granier S, Kobilka BK, and Zare RN (2007) Counting low-copy number proteins in a single cell. Science 315:81–84
70. Mitchell MC, Spikmans V, and de Mello AJ (2001) Microchip-based synthesis and analysis: control of multicomponent reaction products and intermediates. Analyst. 126:24–27
71. Choi CJ and Cunningham BT (2007) A 96-well microplate incorporating a replica molded microfluidic network integrated with photonic crystal biosensors for high throughput kinetic biomolecular interaction analysis. Lab Chip. 7:550–556
72. McAllister DV, Wang PM, Davis SP, Park JH, Canatella PJ, Allen MG, and Prausnitz MR (2003) Microfabricated needles for transdermal delivery of macromolecules and nanoparticles: Fabrication methods and transport studies. Proc. Natl. Acad. Sci. USA 100:13755–13760
73. Hilt JZ and Peppas NA (2005) Microfabricated drug delivery devices. Int. J. Pharmaceutics 306:15–23
74. Chakraborty S and Tsuchiya K (2008) Development and fluidic simulation of microneedles for painless pathological interfacing with living systems. J. Appl. Phys. 103:114701

75. Wiggins S and Ottino JM (2004) Foundations of chaotic mixing. Phil. Trans. R. Soc. Lond. A 362:937–970
76. Campbell CJ and Grzybowski BA (2004) Microfluidic mixers: From microfabricated to self-assembling devices. Phil. Trans. R. Soc. Lond. A 362:1069–1086
77. Di Carlo D, Irimia D, Tompkins RG, and Toner M (2007) Continuous inertial focusing, ordering, and separation of particles in microchannels. Proc. Natl. Acad. Sci. USA 104:18892–18897
78. Dutta D, Ramachandran A, and Leighton Jr DT (2006) Effect of channel geometry on solute dispersion in pressure-driven microfluidic systems. Microfluid. Nanofluid. 2: 275–290
79. Huiqian Y, Nguyen N-T, and Huang X (2006) Micromixer Based on Taylor Dispersion. J. Phys.: Conference Series. 34:136–141
80. Hardt S, Drese KS, Hessel V, and Schönfeld F (2005) Passive micromixers for applications in the microreactor and μTAS fields. Microfluid. Nanofluid. 1:108–118
81. Stroock AD, Dertinger SKW, Ajdari A, Mezić I, Stone HA, and Whitesides GM (2002) Chaotic mixer for microchannnels. Science. 295:647–651
82. Lin J-L, Lee K-H, and Lee G-B (2005) Active mixing inside microchannels utilizing dynamic variation of gradient zeta potentials. Electrophoresis 26:4605–4615
83. Giddings JC, Yang FJF, and Myers MN (1976) Flow Field-Flow Fractionation: A Versatile New Separation Method. Science 193:1244–1245
84. Liu G, and Giddings JC (1991) Separation of particles in nonaqueous Suspensions by thermal-electrical field-flow fractionation. Anal. Chem. 63:296–299
85. Sant HJ and Gale BK (2006) Geometric scaling effects on instrumental plate height in field flow fractionation. J. Chromatograph. A 1104:282–290
86. Chon K, Moon J, Kim S, Kim S-D, and Cho J (2006) Bio-particle separation using microfluidic porous plug for environmental monitoring. Desalination. 202:215–223
87. Gale BK, Caldwell KD, and Frazier AB (1998) A micromachined electrical field-flow fractionation (μ-EFFF) system. IEEE Trans. Biomed. Eng. 45:1459–1469 (1998)
88. Edwards TL, Gale BK, and Frazier AB (2002) A microfabricated thermal field-flow fractionation system. Anal. Chem. 74:1211–1216
89. Ristenpart WD, Wan J, and Stone HA (2008) Enzymatic reactions in microfluidic devices: Michaelis-Menten kinetics. Anal. Chem. 80:3270–3276
90. Gomez FA (2008) Biological Applications of Microfluidics, Wiley-Interscience, New York
91. Starkey DE, Han A, Bao JJ, Ahn CH, Wehmeyer KR, Prenger MC, Halsall HB, and Heineman WR (2001) Fluorogenic assay for β-glucuronidase using microchip-based capillary electrophoresis. J. Chromatograph. B. 762:33–41
92. Schilling EA, Kamholz AE, and Yager P (2002) Cell lysis and protein extraction in a microfluidic device with detection by a fluorogenic enzyme assay. Anal. Chem. 74: 1798–1804
93. Cohen CB, Chin-Dixon E, Jeong S, and Nikiforov TT (1999) A microchip-based enzyme assay for protein kinase A. Anal. Biochem. 273:89–97
94. Yang R-J, Fu L-M, and Lin Y-C (2001) Electroosmotic flow in microchannels. J. Colloid Interf. Sci. 239:98–105
95. Ross D, Johnson TJ, and Locascio LE (2001) Imaging of electroosmotic flow in plastic microchannel. Anal. Chem. 73:2509–2515
96. Fu L-M, Lin J-Y, and Yang R-J (2003) Analysis of electroosmotic flow with step change in zeta potential. J. Colloid Interface Sci. 258:266–275
97. Joo S, Chung TD, and Kim HC (2007) A rapid field-free electroosmotic micropump incorporating charged microchannel surfaces. Sens. Actuators B: Chem. 123:1161–1167
98. Wu H-Y and Liu C-H (2005) A novel electrokinetic micromixer. Sens. Actuators A: Phys. 118:107–115
99. Oddy MH, Santiago JG, and Mikkelsen JC (2001) Electrokinetic instability micromixing. Anal. Chem. 73:5822–5832

100. Studer V, Pépin A, Chen Y, and Ajdari A (2002) Fabrication of microfluidic devices for AC electrokinetic fluid pumping. Microelectronic Eng. 61–62:915–920
101. Kandlikar SG, Garimella S, Li D, Colin S, and King MR (2006) Heat transfer and fluid flow in minichannels and microchannels, Elsevier, Amsterdam
102. Gao J, Xu J, Locascio LE, and Lee CS (2001) Integrated microfluidic system enabling protein digestion, peptide separation, and protein identification. Anal. Chem. 73:2648–2655
103. Doh I and Cho Y-H (2005) A continuous cell separation chip using hydrodynamic dielectrophoresis (DEP) process. Sens. Actuators A: Phys. 121:59–65
104. Kang KH, Xuan X, Kang Y, and Li D (2006) Effects of dc-dielectrophoretic force on particle trajectories in microchannels. J. Appl. Phys. 99:064702
105. Kua CH, Lam YC, Rodriguez I, Yang C, and Youcef-Toumi K (2007) Dynamic cell fractionation and transportation using moving dielectrophoresis. Anal. Chem. 79:6975–6987
106. Cho SK, Moon H, and Kim C-J (2003) Creating, transporting, cutting, and merging liquid droplets by electrowetting-based actuation for digital microfluidic circuits, J. Microelectromech. Syst. 12:70–80
107. Fair RB (2007) Digital microfluidics: is a true lab-on-a-chip possible? Microfluid. Nanofluid. 3:245–281
108. Sigurdson M, Wang D, and Meinhart CD (2005) Electrothermal stirring for heterogeneous immunoassays. Lab Chip. 5:1366–1373
109. Feng JJ, Krishnamoorthy S, and Sundaram S (2007) Numerical analysis of mixing by electrothermal induced flow in microfluidic systems. Biomicrofluidics. 1:024102
110. Feldman HC, Sigurdson M, and Meinhart CD (2007) AC electrothermal enhancement of heterogeneous assays in microfluidics. Lab Chip. 7:1553–1559
111. Ng WY, Goh S, Lam YC, Yang C, and Rodríguez I (2009) DC-biased AC-electroosmotic and AC-electrothermal flow mixing in microchannels. Lab Chip. 9:802–809
112. Pamme N (2006) Magnetism and microfluidics. Lab Chip. 6:24–38
113. Mori S (1986) Magnetic field-flow fractionation using capillary tubing. Chromatographia. 21:642–644
114. Latham AH, Freitas RS, Schiffer P, and Williams ME (2005) Capillary magnetic field flow fractionation and analysis of magnetic nanoparticles. Anal. Chem. 77:5055–5062
115. Doyle PS, Bibette J, Bancaud A, and Viovy J-L (2002) Self-assembled magnetic matrices for DNA separation chips. Science. 295:2237
116. Jang J and Lee SS (2000) Theoretical and experimental study of MHD (magnetohydrodynamic) micropump. Sens. Actuators A: Phys. 80:84–89
117. Hatch A, Kamholz AE, Holman G, Yager P, and Böhringer KF (2001) A ferrofluid magnetic micropump. J. Microelectromech. Syst. 10:215–221
118. Hartshorne H, Backhouse CJ, and Lee WE (2004) Ferrofluid-based microchip pump and valve. Sens. Actuators B: Chem. 99:592–600
119. Yamahata C, Chastellain M, Parashar VK, Petri A, Hofmann H, and Gijs MAM (2005) Plastic Micropump With Ferrofluidic Actuation. J. Microelectromech. Syst. 14: 96–102
120. Atencia J and Beebe DJ (2004) Magnetically-driven biomimetic micro pumping using vortices. Lab Chip. 4:598–602
121. West J, Gleeson JP, John A, Collins JK, and Berney H (2003) Structuring laminar flows using annular magnetohydrodynamic actuation. Sens. Actuators B: Chem. 96:190–199
122. Jackson WC, Tran HD, O'Brien MJ, Rabinovich E, and Lopez GP (2001) Rapid prototyping of active microfluidic components based on magnetically modified elastomeric materials. J. Vac. Sci. Technol. B. 19:596–599
123. Yuen PK, Li GS, Bao YJ, and Muller UR (2003) Microfluidic devices for fluidic circulation and mixing improve hybridization signal intensity on DNA arrays. Lab Chip. 3:46–50
124. Biswal SL and Gast AP (2004) Micromixing with linked chains of paramagnetic particles. Anal Chem. 76:6448–6455
125. Verpoorte E (2003) Beads and chips: new recipes for analysis. Lab Chip. 3:60 N–68 N

126. Hayes MA, Polson NA, Phayre AN, and Garcia AA (2001) Flow based microimmunoassay. Anal. Chem. 73:5896–5902
127. Jiang GF and Harrison DJ (2000) mRNA isolation in a microfluidic device for eventual integration of cDNA library construction. Analyst 125:2176–2179
128. Kwakye S and Baeumner A (2003) A microfluidic biosensor based on nucleic acid sequence recognition. Anal. Bioanal. Chem. 376:1062–1068
129. Furdui VI, Kariuki JK, and Harrison DJ (2003) Microfabricated electrolysis pump system for isolating rare cells in blood. J. Micromech. Microeng. 13:S164–D170
130. Furdui VI and Harrison DJ (2004) Immunomagnetic T cell capture from blood for PCR analysis using microfluidic systems. Lab Chip. 4:614–618
131. Reits EAJ and Neefjes JJ (2001) From fixed to FRAP: measuring protein mobility and activity in living cells. Nat. Cell Biol. 3:E145–E147
132. Lima R, Wada S, Tsubota K, and Yamaguchi T (2006) Confocal micro-PIV measurements of three-dimensional profiles of cell suspension flow in a square microchannel. Meas. Sci. Technol. 17:797–808
133. Chen Y, Lagerholm BC, Yang B, and Jacobson K (2006) Methods to measure the lateral diffusion of membrane lipids and proteins. Methods. 39:147–53
134. Mere L, Bennett T, Coassin P, England P, Hamman B, Rink T, Zimmerman S, and Negulescu P (1999) Miniaturized FRET assays and microfluidics: key components for ultra-high-throughput screening. Drug Discov. Today. 4:363–369
135. Huang P, Guasto JS, and Breuer KS (2006) Direct measurement of slip velocities using three-dimensional total internal reflection velocimetry. J. Fluid Mech. 566:447–464
136. Huang P and Breuer KS (2007) Direct measurement of anisotropic near-wall hindered diffusion using total internal reflection velocimetry. Phys. Rev. E. 76:046307
137. Heng X, Erickson D, Baugh LR, Yaqoob Z, Sternberg PW, Psaltis D, and Yang C (2006) Optofluidic microscopy – a method for implementing a high resolution optical microscope on a chip. Lab Chip. 6:1274–1276
138. Psaltis D, Quake SR, and Yang C (2006) Developing optofluidic technology through the fusion of microfluidics and optics. Nature. 442:381–386
139. Sinton D (2004) Microscale flow visualization. Microfluid. Nanofluid. 1:2–21
140. Santiago JG, Wereley ST, Meinhart CD, Beebe DJ, and Adrian RJ (1998) A particle image velocimetry system for microfluidics. Exp. Fluids. 25:316–319
141. McKnight TE, Culbertson CT, Jacobson SC, and Ramsey JM (2001) Electroosmotically induced hydraulic pumping with integrated electrodes on microfluidic devices. Anal. Chem. 73:4045–4049
142. Minor M, van der Linde AJ, van Leeuwen HP, and Lyklema J (1997) Dynamic aspects of electrophoresis and electroosmosis: A new fast method for measuring particle mobilities. J. Colloid Interface Sci. 189:370–375
143. Wereley ST, Gui L, and Meinhart CD (2002) Advanced algorithms for microscale particle image velocimetry. AIAA J. 40:1047–1055
144. Klank H, Goranovic G, Kutter J, Gjelstrup H, Michelsen J, and Westergaard C (2002) PIV measurements in a microfluidic 3D-sheathing structure with three-dimensional flow behaviour. J. Micromech. Microeng. 12:862–869
145. Kinoshita H, Kaneda S, Fujii T, and Oshima M (2007) Three-dimensional measurement and visualization of internal flow of a moving droplet using confocal micro-PIV. Lab Chip. 7:338–346
146. Anderson MS (2005) Microfluidics and chromatography with an atomic force microscope. Anal. Chem. 77:2907–2911
147. Blanco FJ, Agirregabiria M, Berganzo J, Mayora K, Elizalde J, Calle A, Dominguez C, and Lechuga LM (2006) Microfluidic-optical integratedCMOScompatible devices for label-free biochemical sensing. J. Micromech. Microeng. 16:1006–1016
148. Shipper EF, Brugman AM, Dominguez C, Lechuga LM, Kooyman RPH, and Greve J (1997) The realization of an integrated Mach-Zehnder waveguide immunosensor in silicon technology. Sens. Actuators B: Chem. 40:147–153

Chapter 5
Perspectives of Micro and Nanofabrication of Carbon for Electrochemical and Microfluidic Applications

R. Martinez-Duarte, G. Turon Teixidor, P.P. Mukherjee, Q. Kang, and M.J. Madou

Abstract This chapter focuses on glass-like carbons, their method of micro and nanofabrication and their electrochemical and microfluidic applications. At first, the general properties of this material are exposed, followed by its advantages over other forms of carbon and over other materials. After an overview of the carbonization process of organic polymers we delve into the history of glass-like carbon. The bulk of the chapter deals with different fabrication tools and techniques to pattern polymers. It is shown that when it comes to carbon patterning, it is significantly easier and more convenient to shape an organic polymer and carbonize it than to machine carbon directly. Therefore the quality, dimensions and complexity of the final carbon part greatly depend on the polymer structure acting as a precursor. Current fabrication technologies allow for the patterning of polymers in a wide range of dimensions and with a great variety of tools. Even though several fabrication techniques could be employed such as casting, stamping or even Computer Numerical Controlled (CNC) machining, the focus of this chapter is on photolithography, given its precise control over the fabrication process and its reproducibility. Next Generation Lithography (NGL) tools are also covered as a viable way to achieve nanometer-sized carbon features. These tools include electron beam (e-beam), Focused-ion beam (FIB), Nano Imprint Lithography (NIL) and Step-and-Flash Imprint Lithography (SFIL). At last, the use of glass-like carbon in three applications, related to microfluidics and electrochemistry, is discussed: Dielectrophoresis, Electrochemical sensors, and Fuel Cells. It is exposed how in these applications glass-like carbon offers an advantage over other materials.

Keywords Carbon · Glass-like carbon · Carbon MEMS · SU-8 · Photolithography · Electron-beam · Focused ion beam · Nanoimprint lithography · Microfluidics · Dielectrophoresis · Electrochemistry · Nanoelectrodes · Fractal electrodes · Fuel cells

---

R. Martinez-Duarte (✉)
Department of Mechanical and Aerospace Engineering, University of California, Irvine 4200 Engineering Gateway, Irvine, CA 92697, USA
e-mail: drmartnz@gmail.com

S. Chakraborty (ed.), *Microfluidics and Microfabrication*,
DOI 10.1007/978-1-4419-1543-6_5,

5.1 Introduction

Etymologically, the English name for carbon comes from the Latin name for burnt wood, *carbo*. Historically, carbon has played an important role in the daily lives of humans since prehistoric times when it could be commonly encountered as soot and as main component in blac inks for body painting. The use of carbon for odor control is well known and dates back to the Egyptians. Charcoal derivation as we know it today dates back to the Roman civilization. However, it was not until the late 18th century when carbon began to be treated as a chemical element. In 1722, René A. F. de Réaumur demonstrated that iron could be transformed into steel by the absorption of certain substance, now known to be carbon [1]. In 1772 the French scientist Antoine Lavoisier proved that diamond is a crystalline allotrope of carbon by comparing the results from heat treating carbon and diamond samples. In 1779, following a similar method than the one used by Lavoisier, Carl Scheele determined that graphite, considered at the time a form of lead, was indeed another carbon allotrope. Lavoisier later listed Carbon as a separate element in his 1789 textbook *Traité Élémentaire de Chimie* [2]. Carbon is the chemical element with symbol C and atomic number 6. It is nonmetallic, tetravalent and has an atomic weight of 12.0107. Its electron configuration, 1 $s^2$2 $s^2$2p^2, makes it a member of group 14 and period 2 on the periodic table. Carbon has the highest melting and sublimation point of all elements (3800 K). It forms more compounds than any other element, with almost 10 million pure organic compounds described to date. It is also the fourth most abundant element in the Universe by mass after hydrogen, helium and oxygen. Nevertheless, it is present in all known life forms and in the human body is the second most abundant element by mass (about 18.5%) after oxygen. This abundance and the easiness of carbon to polymerize make this element the chemical basis of all known life on Earth.

5.2 Carbon Allotropes

Several allotropes of carbon exist (see Fig. 5.1 for some examples), resulting in a large variety of molecular configurations for multi-atomic structures. This is partially due to the fact that atomic carbon is a very short-lived species that requires to be promptly stabilized [3, 4]. Carbon allotropes include diamond, lonsdaleite, buckminsterfullerenes, graphene, carbyne [5–11], graphite, carbon nanofoams, diamond-like carbon, amorphous carbon and those carbons derived from the pyrolysis of organic materials, better known as glass-like carbons. It is important to describe these allotropes and show how they compare to glass-like carbon for the reader to understand the advantages and possible disadvantages of the use of the latter in the proposed applications. The physical properties of carbon vary widely with the allotropic form. For example, diamond is highly transparent while graphite is opaque and black; diamond is among the strongest materials while graphite is soft enough to allow its use as solid lubricant; diamond is an excellent electrical insulator while graphite is a good

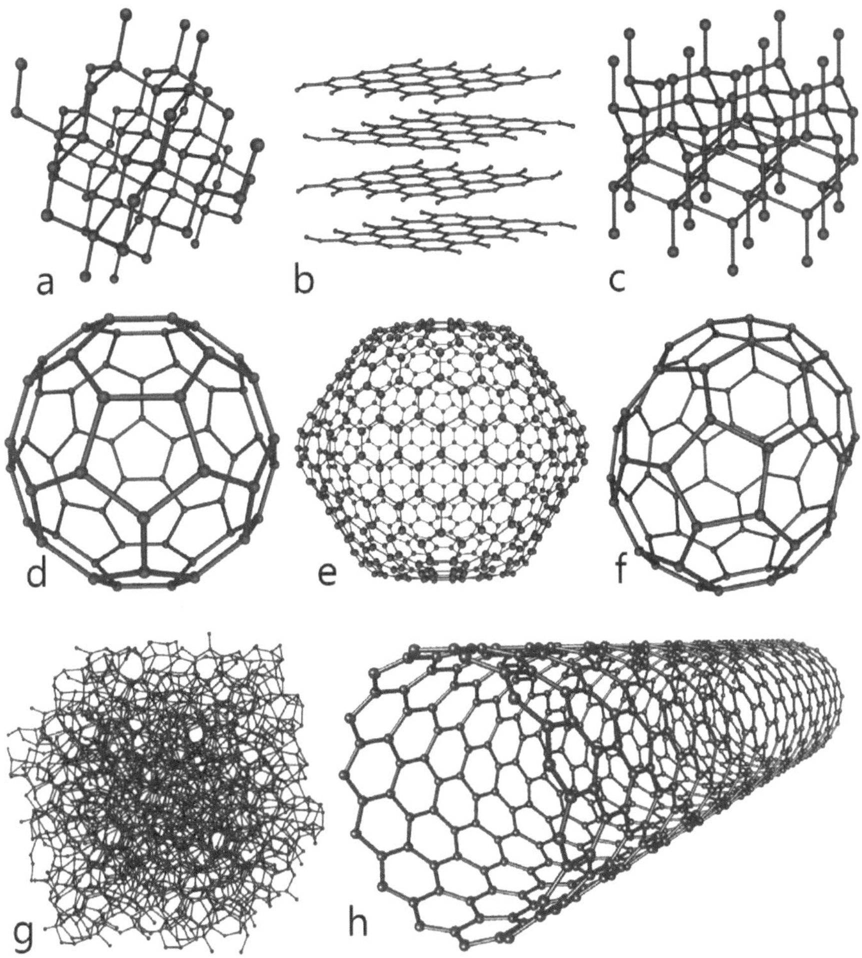

Fig. 5.1 Some carbon allotropes: (**a**) Diamond, (**b**) Graphite and (**c**) Lonsdaleite. Examples of Buckminsterfullerenes: (**d**) C_{60} (buckyball), (**e**) C_{540}, (**f**) C_{70} (**g**) Amorphous carbon, and (**h**) single-walled carbon nanotube. Illustration by Michael Ströck. Reprinted under a GNU free documentation license www.gnu.org

electrical conductor; diamond is the best known naturally occurring thermal conductor but some forms of graphite are used in thermal insulation.

Diamond is an allotrope of carbon that is twice as dense as graphite. It is formed at very high pressures with the conversion of graphite into diamond. The resulting chemical bonding between the carbon atoms is covalent with sp3 hybridization. Diamond has a cubic crystalline structure and is thermodynamically stable at pressures above 6 GPa at room temperature and metastable at atmospheric pressure. At low pressures it turns rapidly into graphite at temperatures above 1900 K in an inert atmosphere.

Lonsdaleite, on the other hand, presents an hexagonal crystalline structure. This carbon allotrope is believed to form from graphite present in meteorites upon their impact on Earth. The sudden great heat and stress of the impact causes graphite to crystallize retaining its hexagonal crystal lattice. With a translucent brownish-yellow color, Lonsdaleite was first discovered in 1967 in the Canyon Diablo meteorite at Barringer Crater (also known as Meteor Crater) in Arizona [12]. Hexagonal diamond was also first lab-synthetized in 1967 by compressing and heating graphite [13].

Buckminsterfullerenes, or fullerenes, are carbon allotropes that have a graphite-like structure but instead of purely hexagonal packing, they also contain pentagonal or even heptagonal carbon rings that bend the sheet into spheres, ellipses or cylinders. Fullerenes are discrete molecular species in contrast to the theoretically infinite lattices of diamond or graphite. Buckyballs, buckytubes (now better known as nanotubes) and nanobuds are all forms of fullerenes. The first fullerene, a soccer-ball-shaped carbon molecule called C_{60} was discovered in 1985 [14]. The name buckminsterfullerenes and its derivatives are given after Richard Buckminster Fuller, the American architect who popularized the use of geodesic domes, which resemble the structure of fullerenes. Interestingly, other names that were considered in the original 1985 publication were ballene, spherene, soccerene and carbosoccer [14].

A basic structural element of some forms of fullerenes (such as nanotubes) is graphene. Graphene is a one-atom-thick sheet of hexagonally arranged carbon atoms. The term graphene appeared in the late 1980s but was not experimentally derived until 2004 [15] and not characterized until 2007 [16]. Graphene is the strongest material known to man, 200 times stronger than steel, and as of 2008, one of the most expensive [17].

Graphite is formed by stacking graphene layers parallel to each other in a three-dimensional, crystalline, long-range order. There are two allotropic forms with different stacking arrangements: hexagonal and rombohedral, but the chemical bonds within the layers are covalent with sp2 hybridization in both cases. Carbon atoms in graphite, as in graphene, are bonded trigonally in a plane composed of fused hexagonal rings. The resulting network is 2-dimensional and the resulting flat sheets are stacked and loosely bound through weak Van der Waals forces. Because of the delocalization of one of the outer electrons of each atom to form a π-cloud, graphite conducts electricity preferentially in the plane parallel to the covalently bonded sheets.

Carbon nanofoam is yet another carbon allotrope discovered as recently as 1997 [18]. It consists of a low-density cluster of carbon atoms strung together in a loose three-dimensional web. Each cluster is about 6 nanometers wide and consists of about 4000 carbon atoms linked in graphite-like sheets. The clusters present a negative curvature by the inclusion of heptagons among the regular hexagonal pattern. Up to this date, it has been synthesized by high-repetition-rate laser ablation of an ultrapure carbon target in Ar environment [18–21]. The large-scale structure of carbon nanofoam is similar to that of an aerogel, yet unlike carbon aerogels, carbon nanofoam has a high electrical resistance. The most unusual feature of this allotrope is its attraction to magnets and its capability to become magnetic below 90 K [20, 21].

Other family of allotropes that feature a certain degree of amorphousness are the diamond-like carbon (DLC) films [22–27]. These films are hard and present a significant fraction of sp3 hybridized carbon atoms that can contain a significant amount of hydrogen atoms. Depending on the deposition conditions, these films can be fully amorphous or contain diamond crystallites. The deposition parameters are (low) total pressure, hydrogen partial pressure, type of precursor molecules, and use (or not) of plasma ionization. High ionization favors amorphous films while high atomic hydrogen contents favor diamond crystallite formation. Because of the confusion about structure generated by the term "diamond-like carbon" films, the term "hard amorphous carbon" has also been suggested as a synonym [28].

Amorphous carbon is an assortment of carbon atoms in a non-crystalline, irregular state that is essentially graphite without a crystalline macrostructure. Short range order exists, but with deviations in the interatomic distances and/or interbonding angles with respect to the graphite and diamond lattices. The term amorphous carbon is restricted to the description of carbon materials with localized π-electrons as described in [29]. Deviations in the bond angles occur in such materials because of the presence of "dangling bonds". This description of amorphous carbon is not applicable to carbon materials with two-dimensional structural elements, such as the ones present in all pyrolysis residues of carbon compounds. Even if coal, soot, carbon black and glass-like carbons are informally called amorphous carbon, they are products of pyrolysis, which does not produce true amorphous carbon under normal conditions [28]. Glass-like carbon is an allotrope with a very high isotropy of its structural and physical properties and with a very low permeability for liquids and gases. The original surfaces and the fracture surfaces have a pseudo-glassy appearance. Glass-like carbons cannot be described as amorphous carbon because they consist of two-dimensional structural elements and do not exhibit "dangling" bonds.

5.3 Glass-Like Carbons

Glass-like carbons are derived through the carbonization, or thermal degradation, of organic polymers in inert atmospheres. The resultant carbon has a glass-like appearance in the sense that is smooth, shiny and exhibits a conchoidal fracture[1] [30]. Because of its appearance, glass-like carbon has also been referred historically as "vitreous carbon" or "glassy carbon". It is impermeable to gases and extremely

---

[1] Some crystals do not usually break in any particular direction, reflecting roughly equal bond strengths throughout the crystal structure. Breakage in such materials is known as fracture. The term conchoidal is used to describe fracture with smooth, curved surfaces that resemble the interior of a seashell; it is commonly observed in quartz and glass. Conchoidal fracture. (2009). In Encyclopaedia Britannica. Retrieved April 08, 2009, from Encyclopaedia Britannica Online: http://www.britannica.com

inert, with a remarkable resistance to chemical attack from strong acids such as nitric, sulfuric, hydrofluoric or chromic and other corrosive agents such as bromine. Even when it does react with oxygen it only does so at high temperatures. Its rates of oxidation in oxygen, carbon dioxide or water vapor are lower than those of any other carbon. It has a hardness of 6–7 on Mohs' scale, a value comparable to that of quartz. Its density ranges from 1.4 to about 1.5 g/cm^{-3}, compared to 2.3 g/cm^{-3} for graphite, which suggests a significant degree of porosity. X-ray diffraction studies have shown that glass-like carbon presents an extremely small pore size of a closed nature, and that has an amorphous structure [31–38]. Glass-like carbon features a coefficient of thermal expansion of 2.2–3.2 $\times$ 10^{-6}/K which is similar to some borosilicate glasses. Its Young Modulus varies between 10 and 40 GPa. Because its thermal conductivity is about a tenth of that of typical graphite, it has been considered as thermally inert [39–44]. Glass-like carbon also has a wider electrochemical stability window than platinum and gold, which makes it ideal in electrochemistry experiments [45]. Even when the overall properties of the resulting carbon depend on the nature of the precursor used, they do not change very significantly [46] and the above values could be employed as an initial reference. Up to this date a consensus on the crystalline structure of glass-like carbon has not been reached. The most widely known and accepted model is the one that considers this type of carbon as made up of tangled and wrinkled aromatic ribbon molecules that are randomly cross-linked by carbon-carbon covalent bonds as shown in Fig. 5.2. The ribbon molecules form a networked structure, the unit of which is a stack of high strained aromatic ribbon molecules. Such structure of crystallites reflects the features of thermosetting resins structure which are commonly used as precursors for glass-like carbons. This model explains the most experimental results obtained so far on glass-like carbons including its impermeability, brittleness and conductivity [44, 46–49]. Other models exist including the "oxygenated Tetrahedral-Graphitic parts" model of Kakinoki [47], the "crumpled sheets" model of Oberlin [38], the "closed pores" model by Shiraishi [50] and the "globular" model by Fedorov [51]. Excellent reviews on the structure of glass-like carbon can be found in [44] and [49]. While the physical structure and chemical properties of glass-like carbons have been

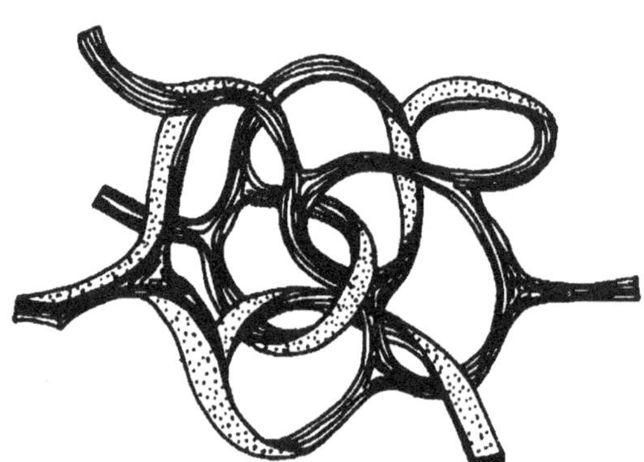

Fig. 5.2 Structural model of glass-like carbon as proposed by Jenkins in 1971. This model is able to explain most of the properties exhibit by glass-like carbon up to this day. Adapted by permission from Macmillan Publishers Ltd: Nature, from Jenkins and Kawamura [48] copyright 1971

extensively studied and debated since the early 1960s it was not until recent years when extended studies on the electric and electronic properties of this carbon appeared [49, 52, 53].

Carbonization is the process by which solid residues with a high content of carbon are obtained from organic materials, usually by pyrolysis in an inert atmosphere [28]. Different precursors to obtain glass-like carbon may be used, including phenolic resins, polyfurfuryl alcohols, cellulose, polyvinyl chloride and polyimides [34, 44, 46, 54–64]. Different degrees of shrinkage and carbon yield (the ratio of the weight of carbon to the weight of the original polymer sample) are obtained during carbonization depending on the precursor used. Phenol-formaldehyde, polyfurfuryl alcohol and polyvinyl alcohol have the highest yields, with an approximate 50% carbon [44]. Novolac resins, or acid-catalyzed phenol formaldehyde resins, are commonly used as photoresists and have recently become the material of choice to derive carbon structures featuring micro and nano dimensions. Their volume shrinkage varies from 50 to 90% [45, 65–67]. As with all pyrolytic reactions, carbonization is a complex process with many reactions taking place concurrently, including dehydrogenation, condensation, hydrogen transfer and isomerization [68–76]. The pyrolysis process of organic compounds can be divided into three major steps: pre-carbonization, carbonization and annealing. During pre-carbonization ($T < 573$ K) molecules of solvent and unreacted monomer are eliminated from the polymeric precursor. The carbonization step can be further divided into two stages. From 573 to 773 K (300 to 500°C), heteroatoms such as oxygen and halogens are eliminated causing a rapid loss of mass while a network of conjugated carbon systems is formed. Hydrogen atoms start being eliminated towards the end of this stage. The second stage of carbonization, from 773 to 1473 K (500 to 1200°C), completely eliminates hydrogen, oxygen and nitrogen atoms and forces the aromatic network to become interconnected. At this point, permeability decreases and density, hardness, Young's modulus and electrical conductivity increase. Sulfur does not evolve until even higher temperatures (> 1800 K), and certain metallic impurities, e.g. iron, may even require a leaching process to be eliminated. The final step, annealing, is carried out at temperatures above 1473 K, allowing to gradually eliminate any structural defects and evolve further impurities [44]. The final pyrolysis temperature determines the degree of carbonization and the residual content of foreign elements. For instance, at T$\sim$1200 K the carbon content of the residue exceeds a mass fraction of 90% in weight, whereas at T$\sim$1600 K more than 99% carbon is found [34, 73, 77]. Glass-like carbon is characterized as a type of Char[2] and is classified as a non-graphitizable, or non-graphitizing, carbon. It is important to mention that glass-like carbon is usually derived from thermosetting resins which do not melt during the carbonization process but rather maintain their shape

---

[2]Char is a solid decomposition product of a natural or synthetic organic material. If the precursor has not passed through a fluid stage, char will retain the characteristic shape of the precursor (although becoming of smaller size). For such materials the term "pseudomorphous" has been used. In contrast, coke is produced by pyrolysis of organic materials that have passed, at least in part, through a liquid or liquid-crystalline state during the carbonization process.

along the process. Glass-like carbon does not graphitize even when heat treated at 3273 K (3000°C). The inability of the graphitic planes to grow and stack parallel to each other even at high temperatures is due to the entangled nature of glass-like carbon structure, as mentioned above. Non-graphitizable carbons are mechanically hard and are sometimes referred as hard carbons [78]. For the sake of clarity, even when the term pyrolytic carbon might suggest a synonym for glass-like carbons, the former term only refers to carbon materials deposited from gaseous hydrocarbon compounds. The term pyrolytic carbon does not describe the large range of carbon materials obtained by thermal degradation (thermolysis, pyrolysis) of organic compounds when they are not formed by chemical vapor deposition. Materials deposited by physical vapor deposition are not covered either by the term pyrolytic carbon. Fig. 5.4 illustrates a variety of carbon products obtained with different processes.

The first derivation of a carbon material from a phenolic resin appears to be from 1915, when Weintraub and Miller in Massachusetts derived disks of a "very bright, shiny looking carbon with hardness equal or greater than 6 on the Mohs mineral scale". These disks were used to improve the useful life of microphones employed in telephone transmitters. In their patent, they describe the slowly heating of a hardened resin to a temperature close to 700°C in about one week and the subsequent firing at temperatures from 800 to 1100°C in a few hours. They also recognized the alternative and easiness of shaping the resist prior to carbonization instead of directly machining carbon [80]. A sustained flow of publications on glass-like carbon did not begin until 1962, when Davidson at the General Electric Co. in Kent, England derived glass-like carbon from cellulose [81] and Yamada and Sato, at the Tokai Electrode Manufacturing Co. in Nagoya, Japan, published preliminary characterization results of carbon derived from organic polymers, which they referred to as glassy carbon [39]. In 1963 Lewis, Redfern and Cowlard postulated a similar glass-like carbon, named "vitreous carbon" by the authors, as an ideal crucible material for semiconductors [40]. Later that year, Redfern disclosed several production processes to derive this vitreous carbon in a patent [82]. Redfern, Lewis and Cowlard worked for the Plessey Company in the United Kingdom. In 1965, X-ray studies of the material were conducted by Bragg at the Lockheed Missiles and Space Co. in Palo Alto, CA [31], while their tensile properties and pore size were characterized by Kotlensky at the Jet Propulsion Laboratory in Pasadena, CA [83] and a preliminary model of the structure was published by Kakinoki in Osaka, Japan [47]. That same year the advantages of glassy carbon electrodes for voltammetry and analytical chemistry were characterized by Zittel and Miller from Oak Ridge National Laboratory using a proprietary glassy carbon from the Tokai Electrode Manufacturing Co. [84] while Lee explored the mechanics of thermal degradation of phenolic condensation polymers [68]. In 1967, Cowlard and Lewis published a detailed description of the properties of vitreous carbon, the fabrication process and its potential applications [41]. In 1968, Rothwell conducted a more detailed study on the small-angle X-ray scattering of glassy carbon upon the suggestion of Bragg in Palo Alto [32]. Starting in 1969, glassy carbon was already receiving mainstream attention and started to be reviewed by different authors [42, 55]. That same year, Halpin and Jenkins studied its interaction with alkali metals and

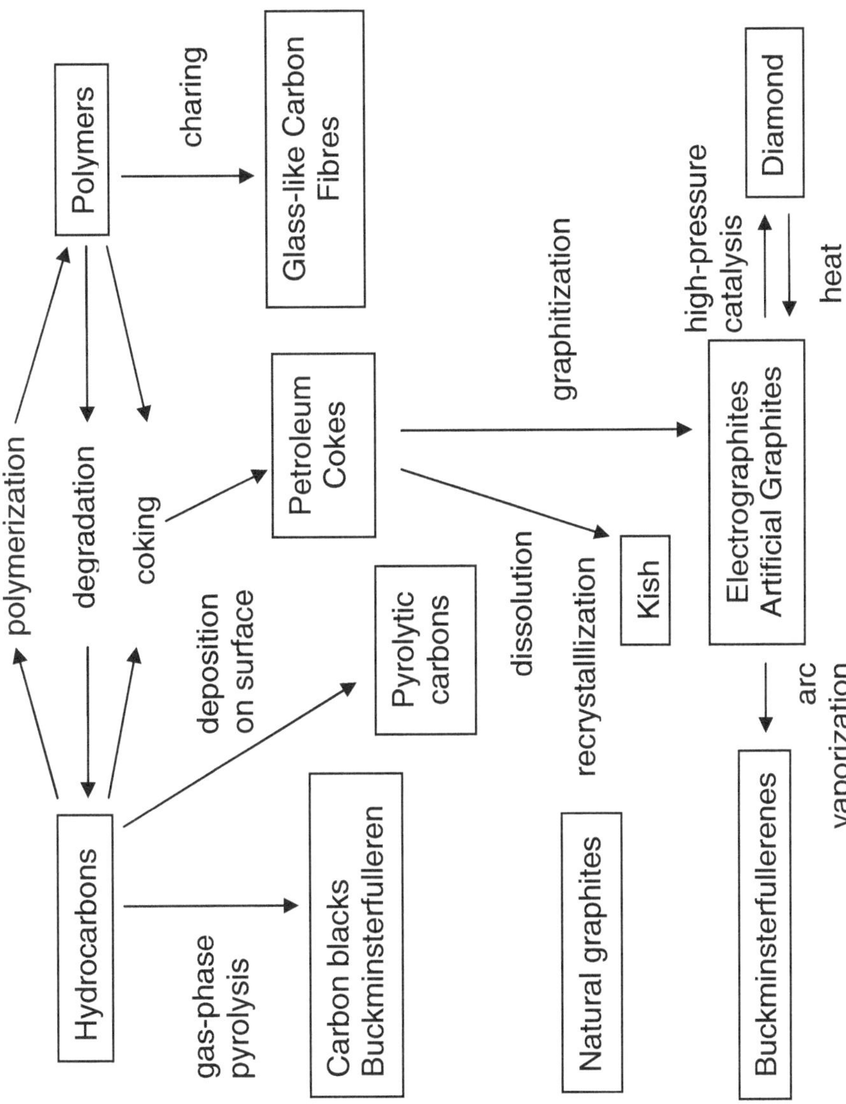

Fig. 5.3 Carbon products obtained by different processes. Adapted from Pocard et al. [79] – Reproduced by permission of The Royal Society of Chemistry

demonstrated potassium intercalation in glassy carbon [43]; Fitzer and Schafer delved on the formation of glasslike carbon by pyrolysis of polyfurfuryl alcohol and phenolic resins [54] and Lewis, Murdoch and Moul published on the heat of combustion of vitreous carbon [85]. In 1970, Fitzer and Schafer further studied the effect of crosslinking on the formation of glasslike carbons from thermosetting resins [69]. At the same time, Benson suggested the use of this carbon as implantable material in the human body [86] and reviewed the advantages of elemental carbon as a biomaterial [87]. The decade of 1970 brought a significant interest on the use of glassy carbon as a material for different implants and biomedical instrumentation. A large volume of publications exist, few examples are [88–95]. In 1971 a new structural model was postulated by Jenkins and Kawamura which was further corroborated in 1972 [46, 48]. This model is up to date the only one capable of explaining most of the experimental results obtained with glass-like carbon in the last 40 years [49]. Other models were later suggested in 1978 [51], 1983 [38], 1984 [50] and 2002 [52]. Further characterization of glass-like carbon was done by Williams, Dobbs and Taylor, who investigated the optical properties [96], fracture behavior [30] and electron-transfer kinetics [97] of glassy carbon respectively. Nathan conducted Raman spectroscopy on glassy carbon in 1974 [98]. The 1970s also witnessed an explosion of the interest on glassy carbon by the analytical and electrochemistry communities which still remains strong [37, 58–60, 79, 99–106]. In 1990, a novel synthesis of glassy carbons at relatively low temperatures (600°C) yielded to homogeneously doped glassy carbon (DGC) materials for electrochemistry [58–60, 79]. Few reviews on the contemporary state-of-the-art of glass-like carbon exist for different years [44, 49, 107–109]. As it can be seen from the historic review above, the same material -carbon derived from organic polymers by pyrolysis in inert atmosphere- was referred to with three different names: vitreous carbon, glassy carbon or glass-like carbon. The term vitreous carbon was first introduced by Redfern, Lewis and Cowlard at the Plessey Company in 1963. Although highly referenced in implant-related publications during the 1970s, the term vitreous carbon started to fall in disuse by the end of that decade. Vitreous carbon is now better identified with Reticulated Vitreous Carbon (RVC), a material introduced in the late 70s by Chemotronics International Inc. from Ann Arbor, MI. The use of RVC in electrochemistry was first documented in 1977 by [101] and was dubbed "optical transparent vitreous carbon". Extensive reviews on RVC are given in [110] and [111] . The term "glassy carbon" was introduced by Yamada and Sato at the Tokai Electrode Manufacturing Corporation in 1962. The commercialization of Tokai's glassy carbon electrodes targeting the electrochemistry market made glassy carbon the term of preference for the electroanalytical chemistry community to refer to glass-like carbon. In 1995 the IUPAC (International Union of Pure and Applied Chemistry) defined glass-like carbon as the material derived by the pyrolysis of organic polymers and recommended that the terms "Glassy carbon" and "Vitreous carbon", which had been introduced as trademarks, should not be used as synonymous for glass-like carbon. From a scientific viewpoint, the terms vitreous and glassy suggest a similarity with the structure of silicate glasses which does not exist in glass-like carbon, except for the pseudo-glassy appearance of the surface

[28]. Since the 1960s the term glass-like carbon has been the term of choice in most of the publications dealing with the mechanisms of thermal degradation and pyrolysis of polymer precursors. The latest review on glass-like carbon was published in 2002 by Pesin [49].

From the microfabrication standpoint, glass-like carbon microstructures were not reported until the late 1990s by Schueller and co-workers at Harvard University. In their process, polydimethylsiloxane (PDMS) molds were fabricated using soft lithography and then used to pattern furfuryl alcohol-modified phenolic resins and phenol-formaldehyde resins, which were subsequently carbonized [61, 112–115]. This fabrication technique allowed them to work with flat or curved surfaces and derive different structures (including free-standing lateral comb drives, diffraction gratings and grids) with dimensions as small as a few micrometers and featuring high aspect ratios if desired [115]. Microelectromechanical functions were also demonstrated [112]. During the same decade, initial work on the derivation of carbon from photoresists emerged in the quest for alternatives to carbon films produced by physical deposition techniques. The interest was driven by the use of pyrolyzed photoresists in batteries, electrochemical sensors, capacitors and MicroElectroMechanical Systems (MEMS). Electrochemical studies of carbon films derived from positive photoresists were conducted in 1998 by Kim et al. in Berkeley, CA [64] and later by Ranganathan et al. [65] at Ohio State University (OSU). By 2000, Kostecki, Song, and Kinoshita patterned these carbon films as microelectrodes and studied the influence of the geometry in their electrochemical response [116]. Pyrolyzed Photoresist Films, or PPF, still derived from positive resists were surface-characterized in 2001 and determined to have a near-atomic flatness [117]. The resultant carbon showed an electrochemical behavior similar to glass-like carbon although with decreased surface roughness. In 2002 the derivation of carbon from negative photoresists was reported by Singh, Jayaram, Madou, and Akbar at OSU. They used SU-8, a relatively new epoxy-based photoresist at that time [118, 119], and polyimide to fabricate circular patterns. The carbon obtained with this precursor showed higher resistivity and vertical shrinkage than the one synthesized from positive resists. Furthermore, the carbon derived from SU-8 showed higher vertical shrinkage and poorer substrate adhesion than that from polyimide. Nevertheless, resistivity from SU-8 carbon was slightly lower than polyimide's [66]. In 2005, structures with aspect ratios higher than 10 were reported by Wang, Jia, Taherabadi, and Madou at the University of California, Irvine (UCI). This achievement was possible thanks to the use of a two-step heating process during pyrolysis. This novel process allowed for the release of residual oxygen contained in the polymer structures that had caused the precursor to burn rather than pyrolyze, even in an oxygen-free atmosphere. A variety of complex high-aspect ratio Carbon-MEMS (C-MEMS) structures, such as posts, suspended carbon wires, bridges, plates, self organized bunched posts and networks, were built in this way. The variation in structure shrinkage depending on the original polymer was also reported [45, 120]. For example, structures with thickness below 10 μm usually shrink approximately by 90%, while hundreds-of-microns thick features shrink approximately 50%. Also in 2005, the electrical properties and shrinkage behavior of both positive and negative resists after pyrolysis were characterized by Park and co-workers at UCI. They corroborated the decrease in resistivity of glass-like carbons as the pyrolysis final temperature increases. They also demonstrated how the largest shrinkage takes place below 873 K (600°C) for all the photoresists tested [121]. Electron beam lithography (EBL) was used in 2006 to fabricate suspended glass-like carbon microstructures [67]. Continuous work on C-MEMS has been conducted since 2004 by the same group at UCI and collaborators in a variety of applications including Lithium-ion batteries [122–125], fuel cells [126, 127], electrochemical sensors [128], cell culturing substrates [129], dielectrophoresis [130–137], micromolding [138] and fractal electrodes [139, 140]. Other groups have also recently integrated carbon structures for their use in gas sensors [141].

The development of novel fabrication techniques keeps reducing the smallest dimensions that can be achieved in carbon. While 40 years ago glass-like carbon was only used in applications requiring large dimensions, such as metallurgy crucibles and laboratory beakers, it can now be employed to fabricate nano-electrodes. Its unique properties have been beneficial to various applications along the years. Initially, its extreme chemical inertness and gas impermeability were exploited to fabricate laboratory equipment such as beakers, basins and boats. Since glass-like carbon is not wetted by a wide range of molten metals, it is an ideal material for the fabrication of crucibles with applications in metallurgical and chemical engineering. Its resistance to erosion and high melting point makes it an ideal material for mandrels, steam, fuel and rocket nozzles and other equipment in mechanical and electrical applications. Moreover, glass-like carbon has also been used in heart valve implants and other biomedical devices thanks to its biocompatibility. "Glassy" carbon electrodes have become so popular that they represent a significant fraction of the multi-billion electrochemistry market.

Carbon-MEMS (or C-MEMS) can be defined as the set of methods that can be used to derive glass-like carbon structures from patterned organic polymers, featuring dimensions ranging from hundreds of micrometers down to tens of nanometers. C-MEMS combines different polymer micro and nanofabrication techniques with pyrolysis or thermal degradation to derive glass-like carbon features. These fabrication techniques include, but are not limited to, stamping, casting, machining and lithography. The choice of each technique is dictated by the quality, complexity and final dimensions of the desired carbon part. In this regard, the incorporation of photolithography (shown in Fig. 5.4) to the C-MEMS toolbox enabled a more precise control on the dimensions and complexity of the precursor polymer structures. The addition of Next-Generation Lithography (NGL) techniques, such as electron beam lithography (EBL), nanoimprint lithography (NIL) and focused-ion beam (FIB), will further reduce the dimensions and greatly increase the intricacy of the resulting carbon structures. Moreover, the existence of commercial high-quality precursors and standardized photolithography tools make the fabrication process and the dimensional control highly reproducible. The fabrication of high-aspect ratio structures [45], nano-electrodes, overhanging motifs [67], free-standing all-carbon micromolds [138] and the integration of carbon electrodes in polymer devices [134, 135, 141] are included in the latest achieved milestones.

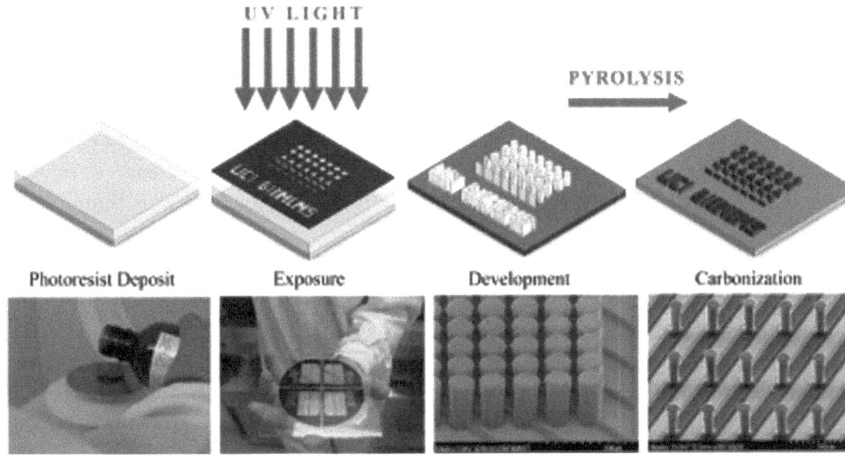

Fig. 5.4 Schematics and figures depicting the C-MEMS process when using photolithography. The case of a negative tone resist is pictured

It is important to differentiate the fabrication of glass-like carbon with nanometer dimensions, as in the case of electrodes derived from polymers patterned with NGL techniques, from the use of Carbon NanoTubes (CNT). One of the main differences is that micro and nanofabrication methods achieve a wider range of length scales, from a few nanometers to several millimeters. In the case of CNT, the dimensions are limited between 1 and 100 nm, depending on the synthesis method being used, the type of CNT (single-walled or multi-walled) [142] and the size of the catalytic particles used to synthesize them [143]. An additional advantage of using NGL methods to pattern carbon is the enhanced placement precision, since the structure geometry is pre-determined using lithography methods. On the other hand, the growth of CNT cannot be pre-patterned, and they need to be positioned after processing when they are to be used individually.

As stated above, the embracing of photolithography in the micro and nanofabrication of glass-like carbon has brought significant advantages and enabled rapid and notable developments. In the following section, the focus is on the derivation of polymer structures using photolithographic methods. Next Generation Lithography technologies are then presented as an alternative to achieve smaller and more complex patterns than those possible with photolithography.

5.4 Photolithography Overview

The most widely used form of lithography is photolithography which is basically the use of light to pattern the substrates. In the Integrated Circuits (IC) industry, pattern transfer from masks onto thin films is accomplished almost exclusively via photolithography. This essentially two-dimensional process has a limited tolerance for non-planar topographies and creates a major constraint for building non-IC miniaturized systems, such as microfluidic devices and polymer precursors for C-MEMS.

Fortunately, research over the last ten years in high-aspect-ratio resists is finally improving dramatically photolithography's capabilities, allowing for it to cover wider ranges of topographies and to resolve ever-smaller features. Performance of a photolithographic process is determined by its resolution (the minimum feature size that can be transferred with high fidelity), the registration (how accurately patterns on successive masks can be aligned), and throughput (the number of parts or devices that can be transferred per hour, a measure of the efficiency of the lithographic process).

Photolithography generally involves a set of basic processing steps: photoresist deposition, soft bake, exposure, post-exposure treatment and developing. Descumming and post-baking might also be part of the process is illustrated in Fig. 5.5. Each one of these steps are detailed below, together with their related topics. For example, alternatives to spin coating for resist deposition are explored, and the properties and types of resists are discussed. Different masking techniques that can be used during the exposure step are also presented.

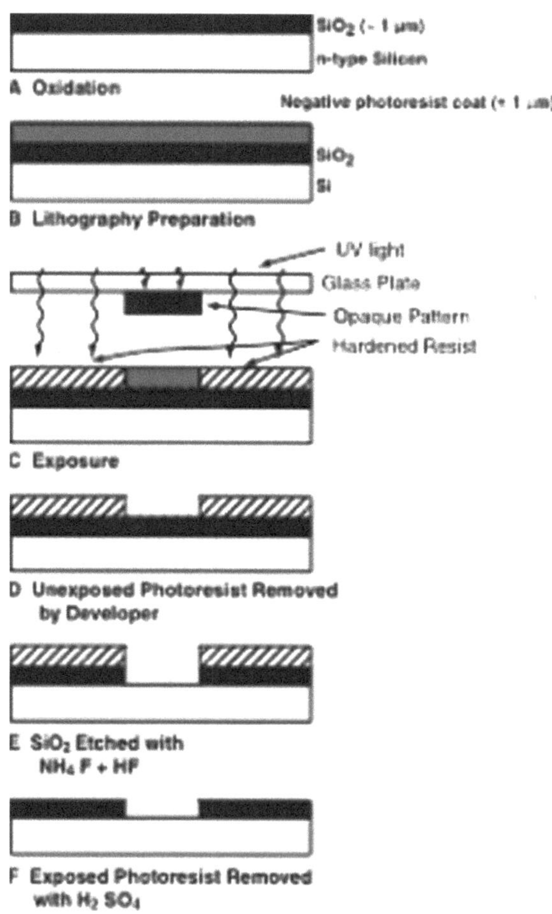

Fig. 5.5 Basic photolithography and pattern transfer. Example uses an oxidized Si wafer and a negative photoresist system. Process steps include exposure, development, oxide etching, and resist stripping. Steps B to D are detailed in the text. For details on steps A, E and F refer to [144]

5.4.1 Substrate Cleaning and the Clean Room

Substrate cleaning is the first and a very important step in any lithographic process, as the adhesion of the resist to the substrate could be severely compromised by the presence of impurities and residual coatings if it is not carried out properly. Contaminants include solvent stains (methyl alcohol, acetone, trichloroethylene, isopropyl alcohol, xylene, etc.), and airborne dust particles from operators, equipment, smoke, etc. In the manufacturing of MEMS and NEMS devices, several substrate choices are available: silicon, glass, quartz, metals and even polymer films such as polyimide (PI) and polyester (PET). Depending on the substrate and the type of contaminants, several cleaning techniques can be used. Wet immersion cleaning might be carried out using diluted hydrofluoric acid, Piranha (a mix of sulfuric acid and hydrogen peroxide at different ratios), RCA (a process involving ammonium hydroxide, hydrogen peroxide, water, hydrofluoric and hydrochloric acids at different stages) or milder, but not as effective, procedures such as DI water rinsing followed by solvent rinse. Other methods include ultrasonic agitation, polishing with abrasive compounds, and supercritical cleaning. Dry methods include vapor cleaning; thermal treatments, for example baking the substrate at 1000°C in vacuum or in oxygen; and plasma or glow discharge techniques, for example in Freons with or without oxygen. In general, vapor phase cleaning methods use significantly less chemicals than wet immersion cleaning. In the case of wet immersion cleaning, dehydration prior to resist deposition is recommended.

In order to increase yield, micro and nanofabrication processes are highly recommended to take place inside a *clean room*, a specially designed area with environmentally controlled airborne particulates, temperature (± 0.1 °F), air pressure, humidity (from 0.5 to 5% RH), vibration, and lighting. Clean rooms are classified based on the maximum particle count per unit volume of air. The size of the particles considered in the count is traditionally of 0.5 μm or larger, and the volume of air is a cubic foot. For example, in a Class 1 clean room, the particle count needs not to exceed one 0.5 μm particle (or larger) per cubic foot. However, the acceptable particle size in IC manufacturing has been decreasing hand in hand with the ever-decreasing feature sizes. With a 64-Kilobyte dynamic random access memory (DRAM) chip, for example, one can tolerate 0.25 μm particles, but for a 4-Megabyte DRAM, one can only tolerate 0.05 μm particles.

5.4.2 Photoresist Deposition

Photoresist deposition is one of the more expensive steps in photolithography, since photoresists may cost as much as $1000 per liter and most of the resist deposition processes waste significant amounts of material. For miniaturized 3D structures, much greater resist thicknesses than those used by the IC industry are often required, and complex topographies might also call for a conformal resist coat over very high aspect ratio features. To deposit photosensitive materials for MEMS and NEMS, spin coating is the most common method for thicknesses smaller than 1 mm. Other coating techniques are also available, such as roller, curtain or extrusion coating, although they are not as efficient when depositing layers thicker than 200 μm. For very thick resist coats (>1 mm), techniques such as casting and the use of thick sheets of dry photoresists replace the use of resist spinners. For conformal coating, resist spraying, or better yet, electrodeposition (ED) of photoresist might be preferable.

Spin Coating. The most common method to deposit photoresists is spin coating. In this method, centrifugal forces causes the resist to flow to the edges, where it builds up until expelled when its surface tension is exceeded. The resulting polymer thickness, T, is a function of spin speed, solution concentration, and molecular weight (measured by intrinsic viscosity). Generally, the photoresist is dispensed onto the substrate, which is held in place by a vacuum-actuated chuck in a resist spinner (see Fig. 5.6) [145]. A rotating speed of about 500 rpm is commonly used during the dispensing step to spread the fluid over the substrate. After the dispensing step, it is common to accelerate to a higher speed to thin down the fluid near to its final desired thickness. Typical spin speeds for this step range from 1500 to 6000 rpm, depending on the properties of the fluid (mostly its viscosity) as well as the substrate. This step can take from a few seconds to several minutes. The combination of spinning speed and time will generally define the final film thickness. An empirical expression to predict the thickness of the spin coated film as a function of its molecular weight and solution concentration is given in [146]

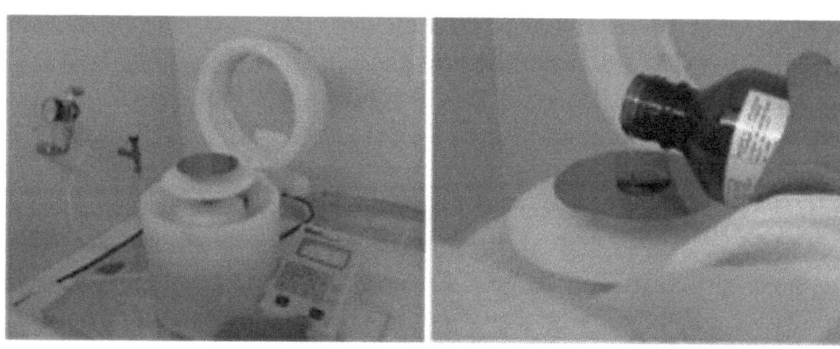

Fig. 5.6 *Left*: A programmable spin coating system with a Si wafer being held in place by a vacuum chuck. *Right*: Manual dispensing of SU-8 photoresist

The photoresist film, after being spin coated into the substrate, must have a uniform thickness and be chemically isotropic so that its response to exposure and development is uniform. The application of too much resist results in edge covering or run-out, hillocks, and ridges, reducing manufacturing yield. Application of too little resist may leave uncovered areas. Optimization of the photoresist coating process in terms of resist dispense rate, dispense volume, spin speed, ambient temperature, venting of the resist spin station and humidity presents a growing challenge.

The need for an alternative photoresist deposition technique arises as the amount of waste material generated by spin coating becomes higher, with most of the resist solution (> 95%) thrown off the substrate during the spin casting process (the wasted resist must be disposed off as a toxic material). Inherent to this process is the formation of edge beads, which might require an additional removal process prior to subsequent processing steps. Oftentimes, the edge of the substrate exhibits resist ridges that are about 10 times higher the mean thickness on the rest of the substrate. Different edge bead removal (EBR) solutions are commercially available, such as AZ EBR Solvent or Microchem's EBR. The main obstacle to overcome when using spin coating in MEMS is caused by varying topography: deeply etched features cause a physical obstruction to the flow of photoresist, preventing complete coverage and often causing striation or resist thickness variation. For example, resist thickness variations can occur on the near and far sides of a cavity, or in cavities at different locations on the substrate. Sizes and shapes of the cavities also have influence on the resist uniformity and coating defects. For substrates with moderate topography, alternative coating techniques, like spray coating, offer better prospects.

5.4.2.1 Alternative Photoresist Deposition Methods

In this section, alternatives to spin coating are compared, closely following James Webster's treatise of the subject [147]. First, the techniques used in conformal coatings are presented. Subsequently, the techniques used to deposit resist thicknesses less than 200 μm are exposed. This sub-section ends with the details of the lamination process for dry resist films.

Spray Coating. In spray coating, the substrates proceed under a spray of photoresist solution. Compared to spin coating, spray coating does not suffer from the variation in resist thickness caused by centrifugal forces, since the droplets of resist stay where they are being deposited. Another major advantage of this method is its ability to uniformly coat over non-uniform surfaces, making the technique appropriate for MEMS processing. More importantly, sprayed coatings do not present the internal stress forces that are common to spin coated films. However, control of the deposited film thickness is not as precise as with spin and extrusion coated substrates (see below), and some waste of photoresist solution occurs as part of this processing technique. It is difficult to deposit layers of resist thicker than 20 μm with spray coating.

Electrostatic Spraying. Electrostatic spraying or electrostatic deposition (ED) is a variant of spray coating. During the atomization of the resist by air or nitrogen pressure, the formed droplets are statically charged by applying a large voltage (e.g., 20 kV). The charge causes the droplets to repel each other, maintaining the integrity of the mist of resist formed.

Electrodeposited (electrophoretic) (ED) [148]. Electrophoretic photoresist deposition is an appropriate technique for the coating of substrates with extreme topography. It uses electric fields to accelerate charged micelles, comprising resist, resist solvent, dye and photoinitiator molecules, towards the substrate to be coated. It is important to emphasize that the solution must be rather resistive for a strong enough electrophoretic field to be established. Typical coating thicknesses, highly dependent on the voltage and the temperature, are in the range of 5–10 microns, but specific resist systems can be deposited up to 35 μm. The main advantage is that it yields a pore-free deposit of the resist, even at very low thicknesses.

Other thin film coating techniques include *Silkscreen Printing, Plasma-Deposited Resist, Meniscus Coating and Dip Coating*. Alternative techniques to spin coating for thicknesses less than 200 μm include:

Roller Coating. Roller, curtain and extrusion coating are all variations of directly casting the coating solution on the substrate. Grooved rubber rolls are used to transfer the liquid resist to the substrate surface. The pitch of the grooves on the roller, the non-volatile content of the resist and the coating roller-to-substrate pressure affect the final thickness of the resist film. Excess resist flows back to a sump for recycling via an automatic viscosity controller and filter unit, thus limiting waste. In general, it is incapable of producing uniform coatings below 5 μm in thickness.

Curtain Coating. In curtain coating the substrate is moved on a conveyor through a sprayed "curtain" of resist. The liquid resist is pumped into a head from which the only exit is a thin nip on the head's underside. The resist forced through this nip forms a curtain of resist through which the substrates to be coated are passed. Typically, thickness in the 25–60 μm range can be obtained with curtain coating. Undeposited material is re-circulated back to the coating head. By carefully controlling the material viscosity, belt speed, and pump speed, reproducible thicknesses can be achieved and maintained over the substrate surface with less than 10% variation in the overall thickness.

Extrusion Coating. In extrusion coating, the extrusion head is positioned at a short, predetermined height above the substrate. A thin curtain of resist falls on the substrate, which is moving horizontally at a controlled rate. Film thicknesses from less than 1 μm to greater than 150 μm in a single coating pass have been demonstrated [147]. Disadvantages of the technique include variation in the substrate surface uniformity and the formation of edge beads along the leading edge of extrusion coated substrates, although not to the extent encountered in spin-coated films. This method requires many controls to achieve good results, and if performed well, can yield very satisfactory coatings.

It is important to mention that there is no forced drying during roller, curtain and extrusion coating other than evaporation. Therefore, the coating material has time to flow and planarize over surface features. The degree of coverage into deep features is highly dependent on the surface wettability and the solution viscosity.

Lamination. Most resists in IC and MEMS fabrication are deposited as liquids, whereas resists used in printed wiring board (PWB) manufacturing are usually rolls of dry film resists that are laminated onto the substrate instead of being spin coated on it. Dry film resist formulations are sandwiched between a polyolefin release sheet and a polyester base, and rolled up onto a support core. These protective covers shield the film from environmental oxygen and facilitate its handling. The dry film resist layers are available in thicknesses ranging from 25 up to 100 μm. Resist thicknesses of 1–1.5 millimeters are common for imaging purposes, and thicker resists (1.5–2.0 mm) are used for plating rather than etch resists. Dry film resists offer advantages such as excellent adhesion on most substrates, no liquid handling since there is no solvent, high processing speeds, excellent thickness uniformity over a whole substrate (even in substrates with holes), facile handling, no formation of edge beads, low exposure energy, low cost, short processing time and near vertical sidewalls. Conformation of the resist to the substrate is achieved by heating under pressure in a hot-roll or cut-sheet laminator. The heat and pressure of the laminating rollers causes the dry film to soften and adapt to surface topologies. The resist is then exposed to a UV light source for patterning. Modern dry film resists are developed in a simple sodium carbonate solution (1–2%) after removal of the top cover layer. The biggest disadvantage of dry resist is its relatively low resolution compared to liquid resists. Two major reasons for this poorer resolution are the thicker resist coating and the fact that the mask is positioned on top of a thick protective cover film. By removing the top cover sheet from the photoresist prior to exposure, higher resolutions are possible. There are a variety of dry film photoresists widely used and commercially available. Examples include Riston®, Ordyl BF 410, Etertec® 5600, DF 4615 and DFR-15. They are all used in the manufacture of circuit boards, can be made quite thick and are all candidates for broad use in MEMS as well. The potential benefits of using dry resist films as permanent components in the mass production of biosensors and microfluidics were recognized and described early on, for example, by Madou et al. [149]. These authors suggest that continuous, web-based manufacturing may finally make ubiquitous, disposable miniaturized devices such as biosensors and microfluidics possible. Dry resist film materials are less expensive than Si and form a convenient substrate. Furthermore, they are available in rolls so that large sheets can be processed. A continuous lithographic process, including exposure and development, taking place between a dry resist supply roll and a pick-up reel has been envisioned by these authors.

5.4.2.2 Resists

In order to understand the photolithographic process better it is necessary to study in detail its basic element: the photoresist. The principal components of a photoresist are the polymer (base resin), a sensitizer, and a casting solvent. The polymer changes its structure when is exposed to electromagnetic radiation; the solvent allows for spin application and the formation of thin layers over the substrate; and sensitizers control the chemical reactions in the polymeric phase. Resists without sensitizers are single-component systems, whereas sensitizer-based resists are two-component systems. Solvents and other potential additives do not directly relate to the photoactivity of the resist. Photoresists must meet several rigorous requirements: good adhesion, high sensitivity, high contrast, good etching resistance (wet or dry etching), good resolution, easy processing, high purity, long shelf life, minimal solvent use, low cost, and a high glass transition temperature, T_g. Most resins used as base for photoresists, such as novolacs, are amorphous polymers that exhibit viscous flow with considerable molecular motion of the polymer chain segments at temperatures above the glass transition. At temperatures below T_g, the motion of the segments is halted, and the polymer behaves as a glass rather than a rubber. If the T_g of a polymer is at or below room temperature, the polymer is considered a rubber; if it lies above room temperature, it is considered to be a glass. For the purpose of deriving glass-like carbon from photopatterned polymers it is of extreme importance to guarantee that the heating rate during pyrolysis is slower than the thermal degradation rate of the resin. In other words, the temperature in the furnace must always be lower than the T_g of the resin and the different compositions it goes through before becoming carbon. If this principle is not followed, the photopatterned polymer melts and flows, resulting in a shapeless carbon piece. In general, polymers that crystallize are not useful as resists because the formation of crystalline segments prevents the formation of uniform high-resolution isotropic films [146].

If the photoresist is of the type called *positive* (also *positive tone*), the photochemical reaction during exposure of a resist weakens the polymer by rupture or scission of the main and side polymer chains, and the exposed resist becomes more soluble in developing solutions. If the photoresist is of the type called *negative* (also *negative tone*), the photochemical reaction strengthens the polymer, by random cross-linkage of main chains or pendant side chains, thus becoming less soluble. Fig. 5.7 illustrates chain scission (positive resists) and cross-linking (negative resists).

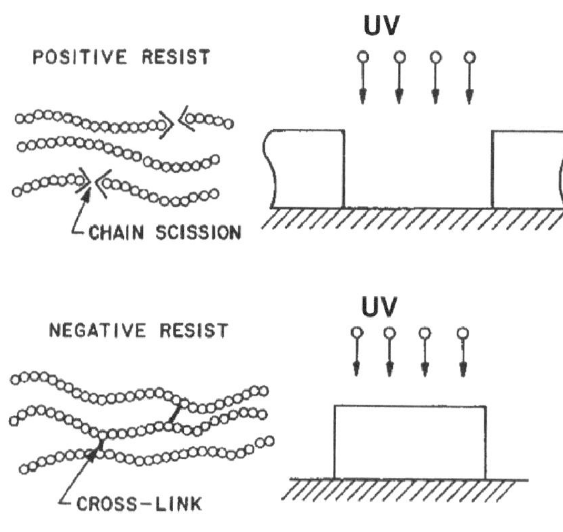

Fig. 5.7 Polymer chain scission (positive resists) and cross-linking (negative resists)

Positive Resists. Two well-known families of positive photoresists are the single component poly(methylmethacrylate) (PMMA)[3] resists and the two-component DNQ resists comprised of a photoactive component (PAC) such as diazonaphtoquinone ester (DNQ) (20–50 wt%) and a phenolic novolac resin.

PMMA becomes soluble through chain scission under Deep UV (DUV) illumination (150 nm < λ < 300 nm). Although the resolution of PMMA is very good, its

[3]Poly(methyl methacrylate) (PMMA) or poly(methyl 2-methylpropenoate) is the synthetic polymer of methyl methacrylate. This thermoplastic and transparent plastic is sold by the trade names Plexiglas, Limacryl, R-Cast, Perspex, Plazcryl, Acrylex, Acrylite, Acrylplast, Altuglas, Polycast and Lucite and is commonly called acrylic glass or simply acrylic.

plasma etch tolerance is very low. Besides traditional photolithography, PMMA is also used in electron beam, ion beam, and X-ray lithography.

The diazonapthoquinone (DQN) resist system is a "workhorse," near-UV (300 nm $< \lambda <$ 400 nm), two-component positive resists, which photochemically transforms into a polar, base-soluble product [150]. The hydrophilic novolac resin (N) is in itself alkali soluble because of the OH groups. Diazonaphthaquinone (DQ) is a hydrophobic and non-ionizable compound and when phenolic resins are impregnated with DQ, they become hydrophobic and are rendered insoluble. The addition of 20–50% in weight of DQ forms a complex with the phenol groups of the novolac resin and reduces the solubility rate of the unexposed resist to less than 1–2 $nm \cdot s^{-1}$. During exposure, DNQ undergoes photolysis, which destroys the inhibitory effect of DQ on film dissolution. The photolysis causes the DNQ to undergo a reaction forming a base-soluble carboxylic acid that can be rapidly developed in aqueous solution of hydroxide ions (e.g. 1% NaOH). In contrast with cross-linked resists, the film solubility is controlled by chemical and polarity differences rather than molecular size. The novolac resin matrix itself is a condensation product of a cresol isomer (para-cresol) and formaldehyde consisting of hydrocarbon rings with two methyl groups and one OH group attached. Phenolic resins are readily cross-linked by thermal activation into rigid forms (Bakelite was the first thermosetting plastic). A novolac resin absorbs light below 300 nm, and the DNQ addition adds an absorption region around 400 nm. The 365, 405, and 435 nm mercury lines can all be used for exposure of DNQ. The intense absorption of aromatic molecules prevents the use of this resist at exposing wavelengths less than about 300 nm; at those shorter wavelengths, linear acrylate and methacrylate copolymers have the advantage. Positive attributes of novolac-based resist are that the unexposed areas are essentially unchanged by the presence of the developer. Thus, line width and shape of a pattern is precisely retained. Most positive resists are soluble in strongly alkaline solutions and develop in mildly alkaline ones.

Negative Resists. The first negative photoresists were based on free-radical-initiated photo-cross-linking processes of main or pendant polymer side chains, rendering the exposed parts insoluble. They were the very first types of resists used to pattern semiconductor devices and still comprise the largest segment of the overall photoresist industry, being widely used to define circuitry in printed wiring boards (PWBs) [151]. A negative photoresist becomes insoluble in organic (more traditional negative resists) or water-based developers (newer negative resist systems) upon exposure to UV radiation. The insoluble layer forms a "negative" pattern that is used as a stencil (usually temporarily) to delineate many levels of circuitry in semiconductors, microelectromechanical systems (MEMS), and printed wiring boards (PWBs). In latest years, features fabricated from negative resist also act as structural elements. The insolubilization of radiated negative resists can be achieved in one of two ways: the negative resist material increases in molecular weight through UV exposure (traditional negative resists), or it is photochemically transformed to form new insoluble products (newer negative resist products).

A disadvantage of negative resists is that the resolution is limited by film thickness. The cross-linking process starts topside, where the light hits the resist first.

Consequently, overexposure is needed to render the resist insoluble at the substrate interface. The greater the desired resist thickness, the greater the overdose needed for complete polymerization and the larger the scattered radiation. Scattered radiation at the resist/substrate interface in turn reduces the obtainable resolution. To improve the resolution of a negative resist, thinner resist layers can be used; however, when using thin layers of negative resist, pinholes become problematic. Negative photoresists, in general, adhere very well to the substrate, and a vast amount of compositions are available. They are highly resistant to acid and alkaline aqueous solutions as well as to oxidizing agents. As a consequence, a given thickness of negative resist is more chemically resistant than a corresponding thickness of positive resist. This chemical resistance ensures better retention of resist features even during a long, aggressive wet or dry etch. Negative resists are also more sensitive than positive resists but exhibit a lower contrast. A good and very practical example of negative resists is that of SU-8.

SU-8 is an acid-catalyzed negative photoresist, made by dissolving EPON®-SU-8 resin (a registered trademark of Shell Chemical Company) in an organic solvent such as cyclopentanone or GBL (gamma-butyloractone) and adding a photoinitiator. The viscosity and hence the range of thicknesses accessible, is determined by the ratio of solvent to resin. The EPON resist is a multifunctional, highly-branched epoxy derivative that consists of bisphenol-A novolac glycidyl ether. On average, a single molecule contains eight epoxy groups which explain the eight in the name SU-8. The material has become a major workhorse in miniaturization science. In a chemically amplified resist like SU-8, one photon produces a photoproduct that in turn causes hundreds of reactions to change the solubility of the film. Since each photolytic reaction results in a "amplification" via catalysis, this concept is dubbed "chemical amplification" [152].

Scientists at IBM discovered that certain photo-initiators, such as onium salts, polymerize low-cost epoxy resins such as EPON®-SU-8. Compositions of SU-8 photoresist were patented by IBM as far back as 1989 [153] and 1992 [154]. Original compositions were intended for printed circuit board and e-beam lithograpy. SU-8 photoresists became commercially available in 1996. Because of its aromatic functionality and highly cross-linked matrix, the SU-8 resist is thermally stable and chemically very inert. After a hard bake, it withstands nitric acid, acetone, and even NaOH at 90°C and it is more resistant to prolonged plasma etching and better suited as a mold for electroplating than Poly(methyl methacrylate) (PMMA) [155]. The low molecular weight [$\sim$ 7000 $\pm$ (1000) Da] and multifunctional nature of the epoxy gives it the high cross-linking propensity, which also reduces the solvent-induced swelling typically associated with negative resists. As a result, very fine feature resolution, unprecedented for negative resists, has been obtained and epoxy-based formulations are now used in high-resolution semiconductor devices. Low molecular weight characteristics also translate into high contrast and high solubility. Because of its high solubility very concentrated resist casting formulations can be prepared. The increased concentration benefits thick film deposition (up to 500 μm in one coat) and planarization of extreme topographies. The high epoxy content promotes strong SU-8 adhesion to many types of substrates and makes the material highly sensitive to UV exposure. On the negative side, strong adhesion makes stripping of the exposed SU-8 material currently one of the most problematic aspects in those applications where the resist must be removed such as in the IC industry. Stripping of SU-8 may be carried out with hot NMP (1-methyl-2-pyrrolidon), plasma or laser ablation. There are other issues to be resolved with this resist; for example, thermal mismatch of SU-8 on a Si substrate (the thermal expansion coefficient for Si is 2.361 10^{-6}/K versus 21–52 10^{-6}/K for SU-8) produces stress and may cause film cracking. Moreover, the absorption spectrum of SU-8 shows much higher absorption coefficients at shorter wavelengths. As a result lithography using a broadband light source tends to result in over-exposure at the surface of the resist layer and under-exposure at the bottom. The resulting developed photoresist tends to have a negative slope, which is not good for mold applications: the mold sidewall should have a positive or at least a vertical slope for easy release of the molded part from the mold. The exaggerated negative slope at the top of the resist structure surface is often called *T-topping* (See Fig. 5.8). UV light shorter than 350 nm is strongly absorbed near the surface creating locally more acid that diffuses sideways along the top surface. Selective filtration of the light source is often used to eliminate these undesirable shorter wavelengths (below 350 nm) and thus obtain better lithography results. For example, Reznikova et al. used a 100 μm thick SU-8 resist layer to filter exposure radiation at 334 nm [156] and Lee et al. reported using a Hoya UV-34 filter to eliminate the T-top (over-exposed top part) [157]. Nearly vertical sidewalls can be achieved using a Hoya UV-34 filter. Aspect ratios up to ~25 for lines and trenches have been demonstrated in SU-8-based contact lithography. When patterned at 365 nm, the wavelength at which the photoresist is the most sensitive, total absorption of the incident light in SU-8 is reached at a depth of 2 mm. In principle, resist layers up to 2 mm thick can be structured [158]. Yang and Wang recently confirmed this astounding potential experimentally [159]. This group at Louisiana State University (LSU), using both wavelength optimization by patterning using a filtered i-line (365 nm) and air gap compensation (with glycerin or a Cargille refractive index matching fluid), demonstrated aspect ratios above 190

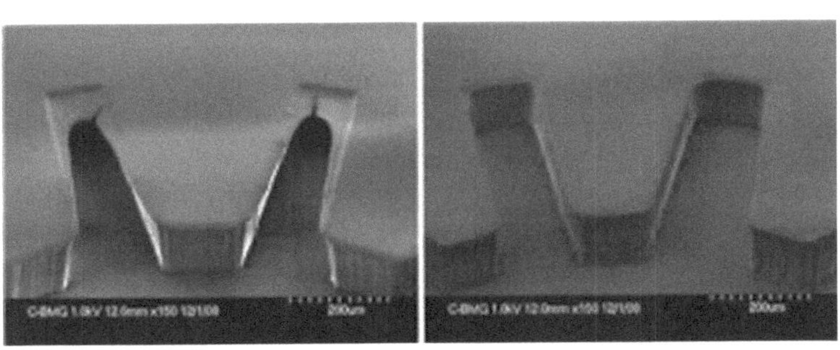

Fig. 5.8 The effect of T-topping. *Left*: Gear pattern showing T-topping. *Right*: Same gear pattern but now T-topping is minimized by the use of an in-house fabricated filter (A 50 μm layer of SU-8 on quartz)

(for a feature with a 6 μm thickness and a height of 1150 μm) and structures as high as 2 mm. Several publications exist on SU-8, a recent review is [160].

A comparison of negative and positive photoresist features is presented in Table 5.1. This table is not exhaustive and is meant only as a practical guide for selection of a resist tone. The choice of whether to use a negative or a positive resist system depends upon the needs of the specific application, such as resolution, ease

Table 5.1 Comparison of traditional negative and positive photoresists

	Resist type	
Characteristic	Positive resist	Negative resist
Adhesion to Si	Fair (priming required)	Excellent (priming not required)
Available compositions	Many	Vast
Baking	In air (+)	In Nitrogen (−) or Air (+)
Contrast γ	Higher, e.g., 2.2	Lower, e.g., 1.5
Cost	More expensive	Less expensive
Developer	Temperature sensitive (−) and aqueous based (Ecologically sound)	Temperature insensitive (+) and organic solvent (−), aqueous based has been introduced.
Developer process window	Small	Wide, relatively insensitive to overdeveloping
Influence of oxygen	No (+)	Yes (−). Minimized in the case of SU-8.
Lift-off	Yes, usually with multiple layer resist (MLR)	Yes, even in single layer resist (SLR)
Mask type	Dark-field: lower-defects	Clear-field: higher-defects
Opaque dirt on clear portion of mask	Not very sensitive to it	Causes printing of pinholes
Photospeed	Slower	Faster
Pinhole count	Higher	Lower
Pinholes in mask	Prints mask pinholes	Not so sensitive to mask pinholes
Plasma etch resistance	Not very good	Very good
Proximity effect	Prints isolated holes or trenches better	Prints isolated lines better
Residue after development	Mostly at < 1 μm and high aspect ratio	Often a problem
Resolution	High	Low (>1 μm)
Sensitizer quantum yield ϕ	0.2–0.3	0.5–1
Step coverage	Better	Lower, can be higher with diluted solutions such as SU-8 2000.5 or 2002.
Strippers of resist over		
Oxide steps	Acid	Acid
Metal steps	Simple solvents	Chlorinated solvent compounds
Swelling in developer	No	Yes
Thermal stability	Good	Fair, good with SU-8.
Wet chemical resistance	Fair	Excellent

of processing, speed, and cost. If the objective is to obtain high-aspect ratios, the use of negative resist is preferred.

5.4.3 Soft Baking or Prebaking

After resist coating, the resist still contains up to 15% solvent and may contain built-in stresses. The photoresist is therefore soft baked (also named pre-exposure baked or prebaked) for a given time in an oven or hotplate at temperatures ranging from 70 to 100°C to remove solvents and stress and to promote adhesion of the resist layer to the substrate. This is a critical step in that failure to sufficiently remove the solvent will affect the resist profile. Excessive baking destroys the photoactive compound and reduces sensitivity. Thick resists may benefit from a longer bake time. The resist thickness, for both negative and positive resists, is typically reduced by 10–25% during soft baking. Hot plating the resist is faster, more controllable, and does not trap solvent like convection oven baking does. In convection ovens the solvent at the surface of the resist is evaporated first, and this can cause an impermeable resist skin, trapping the remaining solvent inside. Commercially, microwave heating or IR lamps are also used in production lines. The optimization of the prebaking step may substantially increase device yield.

5.4.4 Exposure

After soft baking, the resist-coated substrates are transferred to an illumination or exposure system where they are aligned with the features on a mask. For any lithographic technique to be of value, it must provide an alignment technique capable of a precise superposing of mask and substrate that is a small fraction of the minimum feature size of the devices under construction. In the simplest case, an exposure system consists of a UV lamp illuminating the resist-coated substrate through a mask without any lenses between the two. The purpose of the illumination systems is to deliver light with the proper intensity, directionality, spectral characteristics, and uniformity across the wafer, allowing a nearly perfect transfer or printing of the mask image onto the resist in the form of a latent image. The incident light intensity (in W/cm^2) multiplied by the exposure time (in seconds) gives the incident energy (J/cm^2) or dose, D, across the surface of the resist film. Radiation induces a chemical reaction in the exposed areas of the photoresist, altering the solubility of the resist in a solvent either directly or indirectly via a sensitizer. The smaller the dose needed to "write" or "print" the mask features onto the resist layer with good resolution, the better the lithographic sensitivity of the resist.

The absolute size of a minimum feature in an IC or a miniature device, whether it involves a line-width, spacing, or contact dimension, is called the *critical dimension* (CD). The overall resolution of a process describes the consistent ability to print a minimum size image, a critical dimension, under conditions of reasonable manufacturing variation [161]. Many aspects of the process, including hardware, materials, and processing considerations can limit the resolution of lithography. Hardware limitations include diffraction of light or scattering of charged particles (in the case of charged-particle lithography or hard X-rays), lens aberrations and mechanical stability of the system. The resist material properties that impact resolution include contrast, swelling behavior, thermal flow, and chemical etch resistance. The most important process-related resist variables include swelling (during development) and stability (during etching and baking steps).

In photolithography, wavelengths of the light source used for exposure of the resist-coated wafer range from the very short wavelengths of extreme ultraviolet (EUV) (10–14 nm) to deep ultraviolet (DUV) (150–300 nm) to near ultraviolet (UV) (350–500 nm). In near UV, one typically uses the g-line (435 nm) or i-line (365 nm) of a mercury lamp. The brightness of shorter-wavelength sources is severely reduced compared to that of longer-wavelength sources, and the addition of lenses further reduces the efficiency of the exposure system. As a consequence, with shorter wavelengths, higher resist sensitivity is required, and newer DUV sources that produce a higher flux of DUV radiation must be used. In general, the smallest feature that can be printed using projection lithography is roughly equal to the wavelength of the exposure source. The current generation of lithography is using 193 nm light [150] from ArF lasers. Step and scan printing Extreme Ultra-Violet (EUV) systems are expected to come on-line by the end of this decade. In order to stretch the lower limits of photolithography, sophisticated Resolution-Enhancing Techniques (RETs) are employed. RET methods enable one to go quite a bit beyond the conventional Rayleigh diffraction limit and may be used to produce features of 160 nm and below. These methods are classified depending on the element or part of the process they enhance: resist, mask or exposure procedure. RETs improving the resist include chemically amplified resists and anti-reflection coatings – thin film interference effects. Those enhancing the mask are Phase Shifting Masks (PSM) and Optical Proximity Correction (OPC). Enhancements to the exposure procedure include Off-Axis Illumination (OAI), Kohler Illumination and Immersion Lithography. RETs are beyond the scope of this work and the reader is referred to [144].

5.4.4.1 Masks and Grayscale Lithography

Standard Photolithography Masks. The stencil used to repeatedly generate a desired pattern on resist-coated wafers is called a mask. In typical use, a photomask – a nearly optically flat glass (transparent to near ultraviolet [UV]) or quartz plate (transparent to deep UV) with an absorber pattern metal layer (e.g., an 800 Å thick chromium layer) – is placed above the photoresist-coated surface, and the mask/substrate system is exposed to UV radiation. The absorber pattern on the mask is generated by e-beam lithography, a technique that yields higher resolution than photolithography (see below). Like resists, masks can be positive or negative. A positive or dark field mask is a mask on which the pattern is clear with the background dark. A negative or clear field mask is a mask on which the pattern is dark with the background clear. A light field or dark field image, known as mask polarity, is then transferred to the surface. This procedure results in a 1:1 image of the entire mask onto the substrate.

Different types of exposure masks are shown in Fig. 5.9. Masks making direct physical contact (also referred to as hard contact) with the substrate, are called contact masks. Unfortunately, these masks degrade faster due to wear than noncontact, proximity masks (also referred to as soft contact masks), which are slightly raised, say 10–20 μm, above the substrate. However, diffraction effects are minimized with the use of hard contact masks. Hard and soft contact masks are still in use in R&D, in mask making itself, and for prototyping (see Fig. 5.10). Contact mask and proximity mask printing are collectively known as shadow printing. A more reliable method of masking is projection printing where, rather than placing a mask in direct contact with (or in proximity of) a substrate, the photomask is imaged by a high-resolution lens system onto the resist-coated substrate. In projection printing, the only limit to the mask lifetime results from operator handling. The imaging lens can reduce the mask pattern by 1:5 or 1:10, making mask fabrication less challenging. Diffraction can be minimized by the use of highly collimated light sources and/or collimator lenses. Projection printing is the masking method employed in Very Large Scale Integration (VLSI)-based devices such as ICs.

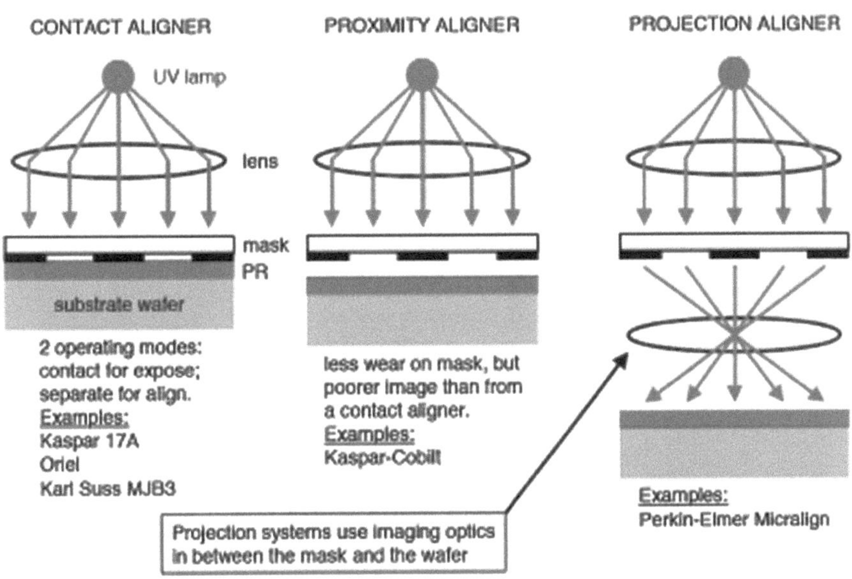

Fig. 5.9 Contact printing, proximity printing and projection printing

In miniaturization science, one often is looking for low-cost and fast-turnaround methods to fabricate masks. This may involve in-house fabricated masks by manually drawing patterns on cut-and-peel masking films and photo reducing them. Alternatively, it may involve direct writing on a photoresist-coated plate with a laser-plotter (~ 2 μm resolution) [162]. Simpler yet, using a drawing program such as Canvas® (ACD Systems, Ltd.), Freehand®, Illustrator® (Adobe Systems, Inc.), or

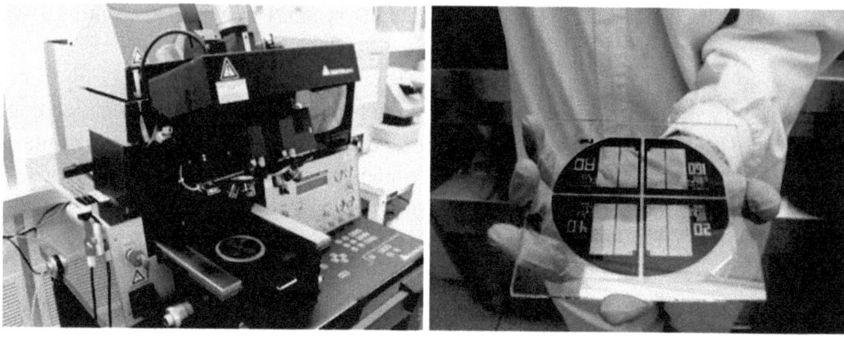

Fig. 5.10 *Left*: A SUSS MicroTec MA/BA6 mask aligner featuring bottom and top-side-alignment microscopes (Integrated Nanosystems Research Facility, UC Irvine). *Right*: A transparency mask taped to a low UV absorption glass holder

L-Edit® (Tanner Research, Inc.) a mask design can be created on a computer and saved as a Postscript® or Gerber® file to be printed with a high-resolution printer on a transparency [163]. The transparency with the printed image may then be clamped between a pre-sensitized chrome-covered mask plate and a blank plate to make a traditional photomask from it. After exposure and development, the exposed plate is put in a chrome etch for a few minutes to generate the desired metal pattern, and the remaining resist is stripped off. Simpler yet, the printed transparency may be attached to a quartz plate to be used as mask directly (See Fig. 5.10). Resolution with these methods is currently in the order of few micrometers and is highly dependent on the photo-plotter used to print the transparency. Even when these masks are significantly less expensive and have a fast-turnaround time, the quality of the exposed patterns, i.e. wall roughness, is obviously less than that obtained with photomasks patterned with e-beam lithography (where resolution can be less than 50 nm) (see Fig. 5.11).

Grayscale Lithography. Photolithography, as described so far, constitutes a binary image transfer process – the developed pattern consists of regions with resist (1) and regions without resist (0). In contrast, in grayscale lithography, the partial exposure of a photoresist renders it soluble to a developer in proportion to the local exposure dose and as a consequence, after development, the resist exhibits a surface relief. Grayscale lithography has a great potential use in miniaturization science as it allows for the mass production of micromachines with varying topography. The possibility of creating profiled micro 3D structures offers tremendous additional flexibility in the design of microelectronic, optoelectronic and micromechanical components [164]. A key part in the development of a grayscale process is the characterization of the resist thickness as a function of the optical density in the mask for a given lithographic process. It is also desirable to use photoresists that exhibit a low contrast in order to achieve a wide process window. Ideally, the resist response can be linearized to the optical density within the mask. Grayscale lithography can be achieved without the use of physical masks (software masks) or by employing Gray-tone masks (GTMs).

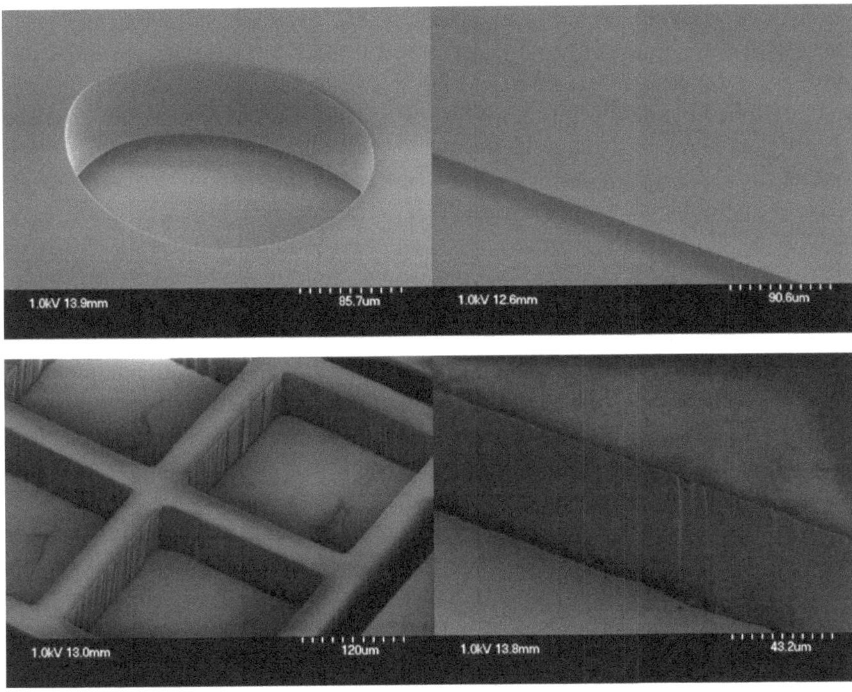

Fig. 5.11 The quality of a pattern is obviously far superior when using photomasks patterned with e-beam lithography (*top*) than when using transparency masks printed with a commercial photo-plotter (*bottom*)

Possible methods for making GTMs, or variable transmission masks, include magnetron sputtering of amorphous carbon (a-C) onto a quartz substrate. Essentially any transmittance (T) desired in the $0 < T < 100\%$ range can be achieved by controlling the film thickness (t) in the $200 > t > 0$ nm range with subnanometer precision [165]. Perhaps more elegantly, gray levels may be created by the density of dots that will appear as transparent holes in a chromium mask. These dots are small enough not to be transferred onto the wafer because they are below the resolution limit of the exposure tool. Another attractive way to fabricate a GTM is with High Energy Beam Sensitive (HEBS) glass. HEBS-glass turns dark upon exposure to an electron beam; the higher the electron dosage, the darker the glass turns. In HEBS-glass, a top layer, a couple of microns thick, contains silver ions in the form of silver-alkali-halide $(AgX)_m (MX)_n$ complex nanocrystallites that are about 10 nm or less in size, and are dispersed within cavities of the glass SiO_4 tetrahedron network. Chemical reduction of the silver ions produces opaque specks of silver atoms upon exposure to a high-energy electron beam (> 10 kV) [164].

Due to problems caused by the use of masks, such as expense and time consumed in fabricating them, contamination introduced by them, their disposal, and the difficulties in their alignment, research into Maskless Optical Projection Lithography

(MOPL) is growing rapidly and broadly. One approach to make multi-level photoresist patterns directly, without a physical mask but instead with a *software mask*, is by variable-dose e-beam writing, in which the electron dosage (the current multiplied by the dwell time) is varied across the resist surface [166]. A laser writer can produce the same result but at a lower resolution. However, variable-dose e-beam and laser writing are serial, slow and costly thus making gray-tone masks (GTMs) a desirable alternative, especially if high throughput production is required. Another MOPL or *software mask* approach already on the market (by Intelligent Micro Patterning LLC), is based on the Digital Micromirror Device (DMD) chip from Texas Instruments Inc. (TI), and relies on the same spatial and temporal light modulation technology used in DLP (Digital Light Processing) projectors and HDTVs (high definition televisions). Enormous simplification of lithography hardware is feasible by using the movable mirror arrays in a DMD chip to project images on the photoresist. This technique is capable of fabricating micromachined elements with any surface topography, and can, just like e-beam lithography or laser writing, be used for implementing maskless binary and grayscale lithography. The maximum resolution of DMD-based maskless photolithography (currently about 1 μm) is less than with e-beam lithography (less than 50 nm) or laser writers (<1.0 μm), but it is a parallel technique and for many applications, i.e. in microfluidics, the lower resolution might not be an obstacle. The unique capability of representing a gray scale is probably the most essential merit of this type of maskless lithography. When a mirror is switched on more frequently than off, it reflects a light gray pixel; a mirror that is switched off more frequently reflects a darker gray pixel. In this way, the mirrors in a DMD system can reflect pixels in up to 1,024 shades of gray to convert the video or graphic signal entering the DMD chip into a highly detailed grayscale image. Examples of grayscale features obtained with a SF-100 are shown in Fig. 5.12.

5.4.5 Post Exposure Treatment

A post-exposure treatment of the exposed photoresist is often desired because the reactions initiated during exposure might not have run to completion. To halt the reactions or to induce new ones, several post-exposure treatments are common: post-exposure baking (PEB), flood exposure with other types of radiation, treatment with a reactive gas, and vacuum treatment. Post-exposure baking (sometimes in vacuum) and treatment with reactive gas are used in image reversal and dry resist development. In the case of a chemically amplified resist, such as SU-8, the post-exposure bake is most critical. Although reactions induced by the catalyst that forms during exposure take place at room temperature, their rate is greatly increased by baking at 60–100°C. The precise control of PEB times and temperatures critically determines the subsequent development and the quality of the final features. Extended PEB times will introduce significant amounts of stress in the polymer that will most likely cause structure bending and peeling from the substrate; especially in

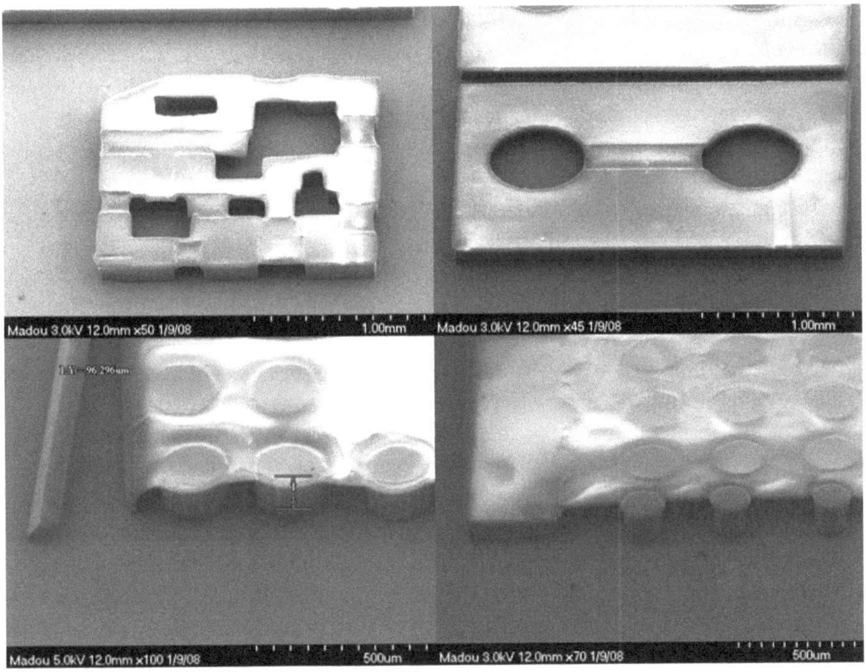

Fig. 5.12 Examples of grayscale lithography obtained with the SF-100 (Intelligent Micro Patterning, LLC) using SU-8 as resist. These features were fabricated by one of the authors in the Indian Institute of Technology in Kanpur, India

extended, large surface area features. Reduced times will yield structures that are not completely cross-linked and can be attacked by the developer. This causes extremely high surface roughness or even complete dissolution. An optimal PEB improves adhesion, reduces scumming (resist left behind after development), increases contrast and resist profile (higher edge-wall angle) and reduces the effects of standing waves in a regular resist.

5.4.6 Development

Development is the dissolution of un-polymerized resist that transforms the latent resist image, formed during exposure, into a relief image that will serve as a mask for further subtractive and additive steps, as a permanent structural element, or as a precursor for carbonization as in the case of C-MEMS. During the development of an exposed resist, selective dissolving takes place. Two main technologies are available for development: wet development, which is widely used in circuit and miniaturization manufacture in general, and dry development, which is starting to replace wet development for some of the ultimate line-width resolution applications. Wet development by solvents can be based on at least three different types of exposure-induced changes: variation in molecular weight of the polymers (by cross-linking or by chain scission), reactivity change, and polarity change [167]. Two main types of wet development setups are used: immersion and spray developers. During batch immersion developing, substrates are batch-immersed for a timed period in a developer bath and agitated at a specific temperature.

During batch spray development, fan-type sprayers direct fresh developing solution across wafer surfaces. Positive resists are typically developed in aqueous alkaline solutions, and negative resists in organic ones. Each developer has a different dilution, and some require longer development times than others. They are generally matched to a type of photoresist. Though they may be interchangeable to some extent, changing the type of developer used in a process will usually change the development time necessary to resolve the pattern.

The use of organic solvents leads to some swelling of the resist (especially for negative resists) and loss of adhesion of the resist to the substrate. Dry development overcomes these problems, as it is based either on a vapor phase process or a plasma [168]. In the latter, oxygen-reactive ion etching (O_2-RIE) is used to develop the latent image. The image formed during exposure exhibits a differential etch rate to O_2-RIE rather than differential solubility to a solvent [146]. Dry developed resists should not be confused with dry film resists, which are resists that come in film form and are laminated onto a substrate rather than spin coated. With the continued pressure by the U.S. Environmental Protection Agency (EPA) for a cleaner environment, dry development and dry etching are becoming the predominant technologies to use.

5.4.7 De-Scumming and Post-Baking

A mild oxygen plasma treatment, so-called de-scumming, removes unwanted resist left behind after development. Negative, and to a lesser degree positive, resists leave a thin polymer film at the resist/substrate interface.

Post-baking or hard baking removes residual coating solvent and developer, and anneals the film to promote interfacial adhesion of the resist that has been weakened either by developer penetration along the resist/substrate interface or by swelling of the resist (mainly for negative resists). Unfortunately, post-bake does induce some stress and resist shrinkage. Nevertheless, hard baking improves the hardness of the film and avoids solvent bursts during vacuum processing. Improved hardness increases the resistance of the resist to subsequent etching steps. Post-baking frequently occurs at higher temperatures (120°C) and for longer times than soft baking. The major limitation for heat application is excessive flow or melt, which degrades wall profile angles and enables impurities to be easily incorporated into the polymer matrix due to the plastic flow of the resist. Special care needs to be taken to prevent the post-baking temperature from exceeding the glass transition temperature, T_g, of the developed resist. However, resist reflow may be used for tailoring resist sidewalls.

5.4.8 Resist Profiles – An overview

Manipulation of resist profiles is one of the most important concerns of a lithography engineer. Depending on the final objective, one of the three resist profiles shown in Fig. 5.13 is attempted. A reentrant, undercut or a reverse resist profile (resist sidewall > 90°) is required for metal lift-off. Some authors confusingly call slopes > 90° *overcut* [168]; most, including these authors, refer to this type of resist profile as an *undercut.* Shallow resist angles (< 90°) enable continuous deposition of thin films over the resist sidewalls. A vertical (90° resist sidewall angle) slope is desirable when the resist is intended to act as a permanent structural element such as in microfludics and molding applications. Vertical walls are also usually desirable in C-MEMS applications. For more details refer to [144].

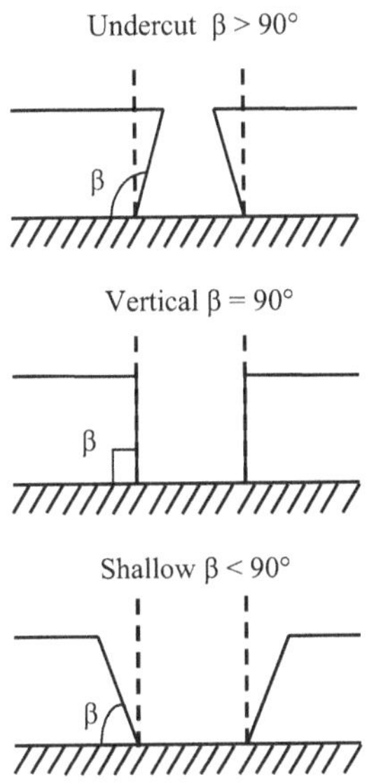

Fig. 5.13 The three important resist profiles. Top: reentrant, undercut or a reverse resist profile (resist sidewall β > 90°) is required for metal lift-off. A vertical (β = 75–90° resist sidewall angle) slope is desirable for a perfect fidelity transfer of the image on the mask to the resist. Shallow resist angles (45° < β < 90°) enable continuous deposition of thin films over the resist sidewalls

5.5 Next Generation Lithography (NGL)

We now introduce the concept of Next Generation Lithography (NGL). This group of techniques are the ones regarded today as sufficiently developed to be postulated as replacements to traditional photolithography in the everlasting race to shrink feature dimensions. Techniques classified now as NGL include: extreme ultraviolet lithography (EUVL), X-ray lithography, charged particle beam lithography based on electrons and ions (such as electron and ion projection techniques) and imprint lithography [Nanoimprint lithography (NIL) and Step-and-Flash Imprint

Lithography (SFIL)]. IC and miniaturization science are taking increasingly separate paths in adopting preferred lithography strategies. For ICs, throughput and finer geometries are needed and batch processing is a prerequisite. In miniaturization science, modularity, good depth-of -focus (DOF), extension of the z-direction, that is, the height of features (skyscraper-type structures), incorporating nontraditional materials (e.g., gas-sensitive ceramic layers, polymers), and replication methods catch the spotlight and batch fabrication is not always a prerequisite. We emphasize on charged-particle-based and imprint lithographies because they represent a more affordable and readily available way to obtain nano-sized structures for Carbon NEMS or nanofluidics applications. EUVL and X-ray lithography may also be used but they are more expensive and are not readily available due to their required infrastructure. EUVL and X-ray are touted by the IC industry as strong candidates to replace DUV lithography in the quest to achieve the 32 nm node although no clear solution has been envisioned yet (International Technology Roadmap for Semiconductors). EUV is actively being promoted by the Extreme Ultraviolet Limited Liability Company (EUV LLC) which is a consortium lead by Intel Corporation with partners such as Motorola Corporation, Advanced Micro Devices Corporation, IBM, Infineon, Micron Technology and the Sandia and Lawrence Livermore National Laboratories. X-ray lithography is being championed by IBM in the US and by NTT, Toshiba, Mitsubishi and NEC in Japan [169]. Other techniques currently in the R&D stage but that could emerge as serious NGLs in the coming years include lithography based on very thin resist layers and block copolymers, zone plate array lithography (ZPAL), quantum lithography (two-photon lithography) and proximal probe based techniques such as atomic force microscopy (AFM), scanning tunneling microscopy (STM), dip-pen lithography (DPL) and near-field scanning optical microscopy (NSOM), and apertureless near-field scanning optical microscopy (ANSOM). In this section we will only explore electron beam, ion beam and nanoimprint lithography. An extensive review on the other techniques mentioned can be found in [144].

Before moving forward it is worth mentioning the LIGA (a German acronym for Lithographie, Galvanoformung, Abformung) technique as an alternative to fabricate very high aspect ratio structures using X-ray lithography. The LIGA technique was invented about 20 years ago [170]. The process, illustrated in Fig. 5.14, involves a thick layer of resist (from micrometers to centimeters), high-energy X-ray radiation, and resist development to make a resist mold. By applying galvanizing techniques, the mold is filled with a metal. The resist structure is removed, and metal products result. Alternatively, the metal part can serve as a mold itself for precision plastic injection molding. Several types of plastic molding processes have been tested, including reaction injection molding, thermoplastic injection molding, and hot embossing. The so-formed plastic part, just like the original resist structure, may also serve as a mold for fast and cheap mass production, since one does not rely on a new X-ray exposure. Of particular interest to miniaturization science is the possibility of creating three-dimensional shapes with slanted sidewalls and step-like structures. The unprecedented precision attainable with LIGA makes this technique stand out against other 3D lithography methods such as Deep UV with SU-8. LIGA

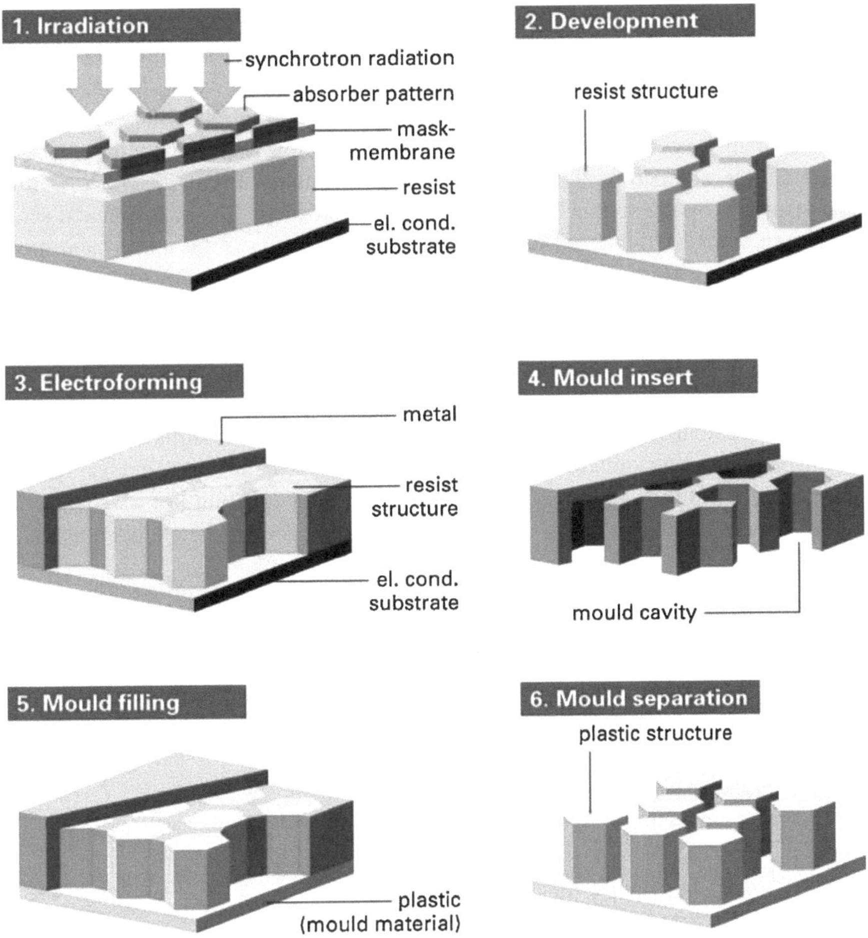

Fig. 5.14 The LIGA process

enables new building materials and a wider dynamic range of dimensions and possible shapes but its wider adoption for commercial and research applications have been halted by the need of an X-ray source (a synchrotron) in the fabrication process. We now proceed to explore charged-particle-beam and imprint lithography techniques.

5.5.1 Charged-Particle-Beam Lithography

Charged-particle-beam lithography includes both narrow beam direct-writing and flood exposure projection systems with electrons and ions. One of the major advantages of direct writing over flood exposure is its independence from a mask and thus a mask fabrication process.

In direct writing systems, the computer-stored pattern is directly converted to address the writing charged particle beam, enabling the pattern to be exposed sequentially, point by point, over the whole wafer. In other words, the mask is a *software mask*. Electron-beam (e-beam) and ion-beam (i-beam) lithographies involve high current density in narrow electron or ion beams. The smaller the beam sizes, the better the resolution, but more time is spent writing the pattern. This sequential (scanning) type system exposes one pattern element or pixel at a time. Within that area, the charged-particle beam delivers maximum current (i), which is limited primarily by the source brightness and column design. The experimental setup imposes a limit on the speed at which the writing beam can be moved and modulated, resulting in a "flash" time in seconds (t). The maximum dose D_{max} (in coulombs per cm^2) deliverable by a particular beam is given by the product of the current and time divided by the pixel area (in cm^2). It will then be necessary to work with resists that react sufficiently fast at D_{max} to produce a lithographically-useful, three-dimensional image (latent or direct image). The continued development of better charged-particle-beam sources keeps widening the possibilities for nanoscale engineering through direct write lithography, etching, depositing, analyzing, and modifying a wide range of materials, well beyond the capability of classical photolithography. Table 5.2 lists some ion-beam and electron-beam applications.

Table 5.2 Electron- and Ion-Beam Applications

Electron-Beam Applications	Ion-Beam Applications
Nanoscale lithography	Micromachining and ion milling
Low-voltage scanning electron microscopy	Microdeposition of metals
Critical dimension measurements	Maskless ion implantation
Electron-beam-induced metal deposition	Microstructure failure analysis
Reflection high-energy electron diffraction (RHEED)	Secondary ion mass spectroscopy
Scanning auger microscopy	–

Flood exposure of a mask in a projection system (that is, parallel exposure of all pattern elements at the same time, as done in Deep UV) is possible with ions and electrons as well. Technologies such as SCattering with Angular Limitation Projection Electron beam Lithography (SCALPEL) and Ion Projection Lithography (IPL) have the potential of making electron-beam and ion-beam high throughput. Exposure masks are fabricated from heavy metals on semi-transparent organic or inorganic membranes. The high cost of mask fabrication and the instability of the mask due to heating have postponed commercial acceptance of these high-energy exposure systems. Moreover, with ion and electron beams, flood exposure is limited to chip-size fields due to difficulties in obtaining broad, collimated, charged-particle beams. The most prevalent use of charged-particle beams remains the narrow beam scanning mode. Because of this reason we emphasize on Direct write Electron-Beam (e-beam) and Focused Ion Beam (FIB) lithography. Details on SCALPEL and IPL can be found elsewhere [144].

5.5.1.1 Direct Write Electron-Beam Lithography

Direct write electron-beam lithography or simply e-beam is a high-resolution patterning technique in which high-energy electrons (10–100 keV) are focused into a narrow beam and are used to expose electron-sensitive resists. There are two basic ways to scan an electron beam. In raster scanning, the patterns are written by an electron beam that moves through a regular pattern. The beam scans sequentially over the entire area and is blanked off where no exposure is required. On the contrary, in vector scanning, the electron beam is directed only to the requested pattern features and hops from features to features. Time is therefore saved in a vector scan system. The direct write technique was first developed in the 1960s using existing SEM technology. As a research solution, several groups modify their standard scanning electron microscopes to create customized electron-beam writing systems. Rosolen, for example, modified a Hitachi S2500 with a purpose built pattern generator and alignment system. The instrument does not require alignment marks on the sample and is able to compensate for positional errors caused by the sample stage and mask tolerances [171]. The e-beam lithography method, like X-ray lithography, does not limit the obtainable feature resolution by diffraction, because the quantum mechanical wavelengths of high-energy electrons are exceedingly small. Advantages and disadvantages of e-beam lithography are shown in Table 5.3. Because of such, the use of electron-beam lithography has been limited to mask making and direct writing on wafers for specialized applications. Usually,

Table 5.3 Advantages and Disadvantages of e-beam lithography

Advantages	Disadvantages
Precise control of the energy and dose delivered to a resist-coated substrate	Proximity effects. Electrons scatter quickly in solids, limiting practical resolution to dimensions greater than 10 nm
Deflection and modulation of electron beams with speed and precision by electrostatic or magnetic fields	Electrons, being charged particles, need to be held in a vacuum, making the apparatus more complex than for photolithography
Imaging of electrons to form a small point of <100 Å, as opposed to a spot of 5000 Å for light in the case of photolithography	The slow exposure speed -an electron beam must be scanned across the entire wafer (for a 4-in wafer with a high feature density this requires ~ 1 h)
No need for a physical mask; only a *software mask* is required	Low throughput. Approximately 5 wafers per hour at less than 0.1 μm resolution.
The ability to register accurately over small areas of a wafer	High system cost
Lower defect densities	–
Large depth of focus because of continuous focusing over varying topography. At 30 keV, electrons will travel on average > 14 μm deep into a PMMA resist layer.	–

this type of slow, expensive fabrication technique prohibits commercial acceptance. Some microstructures, especially intricate microsystems, might be worth the bigger price tag. Nanostructures could also be readily achieved featuring dimensions not possible with photolithography. In those cases, serial microfabrication techniques may not be as prohibitive as they would be in the case of ICs.

Numerous commercial e-beam resists are produced for mask-making and direct write applications. Bombardment of polymers by electrons causes bond breakage and, in principle, any polymer material can function as a resist. However, important considerations include sensitivity, tone, resolution, and etching resistance. PMMA exemplifies an inexpensive positive e-beam resist with a high-resolution capability and a moderate glass transition temperature T_g (114°C). Microposit SAL601 is an often used negative e-beam resist. SAL601, being novolac-based, has much better dry etch resistance than PMMA resists. The same materials act as X-ray resists as well. This is not coincidental, as there is a strong relation between X-ray and e-beam sensitivity. A copolymer of glycidyl methacrylate and ethyl acrylate (COP) is another frequently used negative resist in mask manufacture. This material has (as is typical for acrylates) poor plasma-etching resistance but exhibits good thermal stability.

E-beam has been used by one of the authors on the fabrication of polymer nanostructures [172]. After pyrolysis, the resulting carbon nanoelectrodes are used for the detection of reversible redox species, such as the ferro/ferricyanide couple and dopamine (unpublished results). In the long term, the objective is to take advantage of the biocompatibility of the glass-like carbon and the higher sensitivity of the interdigitated electrode configuration to detect dopamine in living cells in vitro.

5.5.1.2 Focused Ion Beam Lithography

In ion-beam lithography, resists are exposed to energetic ion bombardment in a vacuum. Direct write ion-beam lithography or Focused Ion Beam (FIB) lithography consists of point-by-point exposures with a narrow ion beam generated by a source of liquid metal. Ion-beam lithography uses ions of a kinetic energy from a few keV up to several MeV. For ion-beam construction, liquid metal ion (LMI) sources are becoming the choice for producing high-current-density submicrometer ion beams. With an LMI source, liquid metal (typically gallium) migrates along a needle substrate. By applying an electrical field, a jet-like protrusion of liquid metal forms at the source tip. The gallium-gallium bonds are broken under the influence of the extraction field and are uniformly ionized without droplet or cluster formation. LMI sources hold extremely high brightness levels and a very small energy spread, making them ideal for producing high-current-density submicrometer ion beams. Beam diameters of less than 50 nm and current densities up to 8 A/cm^2 are the norm. In addition to gallium, other pure element sources are available, such as indium and gold. By adopting alloy sources, the list expands to dopant materials such as boron, arsenic, phosphorus, silicon, and beryllium.

Compared to photons (X-rays and DUV light) or electrons, ions chemically react with the substrate, allowing for a greater variety of surface modifications such as patterned doping. The ion-beam spot size is the smallest possible, smaller than UV, X-ray, or electron-beam spots. The smallest FIB spot achieved upto this date is about 8 nm, accomplished by using a two-lens microprobe system and a single-isotope gallium ion source. Ion lithography achieves higher resolution than optical, X-ray, or electron beam techniques because ions undergo almost no diffraction and scatter much less than electrons, since the secondary electrons produced by an ion beam are of lower energy and have a short diffusion range. The total spread including forward and backward scattering of the "stiffer" ion-beams is typically less than 10 nm and they only require about 1–10% of the electron dose to expose a resist. Materials that can be used as resists include an ordinary PMMA resist [173, 174] and SU-8 [175]. FIB shares the same drawbacks with an electron-beam system in that it requires a serially-scanned beam and a high vacuum. Because FIB systems operate in a similar fashion to a scanning electron microscope (SEM) they can be used for imaging (when operated at low beam currents) or for site specific sputtering or milling (when operated at high beam currents).

FIB can be used to perform maskless implantation and metal patterning with submicrometer dimensions. It has also been applied to milling in IC repair, maskless implantation, circuit fault isolation, and failure analysis. FIB systems have been produced commercially for approximately ten years, primarily for large semiconductor manufacturers. As a machining tool, FIB is very slow. Except for research, it may take a long time to become an accepted "micromachining tool." For additional reading on ion-beam lithography in general refer to [176]; for more specific reading on focused ion-beam-induced deposition, see [177].

FIB milling has been employed recently for the patterning of commercial "glassy" carbon. The resultant structures are used for the molding of borosilicate glasses and quartz [178–180]. A comparison between FIB, laser and mechanical milling for the patterning of carbon is given in [181, 182].

5.5.2 Nano Imprint Lithography

Nano Imprint Lithography (NIL) patterns a resist by deforming the resist shape through embossing (with a mold/stamp/template), rather than by altering the resist's chemical structure through radiation (with particle beams for instance). After imprinting the resist, a dry anisotropic etch is used to remove the residual resist layer in the compressed area to expose the substrate underneath. In NIL, a template (the mold/stamp/template) is made of a hard material (usually Ni or Si) and is pressed against a layer of polymer. High temperature and pressure conditions mold and harden the polymer layer. Stephen Chou, now at Princeton University but with the University of Minnesota at the time, invented the technique in 1995 [183, 184] The method relies on the excellent replication fidelity obtained with polymers and combines thermo-plastic molding with common pattern transfer methods. Once a solid stamp with a nano-relief on its surface is fabricated it can be used for the replication of many identical surface patterns. The resolution of the NIL process is a direct function of the resolution of the original template/stamp fabrication process. Electron beam writers that provide high resolution, but lack the throughput required for mass production, are used to make them.

The University of Texas (UT)-Austin developed its version of nanoimprint lithography, step-and-flash imprint lithography (SFIL), in 1998 [185]. In 2001, the SFIL concept was licensed to Molecular Imprints, Inc. (MII), a company that develops SFIL-capable tools (S-FIL® systems) and related processes. The SFIL method is distinct from the original NIL in its use of UV-assisted nanoimprinting that molds photocurable liquids in a step-and-repeat, die-by-die fashion rather than by heat-assisted molding of full, polymer-coated wafers. As shown in Fig. 5.15, in SFIL, a hard but transparent template/stamp/mold (fused silica) is used to mold

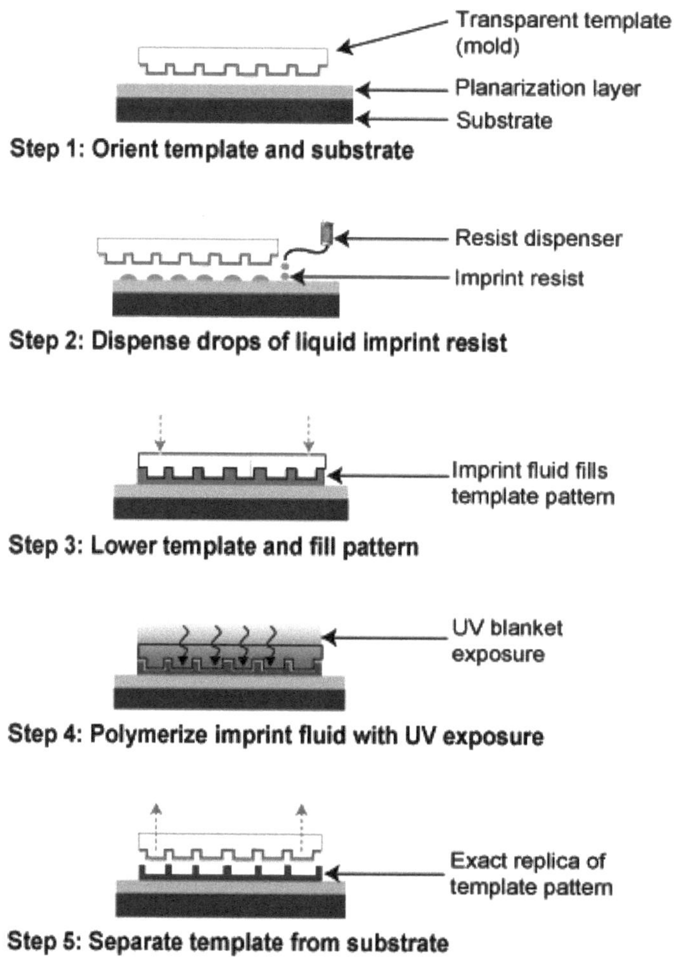

Fig. 5.15 Schematic of step-and-flash imprint lithography (SFIL). Adapted from www.molecularimprints.com

a polymer photoresist layer. The fused silica surface of the stamp, coated with a release layer, is gently pressed into the thin layer of low viscosity photoresist. The photoresist is subsequently exposed to UV light through the transparent template in order to harden it. Upon separation of the fused silica template, one layer of the circuit pattern is left on the wafer surface. Molecular Imprints uses a family of photo-curable, low viscosity materials called MonoMat^{TM} as its imprint resists. MonoMat^{TM} is comprised of an organic monomer that polymerizes in seconds using low cost, broadband UV light sources. Only the template fabrication process, typically accomplished with an e-beam writer, limits the resolution of the features. It has been demonstrated that the mold templates do not deteriorate even after 1500 imprints with sub-100 nm feature sizes.

Imprinting processes are economical due to the simplicity of the equipment involved and the potential for high-throughput. As the rate of improvements in optical lithography decelerates and the costs of manufacturing continue to escalate, there is an increased interest in imprinting and molding as alternative processes for micro and nanofabrication. SFIL is one of the few methods currently available for low-volume prototyping at the 32 nm node. Sub-20 nm features have been made to date that exceed the present requirements in the International Technology Roadmap Semiconductors (ITRS). The SFIL process is now also being explored for manufacturing of several emerging technologies, such as photonic crystals, micro/nano-optical components, and nanopatterned magnetic media for future hard disk drives. The original Thermal NIL using thermoplastic polymer films is focused on applications such as bio-chips, life sciences, storage media and optical devices. Progress in these new areas has been such that Imprint Lithography is likely to find its first commercial manufacturing application in one of these emerging technologies well before it would be required for high-volume, sub-50-nm semiconductor lithography. A notable feature of nanoimprint technologies is their relatively low cost, which allows researchers to explore applications of nanopatterning that would never be economically feasible given the extraordinary cost associated with extreme ultraviolet (EUV) lithography or even current-generation 193 nm steppers. However, the approach faces alignment challenges, and critics say the cost of making the reticles will be high. Unlike conventional lithography, with mask patterns that are 4 $\times$ larger than the printed image, nanoimprint is a 1:1 template technology.

5.6 Microfluidic and Electrochemistry Applications

Micro and nanofabrication techniques like the ones presented above have recently made possible a significant number of breakthroughs in different fields, including environmental, clinical, pharmaceutical and biochemical technologies. The miniaturization of analytical systems offers important advantages such as lower cost, reduced sample and reagent consumption, shorter response time, and greater sensitivity and portability [186–190]. Moreover, quality and performance improvements are making it possible for microanalytical systems to become common aids to many different applications [187, 191, 192].

Although the above-mentioned techniques were described as ways to fabricate polymer precursor structures for carbonization, they can also be utilized to fabricate the polymer designs to be used as fluidic manifolds for a variety of applications. Polymer structures can further serve as master molds for casting of elastomers and other polymers such as in Soft Lithography [193–196]. The reader is encouraged to review further references on polymeric microfluidic devices and their means of fabrication [197–199].

Even when glass-like carbon does not strike as the ideal structural material for microfluidic manifolds, it does offer several advantages in the fabrication and of elements to be incorporated in microfluidic devices, i.e., electrodes. These electrodes can be used for sensing, as in the case of electrochemical detection of glucose or dopamine, or for actuating, as in dielectrophoresis-based particle manipulation. The use of glass-like carbon brings several advantages: (1) Carbon has a wider electrochemical stability window than gold or platinum, so that more voltage can be applied to the carbon electrodes without inducing electrolysis in the solution, (2) it has an excellent biocompatibility, carbon is the building block of nature, (3) carbon is chemically inert to most solvents and (4) it presents excellent mechanical properties. It is important to mention that as with other commonly used materials, polymers in particular, it is possible to change the hydrophobic properties of carbon using surface treatments [200]. This is a very important feature for the fabrication of microfluidic devices, as contact angle plays a very important role in most of them. The following applications detail the use of glass-like carbon elements in microfluidic and electrochemical applications.

5.6.1 Carbon-Electrode Dielectrophoresis (carbon-DEP)

Separation and sorting techniques strive to isolate a targeted particle population from a mixture of various populations by taking advantage of the differences between particle's characteristics. Properties that can be exploited include size, geometry, density, protein affinity and light scattering but additional tags, such as fluorescent or magnetic, might be attached to a particle population to enhance the level of discrimination. These additional tags enable flow cytometry technologies such as FACS^{TM} (Fluorescence Activated Cell Sorting) and other cell sorting technologies like MACS^{TM} (Magnetic Activated Cell Sorting). However, the required linkage of these tags to expensive antibodies increases the cost of the assay per patient.

Dielectrophoresis (DEP) [201, 202] uses only the dielectric properties of the targeted particles for their isolation from a mixture. Different particles present characteristic dielectric properties based on their phenotype, i.e., their membrane structure and surface properties, their internal compartmentalization and other physical characteristics. The collection of these properties gives to each particle a unique dielectric signature and a characteristic dielectrophoretic behavior when under the influence of a non uniform electric field [203–207]. This approach eliminates the use of additional tags and potentially reduces the cost of a given assay.

DEP refers to the induction of a dielectrophoretic force, F_{DEP} [4], on a polar particle immersed in a polar media by a non uniform AC or DC electric field. As the particle interacts with the established electric field gradient, a dipole moment, with a specific magnitude and direction, is induced on it [208, 209]. The magnitude of F_{DEP} is mainly determined by the value of the electric field gradient and the radius of the particle while the direction is given by the relative difference of the dielectric properties between the particles and their surrounding media.[5] DEP is a non-destructive electrokinetic transport mechanism and is highly amenable to miniaturization since the magnitude of the induced force scales favorably with the reduction of the distance between the electrodes [210]. F_{DEP} is also strongly dependent on the voltage applied between electrodes and it gradually vanishes as the distance from an electrode surface is increased and the electric field gradient tends to disappear (see Fig. 5.16).

Fig. 5.16 Finite Element Analysis of the electric field induced by two different electrode geometries: (*left*) 2D and (*right*) 3D. Note how in both cases the F_{DEP}, directly dependent on the induced electric field gradient, gradually vanishes as one moves away from the electrode surface. F_{DEP} is strongest in lighter areas and weakest in darker areas. The darkest square areas in the middle represent the electrodes. Results obtained in collaboration with Tecnológico de Monterrey, Campus Monterrey in Mexico

Due to advances in microfabrication technology achieved in the last 20 years by the IC industry, the traditional method to induce DEP has been based on the use of planar metal electrodes [211–214]. The main advantage of this approach is the

[4] F_{DEP} is given by the equation $2\pi\varepsilon_m r^3 \mathrm{Re}[f_{CM}]\mathrm{grad}(E_{rms}{}^2)$ where ε_m is the permittivity of the medium, r is the particle radius, $\mathrm{Re}[f_{CM}]$ denotes the real part of the Clausius-Mossotti factor and $\mathrm{grad}(E_{rms}{}^2)$ illustrates an electric field gradient. E is given by V/d where V is the voltage applied to the electrodes and d represents the distance between them.

[5] This difference is given by the Clausius-Mossotti factor. This factor is named after the Italian physicist Ottaviano-Fabrizio Mossotti, whose 1850 book analyzed the relationship between the dielectric constants of two different media, and the German physicist Rudolf Clausius, who gave the formula explicitly in his 1879 book in the context not of dielectric constants but of indices of refraction. The Clausius-Mossotti factor is given by $(\varepsilon_p{}^*-\varepsilon_m{}^*)/(\varepsilon_p{}^*+2\varepsilon_m{}^*)$. ε^*denotes complex permittivity and is given by $\varepsilon + (\sigma/i\omega)$ where ε is the permittivity, σ is conductivity, i denotes the square root of -1 and ω is the angular frequency of the applied electric field; p and m denote particle and media respectively.

creation of high electric field gradients employing low applied voltages. However, the use of metal electrodes limits the magnitude of the applied voltage in order to prevent electrolysis. In terms of fabrication, even when the fabrication of planar metal electrodes is relatively easy and readily achievable, the fabrication of three-dimensional electrodes is rather cumbersome. 3D structures require the use of more complicated techniques, such as electroplating, that might result in low fabrication yields. This limitation often results in expensive devices that halt the wider adoption of metal-electrode DEP for commercial applications. An alternative to metal electrodes is the use of insulator-based DEP (iDEP) [215–219]. In this approach, a uniform electric field is applied across an array of insulating structures by a couple of conductive electrodes. The presence of the insulating structures distorts the uniform electric field rendering it non uniform around them. Because the conductive electrodes on either side of the insulating electrode array are separated by a few millimeters, the magnitude of the applied voltage must be significant. One of the primary drawbacks of an iDEP system is precisely the requirement of a high-voltage excitation source to apply the required high electric fields across the insulating array. Another major disadvantage is that such fields can cause significant Joule heating and damage biological particles. iDEP also greatly depends on sample composition. In order to avoid Joule heating, only low-conductivity samples can be used. On the positive side, the possibility of electrolyzing the sample is greatly reduced in iDEP, since the sample is only in contact with an insulating material. Furthermore, insulating structures can be readily fabricated in planar or 3D shapes with fabrication techniques that do not require infrastructure as expensive as metal deposition equipment.

Regardless of the material, the use of 3D electrodes in DEP applications allows for higher separation throughputs than when 2D electrodes are used. This advantage is provided by the fact that, given a channel cross-section, many targeted particles immersed in the bulk volume of the channel flow over the planar electrodes without experiencing any force. This makes it necessary to re-flow the same sample several times, a procedure that negatively impacts biological samples. The use of 3D structures that penetrate the bulk of the channel greatly reduces the mean distance of any particle to its closest electrode surface (see Fig. 5.17).

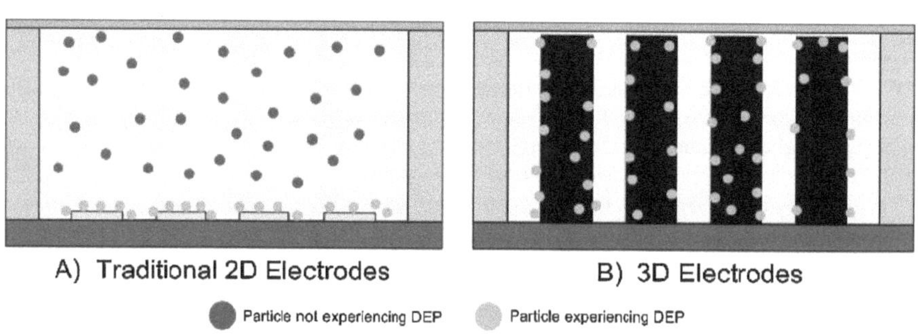

Fig. 5.17 A schematic showing how the use of 3D electrodes is advantageous over 2D ones. Given a channel cross-section, more particles will experience F_{DEP} when using 3D electrodes

DEP has been successfully applied to the manipulation and concentration of a wide array of bioparticles; from proteins [219–221] to microorganisms [222–226], including mammalian cells [214, 227–232] and viruses [205, 233–236]. Several applications have been demonstrated using planar (2D) metal [230, 237, 238] or volumetric (3D) insulator-based electrodes [239, 240], until now the two main trends on the material choice and geometry for electrodes.

Even when the general tendency when employing metals has been to use planar electrodes, few examples of micro fabricated three-dimensional (volumetric) electrodes have been recently implemented. Wang et al. [241] and Voldman et al. [242] used electroplated gold electrodes for particle focusing and for particle trapping for cytometry applications, respectively. Illiescu [243] used complex microfabrication techniques to obtain volumetric doped silicon electrodes in a multi-step process. Another true 3D approach, although not technically microfabricated, is the one described by Fatoyinbo et al. [244], who in an elegant solution used a drilled insulator-conductor sandwich to obtain wells with electrodes all along their walls.

In contrast to the approaches exposed above, carbon-electrode Dielectrophoresis (carbon-DEP) is a technology that employs carbon structures or surfaces as electrodes. The use of carbon as electrode material offers several advantages that make it a more suitable material than those traditionally used in DEP including gold, platinum or other conductive materials such as Indium Tin Oxide (ITO) or doped silicon. Carbon-DEP combines the advantages of metal-based and insulator-based DEP. More specifically, carbon has a wider electrochemical stability window making electrodes less likely to induce electrolysis and less prone to fouling when compared to metal electrodes. Carbon electrodes (planar, volumetric or a combination of both) can also be fabricated relatively inexpensively and with high yields when using fabrication techniques such as C-MEMS. When compared to insulator-based electrodes, or iDEP, the main advantage is that carbon electrodes prevent the use of high voltages and high fields to induce DEP. Carbon is conductive enough to allow for the use of low voltages to apply the required electric fields to induce DEP, thereby minimizing damage to biological particles. Furthermore, it has already been demonstrated that carbon is a highly biocompatible material [86, 129]

The fabrication of carbon-DEP devices, as it is done currently, emanates from the manufacturing of polymer structures. Such structures have commonly been fabricated in SU-8 photoresist following a two-step photolithography process (see Fig. 5.18a). Two steps are needed in order to fabricate the connection leads to

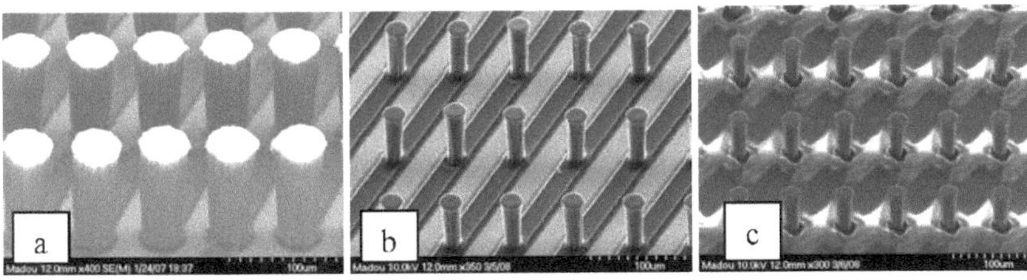

Fig. 5.18 (**a**) SU-8 polymer arrangement featuring posts and connection leads. (**b**) Carbon electrodes obtained after pyrolysis. (**c**) A thin polymeric layer patterned around carbon electrodes

the electrodes and the electrodes themselves since they do not usually feature the same topography. An alternative is to pattern the connection leads out of evaporated metal. The substrates have been limited so far to silicon wafers with a thick layer of silicon dioxide, which acts as the electrical insulator underneath the carbon electrodes. This limitation is mainly due to (1) the high temperatures of the pyrolysis process and (2) the fact that structures will lift-off from substrates that have a coefficient of thermal expansion that does not match the one of the different products being synthesized during pyrolysis, from the precursor polymer to the final glass-like carbon. Carbon and silicon happen to have very similar coefficients of thermal expansion (CTE) during the pyrolysis process; the CTE of glass-like carbon changes from 2.85 to 3.7×10^{-6}/K [245, 246] while that of silicon falls in between 2.6 and 4.442×10^{-6}/K [247]. Quartz wafers have been employed before as substrates with limited success. Titanium and chromium leads have also been fabricated with no better results.

Once the polymer precursor is patterned, it is introduced inside a high temperature furnace for carbonization. This requires the use of an inert atmosphere in the furnace during pyrolysis to prevent combustion of the structures. The carbonization process takes approximately 4 h with an additional 6–8 h for cooling.

An example of the resulting carbon electrodes and their connection leads is shown in Fig. 5.18b. A certain degree of isometric shrinkage with respect to the original polymer structures is always obtained. An optional further step is the patterning of a thin polymeric layer around the electrodes as shown in Fig. 5.18c. This layer serves two purposes: (1) to planarize the channel bottom and (2) to protect the carbon leads and electrodes from lifting-off when immersed in aqueous media.

The next step is to embed the DEP active area, delimited by the carbon electrodes, in a fluidic network. This network includes channels, chambers and fluidic interfaces to the outside world for sample loading and retrieval. Several approaches and materials for the fabrication of the fluidic network have been implemented by the authors, including SU-8, Poly(dimethylsiloxane) (PDMS) and polycarbonate (PC). The polycarbonate approach has yielded the most rapid, affordable and robust solution. An example of a carbon-DEP device and its cross section using PC is shown in Fig. 5.19. For more fabrication details the reader is referred to [134]

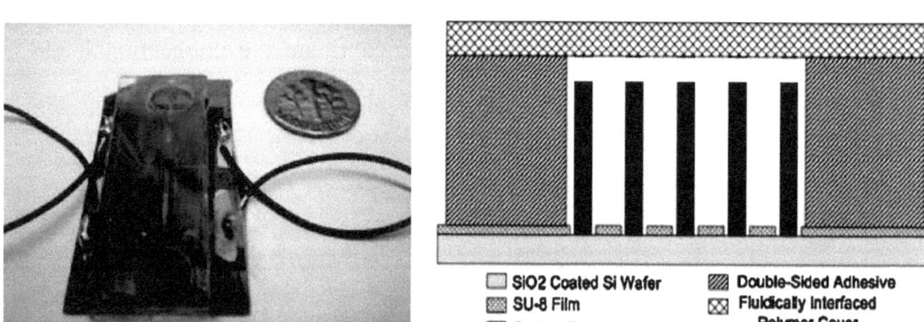

Fig. 5.19 An example and cross-section of a CarbonDEP chip using polycarbonate to fabricate the fluidic network

Fluido-dynamic and electromagnetic modeling and simulation of carbon-DEP devices have already been conducted [121, 130, 133, 134]. Figure 5.20 shows some of these results obtained with Finite Element Analysis. As depicted in the electromagnetic simulation results (Fig. 5.22c and d), an electric field gradient is established in the media with an electric field magnitude that has its maximum at the electrode surfaces and decreases as one moves away from them. This electric field distribution yields two different regimes of DEP: positive DEP (pDEP) and negative DEP (nDEP). The sign, or direction, of F_{DEP} is given by $Re[f_{CM}]$. From the particle's point of view, it will experience pDEP when its permittivity is higher than that of the media ($Re[f_{CM}]$ is positive), and it will undergo nDEP when the media permittivity is higher than its own. In other words, the most polarizable element always gets directed to the regions with highest electric field magnitude. The magnitude of F_{DEP} is given by the magnitude of the established gradient; the sharper

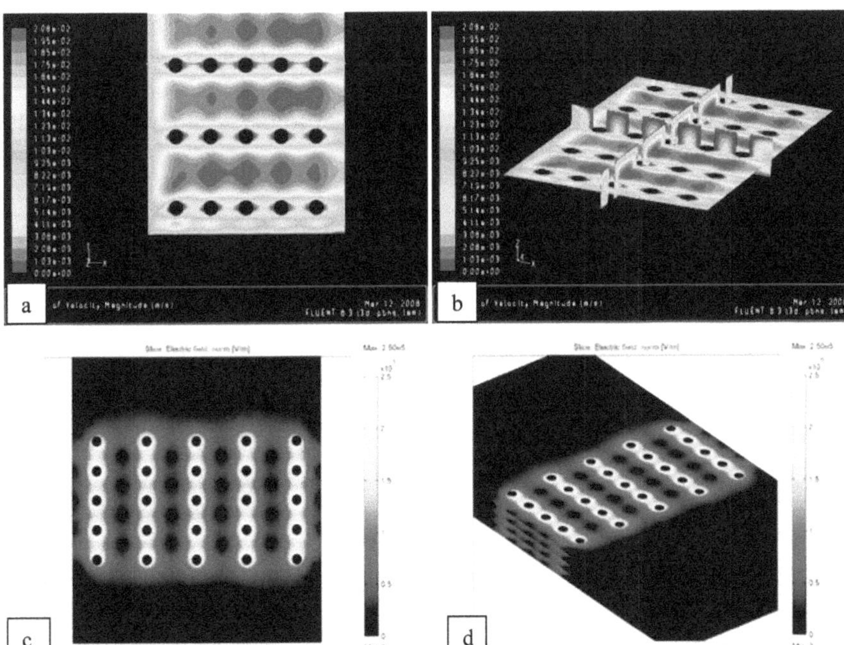

Fig. 5.20 Flow velocity and electric fields induced in a DI water-based conductive media ($\sigma = 10$ mS/m) by a polarized carbon electrode array. (**a**) Top and (**b**) isometric views of the flow velocity field. Flow is in the horizontal direction. Highest velocity is 2 mm/s and is given by red colors in between the posts. Minimum velocity is 0 mm/s, denoted by darker blue colors present on the electrode and channel surfaces due to the no-slip condition. A flow rate of 10 μl/min was simulated. (**c**) Top and (**d**) isometric views of the induced electric field when polarizing electrodes at 10 V_{pp}. Maximum value is 2.5×10^5 V/m, given by lightest colors, and is present on the electrode surfaces. The minimum value of 0 V/m is present in the regions farther away from the electrodes and is illustrated by the dark regions. Simulation results obtained in collaboration with Tecnológico de Monterrey, Campus Monterrey in Mexico and Universitat Rovira i Virgili in Catalonia, Spain

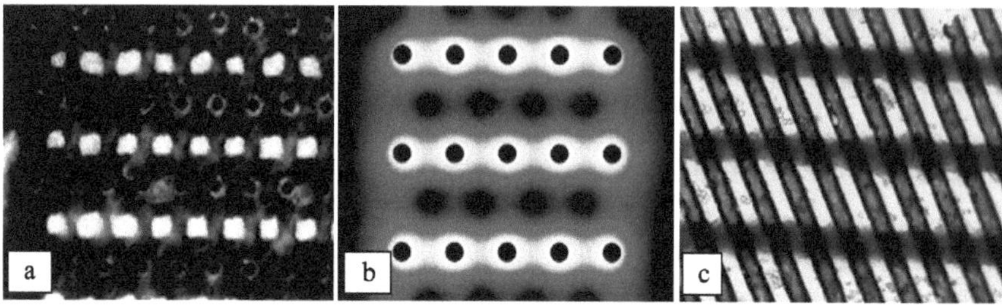

Fig. 5.21 Experimental validation (**a** and **c**) of simulation results (**b**). Latex particles get clustered in nDEP volumes as shown in (**a**) (correlate light areas denoting latex particles with darker areas in b delineating nDEP regions) and yeast cells get trapped in the vicinity of the electrode surfaces by pDEP (correlate darker areas on (**c**) with lighter (**a**) areas of (**b**). Maximum electric field magnitude is 2.5×10^5 V/m and is given in the lightest areas of b. Minimum magnitude is 0 and is given by the darkest areas of (**b**)

the gradient the stronger F_{DEP} is. Figure 5.21 shows the experimental validation of the simulation results: 8 μm latex particles, usually less polar than the media containing them, become clustered by a negative F_{DEP} in nDEP regions, while yeast cells, immersed in a less polar media, get attracted to the electrode surfaces, the pDEP region, by a positive F_{DEP}.

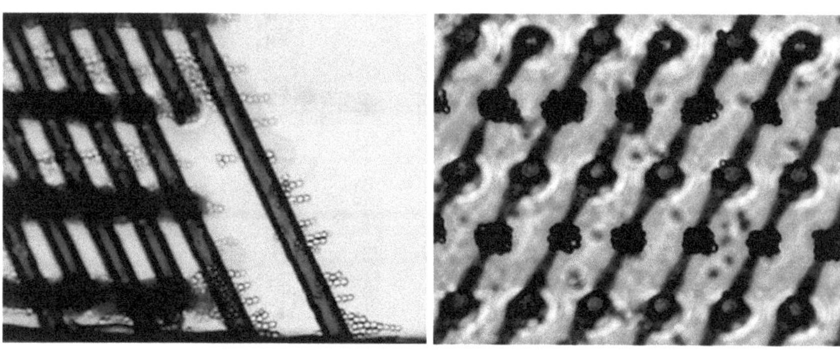

Fig. 5.22 Filtering of yeast cells with pDEP (*left*) and selective positioning of latex particles with nDEP (*right*)

Carbon-DEP has been demonstrated for a variety of functions including positioning [137], filtering [132] and separation of targeted particles [131, 132]. Selective filtering and positioning are shown in Fig. 5.22. The advantages of using 3D electrodes over 2D ones, specifically higher throughput and better efficiency, were demonstrated when using glass-like carbon electrodes to filter viable from non viable yeast cells [131].

A novel multi-stage filter, as depicted in Fig. 5.23, has also been implemented. With this approach it is possible to excite different electrode arrays embedded in a channel using different electric fields. Each one of these fields is optimized in their magnitude and frequency to trap specific particle populations at each array.

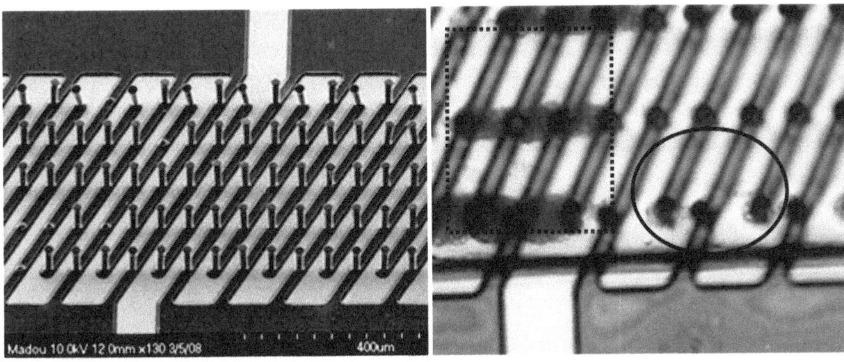

Fig. 5.23 A multi-stage filter featuring two stages. *Left*: fabricated CarbonDEP device. *Right*: Non viable yeast cells are trapped in one array (*dotted square*) while viable yeast get trapped in the other (*solid ellipse*)

By implementing a sequential release protocol one can then retrieve each of the different populations at the end of the channel at different times [132]. Another promising application is the continuous separation of particles by pDEP focusing, as demonstrated in Fig. 5.24. The principle works when the hydrodynamic force overcomes the pDEP trapping force. Since laminar flow is established in the channel, the particles attracted to pDEP regions flush away contained in those streamlines co-linear with the polarized electrode rows (and pDEP areas). Such principle allows continuous separation at higher flow rates than those achieved when implementing separation by trapping, such as in a filter, but requires more complicated

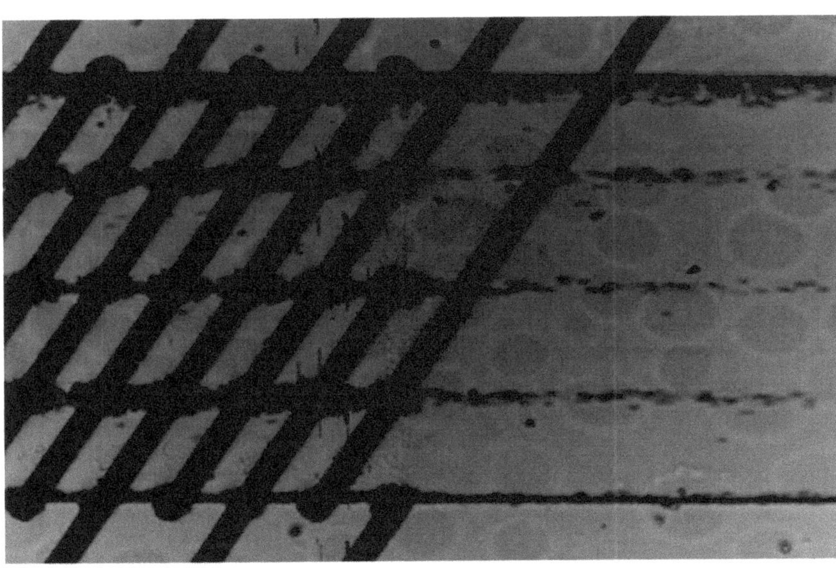

Fig. 5.24 Continuous separation of viable yeast cells using pDEP focusing

geometries for enriched population retrieval. pDEP focusing can be further combined with nDEP focusing to implement a continuous enrichment scheme of two different populations [134]. More recently, carbon-DEP has been integrated in a Compact Disk (CD)-like centrifugal platform (pictured in Fig. 5.25) towards implementing a portable and automated cell separation platform employing DEP [135].

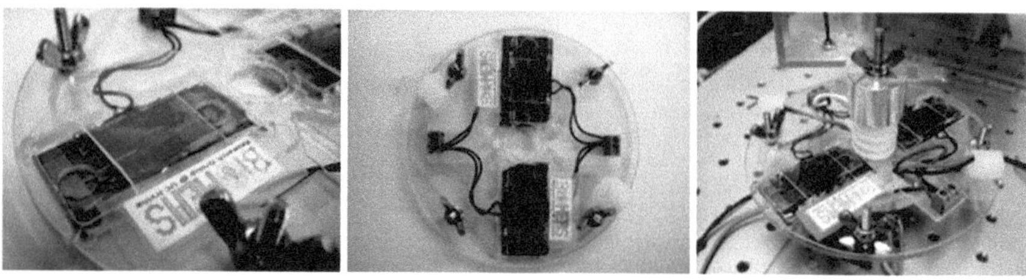

Fig. 5.25 Different views of a SpinDEP platform featuring carbon-DEP devices integrated in a CD-like centrifugal platform

5.6.2 Electrochemical Uses of Carbon in Microfluidic Applications

Previous sections have presented the use of glass-like carbon in dielectrophoresis to manipulate colloids and suspended particles. This subsection is dedicated to discuss the applications of glass-like carbon in electrochemical transducers, specifically sensors and biosensors. We first give a short overview of the specific benefits of using carbon as electrochemical sensors and how photoresist-derived glass-like carbon compares to commercial "glassy" carbon electrodes. Two different sensor geometries, interdigitated electrode arrays and fractal electrodes, are then presented along with their means of fabrication. The subsection ends with an example on the use of carbon electrodes.

Carbon presents a unique combination of properties that make it very suitable for electrochemical applications, namely, good electrical conductivity, acceptable corrosion resistance, availability in high purity, low cost, dimensional and mechanical stability, and ease of fabrication into composite structures [107]. Not surprisingly, the use of carbon in microfluidics is intimately linked to its behavior as electrode material. Regarding the use of carbon as a sensor, an outstanding and very useful property of carbon is that it can be functionalized with surface molecules which allows for the creation of sensors with higher selectivity and lower detection limits. It is important to mention that in order to functionalize its surface, the electrodes first need to be activated. One simple method to make its surface more reactive is by treating it with oxygen or water plasmas [248] and is even possible to selectively functionalize carbon electrodes using variations of the work presented in [249].

The expanding set of techniques for the fabrication of micro- and nanostructures in carbon opens up an entire new world of possibilities for the integration of carbon electrodes with microfluidic systems. Both individual and arrays of micro- and nano-electrodes, created for example with photolithographic methods, enable larger surface to volume ratios and translate into sensors with higher signal-to-noise ratios and enhanced sensitivities [45]. If certain geometries are used, for instance, interdigitated electrode arrays, this sensitivity can be further enhanced using redox cycling mechanisms, as it will be seen further below. The use of micro- and nano-electrodes results in three important consequences for the electrochemical behavior of these electrodes when compared to electrodes at the macro-scale: (1) the mass transport rates to (and from) the electrodes are increased, (2) the double layer capacitance is reduced due to the decrease in surface area, and (3) the ohmic losses (the product of the electrode current and the solution resistance) are reduced due to diminished current [250]. The choice for arrays or individual micro or nano-electrodes will depend on each targeted application, but both types can be successfully combined with microfluidic devices. In this sub-section, only applications related to the use of arrays of microelectrodes will be discussed. Details will be given on the derivation of carbon electrodes from photoresists, such as SU-8, and how its electrochemical behavior compares to the commercial "glassy" carbon.

As mentioned before, several studies have shown that the carbon obtained through pyrolysis of photoresists presents an electrochemical behavior similar to the glass-like carbon derived from other phenolic resins or polyfurfuryl alcohols. Using cyclic voltammetry it can be shown that oxygen reduction at photoresist-derived electrodes begins at about − 0.4 V vs. SCE (Saturated Calomel Electrode), a value similar to that of commercial "glassy" carbon [117, 251]. Moreover, the analysis and comparison of electron-transfer kinetics of various redox systems (such as $Ru(NH_3)_6{}^{3+/2+}$, chlorpromazine, $Fe(CN)_6{}^{3-/4-}$ and dopamine) show well-defined, symmetric voltammograms for all the redox systems under consideration, which indicates that electrodes fabricated through the pyrolysis of photoresist have the same attractive features for electroanalytical applications than their commercial "glassy" counterparts; including low capacitance and weak adsorption properties [65, 252].

We now proceed to detail some possible fabrication techniques and applications for two different sensing arrays: interdigitated and fractal electrodes.

5.6.2.1 Interdigitated Electrode Arrays

Interdigitated electrodes used in electrochemistry typically consist of two comb-like electrodes on a planar surface. A distinct electrochemical advantage of this configuration is that redox compounds can undergo "redox cycling" [253] by oxidizing a compound on one electrode (the generator) and reducing it at the other (the collector) (See Fig. 5.26). The advantages of using interdigitated electrode arrays (IEAs) for the detection of reversible redox species have been studied, both theoretically and experimentally, by several groups [254–256]. One of the pioneering research papers [257] derived the electrical and electrochemical behavior of such electrode arrays and showed that, by decreasing their size, sigmoidal current responses are obtained. These curves are indicative of enhanced mass transport due to nonlinear radial diffusion. Furthermore, reducing the spacing gap between the generators and

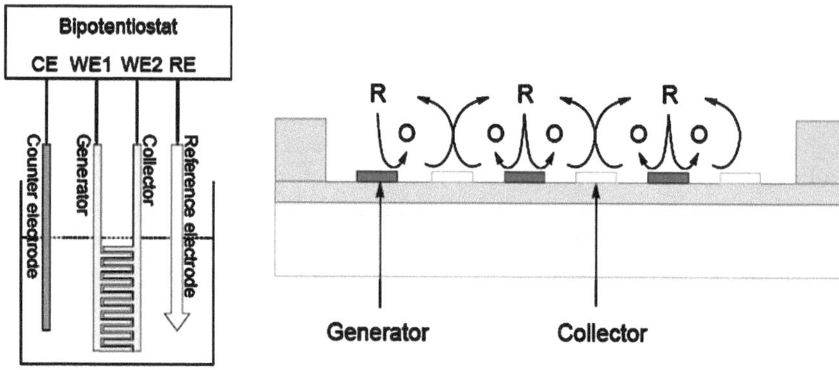

Fig. 5.26 *Left*: Schematic representation of the experimental setup of dual-mode electrochemical experiments using interdigitated electrode arrays. *Right*: Current enhancement mechanism for redox cycling on an interdigitated electrode (R: reduced molecule, O: Oxidized molecule). From [259] Re-printed with permission of BASi Inc

collectors leads to increased collection efficiency, enhanced redox cycling current, and decreased equilibrium time. For instance, by reducing the electrode gap from 1 μm to 50 nm, the collection efficiency increases from 89.40 to 99.95% [258].

Interdigitated electrode arrays are currently microfabricated using noble metals such as gold or platinum, but this results in unstable electrodes if a metallic adhesion layer is placed underneath the noble metal (for instance, Ti under Au), as this creates a galvanic couple that degrades the electrodes in just a few hours of being in contact with the electrolyte [172]. A further disadvantage is that the fabrication process is lengthy as it is based on the lift-off technique [172]. In contrast, the use of photoresist-derived carbon yields much more stable IEAs (as carbon do not degrades when in contact with an electrolyte) and dramatically cuts fabrication time (as it is simplified to a basic photolithography process and pyrolysis). This results in arrays that are simpler to fabricate, easier to miniaturize and less expensive than IEAs made with noble metals.

As mentioned above, interdigitated carbon electrodes can be derived by pyrolyzing polymer microstructures obtained with traditional lithographic methods, but in order to reduce their critical dimensions (gaps and widths), advanced lithographic techniques need to be used. The size reduction of the electrodes in an IEA to a nanoscale yields Interdigitated Nano-Electrode Arrays or INEAs. Two techniques, among others, that allow this size shrinking are electron-beam lithography (EBL) and nano-imprint lithography (NIL). These two methods do not rely on the wavelengths of light in the UV or deep-UV ranges, and therefore allow for much finer patterning of the photoresist precursor. That makes it possible to fabricate electrodes with gaps in the order of a few tens of nanometers. On the commercial side, it is important to note that NIL is expected to be of crucial importance to bring carbon-based INEAs into large-scale production, since it allows for parallel processing of the substrates (see section on NIL). On the other hand, EBL is extremely useful when the objective is to test and optimize different electrode designs, given its higher flexibility in the implementation of electrode geometries (see section on EBL).

Carbon-based INEAs offer high amplification factors that have enabled the monitoring of short-lived chemicals present at very low concentrations, (e.g. exocytosis of catecholamines from neurons) [260, 261]. They have also made possible the sub-femtomole detection of catecholamine in high-performance liquid chromatography (HPLC) [262]. A fabrication technique similar to the C-MEMS one discussed in this chapter was reported by Kostecki in 2000 [116].

5.6.2.2 Fractal Electrodes

Another strategy that can be pursued in order to increase the detection signal from carbon-based sensors is to increase their surface using fractal structures. Fractal-like geometries are ubiquitously found in nature, especially in situations that require minimizing the work lost, due to the transfer network, while maximizing the effective surface area. This is the case in vascular systems and energy-harvesting interfaces, to name a few. Fractals yield advantageous electrochemical characteristics as well [139], and have been proposed as the optimal geometry to be used in the design and fabrication of sensors and energy systems [263, 264].

In order to fabricate carbon fractal electrodes, the general process can be split in two main modules: (1) the backbone and (2) the fractal. The carbon backbone can be conveniently derived from photolitographically defined polymer micro-structures as detailed in the previous sections. The fractal geometries can be built up at the sub-micron and nano-scale using a range of methodologies and fabrication techniques and eventually get combined, if necessary, with the pre-existing carbon backbone architectures to obtain all sorts of fractal-like geometries. Some of the methods and techniques to fabricate fractals and to combine them with the backbone are briefly presented here:

Self-Assembling. Self-assembly is one solution to the problem of synthesizing structures larger than molecules. The stability of covalent bonds enables the synthesis of almost arbitrary configurations of up to 1000 atoms. Larger molecules, molecular aggregates, and forms of organized matter more extensive than molecules cannot be synthesized bond-by-bond. Self-assembly is one strategy for organizing matter on these larger scales, and it offers a route to three-dimensional microsystems [265]. This approach can be used to add pre-fabricated structures, synthesized either from organic precursors or carbonaceous materials, to the Carbon-MEMS (C-MEMS) backbone structures. It is important to note that these C-MEMS structures can already have several levels of “fractality” before being decorated with the self-assembled structures.

Resorcinol-formaldehyde (RF) based carbon xerogel microspheres and high surface area fractal-like structures have been successfully prepared by inverse emulsification of RF sol in cyclohexane in the presence of a non ionic surfactant, followed by pyrolysis at 900°C under inert nitrogen atmosphere. This processing technique leads to either carbon particles or carbon "flowers", depending on the exact processing conditions. Increasing the surfactant concentration leads to smaller and relatively mono-dispersed spherical particles, whereas further increase in surfactant concentration leads to the formation of oval shaped structure and ultimately to high surface area fractal structures (see Fig. 5.27).

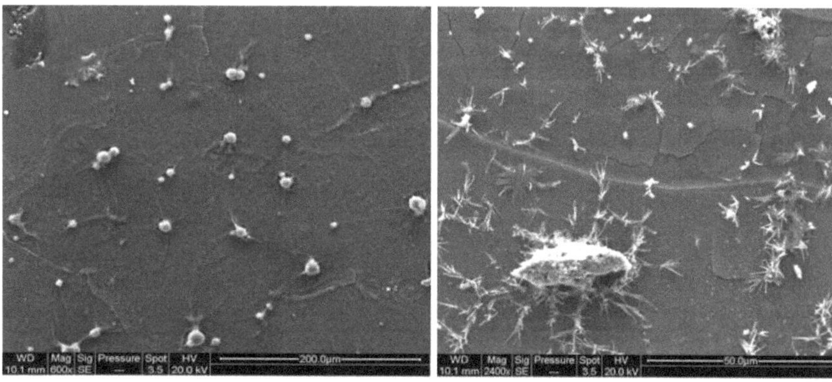

Fig. 5.27 Different geometries based on RF gel processing. Images courtesy of C.S. Sharma from the Indian Institute of Technology, Kanpur

These structures can be easily integrated with C-MEMS structures using different methods. One possible method of integration is described as follows: the RF gel is uniformly distributed over the three dimensional microstructure, for example, by spin coating. After drying, the gel solidifies and covers the three dimensional microstructure. Following pyrolysis, the RF gel is carbonized resulting in a conformal coating of carbon "particles" or "flowers" over the underlying Carbon-MEMS microstructures.

Another possible approach is to use commercially available carbon particles, such as Meso-Carbon-Micro-Beads (MCMB, Osaka Gas Inc., Japan). In this case, carbon particles are uniformly dispersed in a solvent, for example, N-Methyl-2-Pyrrolidone (NMP). The colloidal solution is then uniformly distributed over the three dimensional microstructure, for example, using spin coating. The surface tension of NPM drags the particles to the polymer microstructures during evaporation. Once the solvent evaporates, the carbon particles remain adhered to the microstructures by Van der Waals forces and capillary forces (see Fig. 5.28).

Electrodeposition and Templating. The appearance of fractal interfaces is a common problem in electroplating. Fractal structures can be purposely created when electrodepositing a polymer by controlling the composition of the deposition solution and by controlling the voltage and current [266]. These polymer structures can then be carbonized to yield a carbon fractal. In addition, cracks and other failures in materials often have a fractal nature. These cracks and crevices can be used as templates for electrodeposition.

Chemical Vapor Deposition (CVD). In this method, a standard photolithography process is followed using SU-8 photoresist premixed with carbon nanofibers

Fig. 5.28 MCMB particles adhered to microstructures after solvent drying

(CNFs). The CNFs are mixed with the SU-8 resin (1% in weight) prior to the process. This doped photoresist is then spin-coated onto a silicon wafer to an approximate thickness of 100 μm. A standard photolithographic process is then conducted. The resulting structures exhibit some clumps of CNFs that are partially embedded inside SU-8 posts. The carbon nanofibers can withstand the pyrolysis process and stay attached to the carbon post [267]. The resulting nanostructures are electrically connected to the carbon microstructure.

It is also possible to use CVD techniques to deposit Multi-Walled Carbon Nano-Tubes (MWCNTs) on the Carbon-MEMS structures [120]. MWCNTs are usually grown on planar substrates; therefore it is quite challenging to grow carbon nanotubes on 3D surfaces because of the difficulty to coat the required catalyst from where the CNT grow uniformly all over a 3D surface. To achieve this, thin Fe films were coated onto the carbon structures through pulsed laser deposition (PLD) on the rotating 3D C-MEMS substrates. The coated C-MEMS substrates were then used for growing MWCNTs in a thermal CVD system.

5.6.2.3 Glucose Sensors

An example of the use of carbon structures as sensors is that of enzyme-based glucose sensors. In enzymatic glucose sensors, glucose oxidase (GOx) is typically used as the biological enzyme to form the electrochemical transducer. In the presence of oxygen, GOx catalyzes the electro-oxidation of glucose, giving as result gluconic acid and oxygen peroxide. Then, the peroxide present at the surface of the carbon electrode is further oxidized to one molecule of oxygen and two protons, with the release of two electrons. The electrochemical current produced in this overall reaction is proportional to the concentration of glucose in the solution, and it can be used as a sensing principle. The complete reaction is represented below:

$$\text{Glucose} + O_2 \rightarrow \text{gluconicacid} + H_2O_2$$
$$H_2O_2 \rightarrow O_2 + 2H^+ + 2e^-$$

The most common materials for the immobilization of GOx include noble metals, such as platinum or gold [268, 269]; and carbon, both graphitic and glass-like carbon [270]. By using novel microfabrication approaches, it is possible to increase the specific surface area of the electrodes to immobilize more glucose oxidase per footprint unit and achieve a sensor with a larger signal per unit of surface [128].

The immobilization of the enzymes on the surface of the carbon electrodes can be done with two main techniques: enzyme [271, 272] and co-deposition [273, 274]. Redox enzymes are incapable of direct contact with the electrodes since their redox centers are insulated from the conductive support by protein matrices [275]. In order to contact these enzymes with the electrode, mediators, which are dependent on the specific enzyme being used, are utilized. Several types of schemes have been developed to enable this electric contact. Examples of such efforts include the development of high-surface-area electrodes by graft polymerization of a redox polymer entrapping the enzyme [276], the utilization of carbon black [277], and the fabrication of cylindrical and porous carbon tubes electrodes with higher surface area [278]. In addition, the utilization of a higher number of smaller electrodes to reduce the thickness of the diffusion layer has also been reported [279]. The use of glass-like carbon micro electrodes, fabricated with the C-MEMS technique, as glucose sensors is demonstrated in [128].

5.6.3 Energy

The world demand for energy is projected to more than double by 2050 and to more than triple by the end of the century [280]. In the portable energy arena, recent years have witnessed an explosion of demand for power-hungry portable electronics like laptop computers and mobile phones. As a direct consequence, the need for portable energy sources is expected to increase dramatically as the market continuously commands the integration of a multitude of functionalities and better data transfer capabilities into portable devices. Even when lithium-based technology is currently the alternative of choice, incremental improvements on existing technologies will not be adequate to supply this demand in a sustainable way. Lithium-ion battery technology requires significant breakthroughs in enhancing battery life and reliability to meet the increasing power demands of portable devices [281].

In the present scenario of a global quest toward a clean and sustainable energy future, fuel cells are perceived to play a key role owing to their high energy efficiency, theoretical energy density (higher than batteries), environmental friendliness and minimal noise. Fuel cells link hydrogen and electricity, two highly compatible energy carriers that embody the ideals of a sustainable energy economy: they are clean, abundant, and adapt flexibly to many sources of fuel production and to many end uses.

However, conventional fuel cells consist of structural components, including gaskets, fasteners and end plates that make the assembly complicated. These structural components act as impediments toward reducing the size of the fuel cells. Therefore, novel design and manufacturing approaches are required to realize component miniaturization and assembly simplification of fuel cells [282]. Recent studies [283–292] have suggested that, in order to comply with the electromechanical integration of the fuel cell structure with high precision, repeatability and productivity, micro electromechanical systems (MEMS) technology is the most attractive fabrication process. MEMS offer a better alternative for small fuel cells for mobile applications, compared to other conventional techniques like machining, molding and fastening. MEMS-based techniques are expected to enable simple and mass-producible fuel cells with uniform specifications, in the same way integrated circuits did it for the electronics industry some years ago [293]. MEMS technology is able to provide the following improvements in the fuel cell [294]: (1) significant reduction of precious (typically Pt) catalyst loading and higher power output due to the controlled microstructure of the three phase boundary for the electrochemical reaction; (2) lower contact resistance at the layer interface and controlled gas permeable structure due to electromechanically integrated fabrication; and (3) flexible connection design of multiple cells. In addition, by using bonding technologies employed in MEMS devices, such as anodic and eutectic bonding, the fuel cell components can be monolithically integrated. Localized bonding at a low temperature, along with a solid polymer electrolyte, could prove beneficial in minimizing the required bonding surface and maximizing the electrochemically active area available in the fuel cell [294]. As a result, components such as gaskets, fasteners and end plates could be eliminated. Furthermore, by integrating auxiliary devices on chip, including micro-pumps, valves, connectors and controllers, small fuel cell systems, or a power-plant-on-a-chip, with high power density could be realized. In addition to the required compactness, portable fuel cells must also feature low manufacturing costs to facilitate their commercial implementation [282, 295].

This section explores the use of microfabricated fuel cells as energy sources for mobile applications. Having introduced the motivation for portable, compact fuel cells, we now proceed to list some of the most important advantages of microfabricated fuel cells, the different types and means of integration. The section closes with an overview of micro PEM fuel cells and enzyme-based biofuel cells featuring carbon-based components.

In general, microfabricated fuel cells offer unique advantages [294, 296]:

(1) As a device is miniaturized, the surface-to-volume ratio of the entire device increases (typically, area scales with l^2 while volume scales with l^3). Thus, the surface-to-volume ratio, in general, scales with proportion to $l^2/l^3 = 1/l$. This has a direct implication in improved heat and mass transport phenomena present in a typical low-temperature polymer-electrolyte membrane-based fuel cell for portable power applications.
(2) Increased power density due to higher surface-to-volume ratio and improved energy density due to the use of liquid fuel and reduced balance of plant.

(3) Better control of fuel cell component fabrication and integration leading to increased catalyst utilization in the electrodes and reduced system complexity.

Among the different types of miniature fuel cells, direct liquid-feed methanol fuel cells (DMFCs) have received widespread acceptance as the potential power source of choice for portable devices, given its enhanced power and energy density. Comprehensive overviews of the progress and challenges in DMFCs are provided in recent reviews [297–304]. Other miniature fuel cell architectures include microfluidic fuel cells [305–308], air-breathing fuel cells [309–311] and biological fuel cells [312–316]. Inherent to these miniature fuel cell architectures is the development, adoption and deployment of suitable microfabrication techniques for component synthesis and compact system integration. The fabrication process also holds the key toward successful commercialization of the fuel cell technology for portable power devices. Effective miniaturization to build portable fuel cells is a challenging task.

There are currently two different approaches for the fabrication and integration of miniaturized fuel cells [293]. A first approach is to follow the design principles of larger fuel cells and stack several cells on top of each other. Vertical stacking is then used to connect the cells in series. In contrast, a second approach avoids the bipolar approach of conventional fuel cells and follows a monolithic, or monopolar, fuel cell design that leverages the surface machining aspect of microfabrication processes. Approaches such as the "flip-flop" interconnect method [289, 317] are used instead of vertical stacking to connect the cells in series. The current trend is toward monopolar planar designs [296]. For further details, the reader is referred to Meyers et al. [293] who compared bipolar and monolithic MEMS fuel cell designs based on silicon technology.

5.6.3.1 Polymer Electrolyte Membrane (PEM) Fuel Cells

The polymer electrolyte membrane (PEM) fuel cell is arguably the front runner among the different types of fuel cells and is poised to cater clean electrical energy to automotive, portable and stationary power devices. A typical PEM fuel cell, shown in Fig. 5.29, exhibits a layered architecture with a proton conducting polymer membrane, typically Nafion®, separating the anode and cathode compartments. The anode and cathode sides each comprises of gas channel, gas diffusion layer (GDL) and catalyst layer (CL). Hydrogen is the fuel in the H_2/air PEM fuel cells for automotive applications, while aqueous methanol is the fuel for direct methanol fuel cells (DMFC) for portable applications. Usually, two thin catalyst layers are coated on both sides of the membrane, forming a membrane-electrode assembly (MEA). Protons, electrons and oxygen combine electrochemically within the active catalyst layer to produce electricity, water and waste heat.

The conventional PEM fuel cell (H_2/air and DMFC) features components fabricated from primarily carbon-based materials for the underlying physico-electrochemical processes to occur. The state-of-the-art catalyst layer, with thickness $\sim$ 10–15 μm, in a PEM fuel cell is a three-phase composite, shown in a high

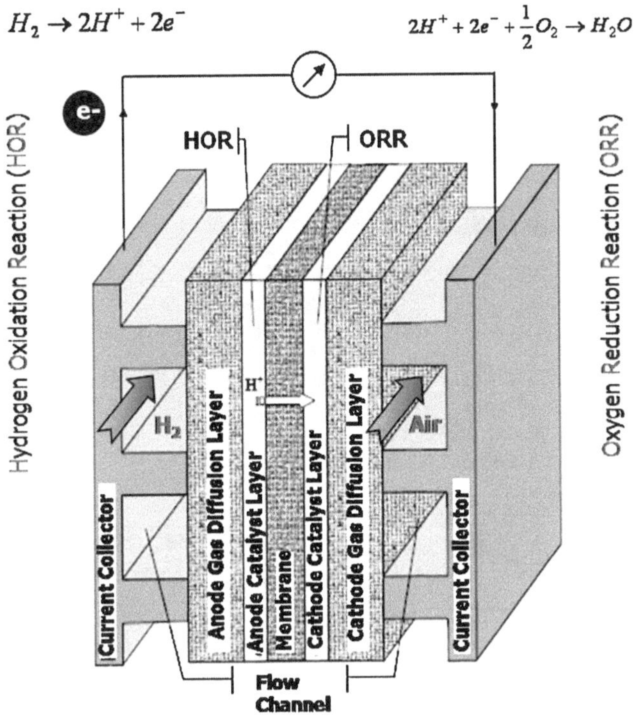

Fig. 5.29 Schematic diagram of a polymer electrolyte fuel cell

resolution transmission electron microscope (TEM) micrograph of an actual catalyst later in Fig. 5.30 [318], and consists of: (1) ionomer, i.e., the ionic phase which is typically Nafion® to provide a passage for protons to be transported in or out, (2) metal (Pt) catalysts supported on carbon, i.e., the electronic phase for electron conduction, and (3) pores for the oxygen gas to be transferred in and product water out. Gottesfeld and Zawodzinski [319], and more recently Eikerling and co-workers [320, 321], have provided good overviews of the catalyst layer structure and functions. Specific properties of the catalyst layer include high catalyst mass activity, high catalyst utilization at all current densities, low mass transport losses, high tolerance to multiple surface area loss mechanisms, tolerance of a wide humidity and temperature operating window, cold start and freeze tolerance, and the ability to be fabricated by robust high volume-compatible, low cost processes. Conventional carbon-supported finely-dispersed electrocatalysts rely on high surface area carbons (carbon blacks or graphitized carbon) for electrical conductivity and on 2–3 nm-sized catalyst particles on those carbon surfaces for high levels of catalyst activity. In an effort to reduce Pt catalyst loading and improve catalyst utilization, recent advances in catalyst layer development include the 3 M® nanostructured thin film (NSTF) catalyst which obviates the use of carbon and ionomer [322, 323]. Catalyst-coated organic nano-whiskers, as shown in Fig. 5.31 [323], form the thin catalyst

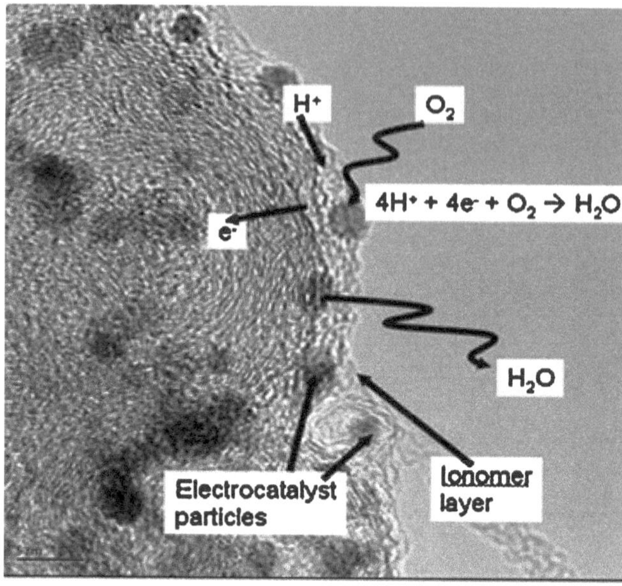

Fig. 5.30 High resolution TEM image of an actual PEM fuel cell catalyst layer illustrating the three-phase interface. Reprinted from [318] with permission from Karren L. More, Oak Ridge National Laboratory, TN, USA

layer (~ 1 μm) and the electrochemically active area is instead a two-phase interface for the electrochemical reaction to occur. The multi-faceted functionality of the gas diffusion layer includes reactant distribution, liquid water transport, electron transport, heat conduction and mechanical support to the MEA. Carbon-fiber-based porous materials, namely non-woven carbon paper and woven carbon cloth, have received wide acceptance as materials of choice for the PEM fuel cell GDL owing to their high porosity (~ 70% or higher) and good electrical/thermal conductivity. The thickness of typical GDL structures range between 200 and 300 μm. SEM micrographs along with 3D representative microstructures of typical non-woven carbon paper and woven carbon cloth GDL are shown in Fig. 5.32 [324]. Mathias et al. provided a comprehensive overview of GDL structure and functions in [325]. In order to facilitate removal of liquid water from GDL and avoid flooding, the GDL is treated with PTFE (polytetrafluoroethylene) with loading varying from 5–30 weight % in order to induce and/or enhance hydrophobicity [325]. The graphite bipolar plate houses the flow channels and consists of solid and porous plate architectures [326]. The bipolar plate properties include high electronic and thermal conductivities and sufficient mechanical/chemical stability. These carbon based components contribute to the effective water and thermal management, described as maintaining a delicate balance between oxygen and liquid water transport as well as heat dissipation in the fuel cell [327, 328].

The fabrication of micro- and nano-sized carbon components holds great potential in future PEM fuel cell components (e.g. gas diffusion layer, catalyst layer) with

Fig. 5.31 Scanning electron micrographs of typical NSTF catalysts as fabricated on a microstructured catalyst transfer substrate, seen (*top*) in cross-section with original magnification of × 10,000, and (*bottom*) in plan view with original magnification of × 50,000. The dotted scalebar is shown in each micrograph. Reprinted from [323] with permission from Elsevier

improved water and thermal management characteristics. It is widely recognized that the performance degradation and the limiting current behavior in the PEM fuel cell are mainly attributed to the excessive build up of liquid water in the cathode side and the resulting flooding phenomena [327, 329, 330]. Liquid water blocks the open pore space in the CL and the GDL leading to hindered oxygen transport and covers the electrochemically active sites in the CL thereby rendering reduced

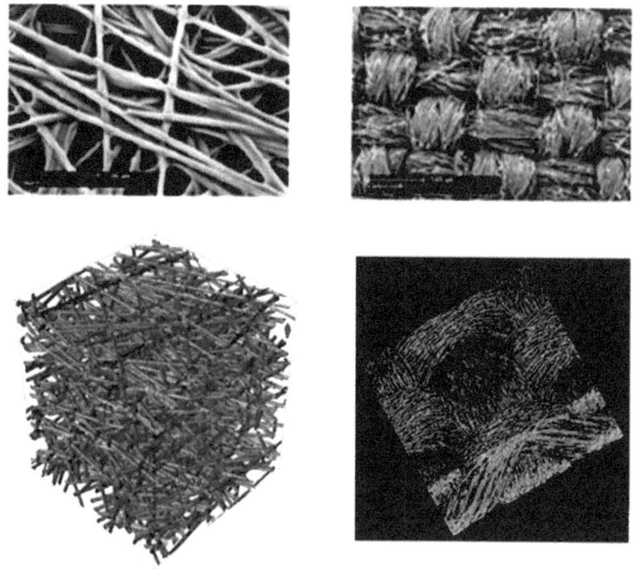

Fig. 5.32 SEM micrographs and representative 3D microstructures of carbon paper and carbon cloth GDL

catalytic activity. The catalyst layer and gas diffusion layer, therefore, play a crucial role in the water management [327] aimed at maintaining a delicate balance between reactant transport from the gas channels and water removal from the electrochemically active sites. In this regard, the capability of the CL as the primary component in the entire fuel cell assembly in generating heat and the impact of the thermal effects on liquid water accumulation and cell performance are of profound importance. Mukherjee and co-workers [324, 330, 331] have recently developed a theoretical framework to study the impact of the evaporative capability due to heat generation in the CL on the water and thermal management and cell performance. Based on a physical description of heat and water balance, they have defined "heat partition factor (β)" which corresponds to the fraction of the total heat generation rate actually available for the evaporation of liquid water in the CL. Considering the heat balance in the CL and GDL as well as the vapor diffusion through the GDL, the heat partition factor can be uniquely defined in terms of the GDL thermal conductivity and the saturation vapor concentration depending on the fuel cell operating temperature. The GDL thermal conductivity and the fuel cell operating temperature through its strong influence on the saturated vapor pressure exhibit profound influence on the heat partition factor which further dictates the net liquid water evaporation and hence the liquid water saturation distribution inside the CL. The variation of the heat partition factor with temperature is shown in Fig. 5.33 for

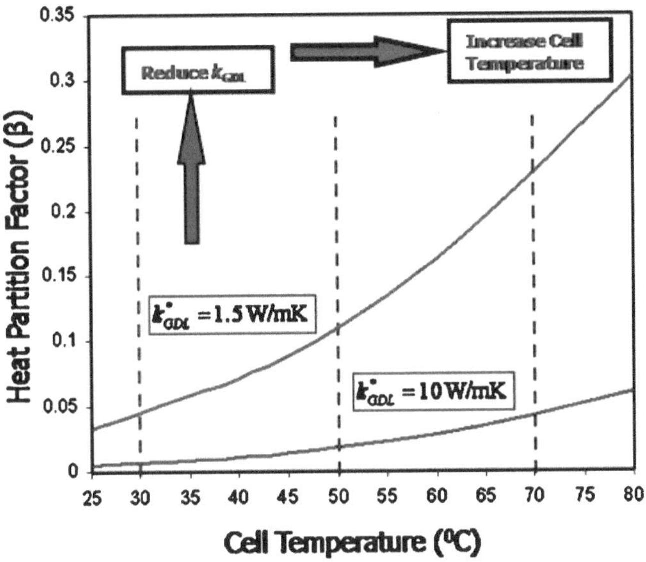

Fig. 5.33 Variation of the heat partition factor with fuel cell operating temperature for two different GDL thermal conductivity values

two different GDL thermal conductivity values. From the figure it can be observed that higher cell operating temperatures and low GDL thermal conductivities promote evaporation in the CL and hence will lower the liquid water saturation level leading to improved performance. The impact of GDL thermal conductivity and operating temperature on the average liquid water saturation level in the CL and electrode performance in terms of the polarization characteristics are shown in Fig. 5.34. The effective thermal conductivity value of 1.5 W/mK is representative of the typical carbon paper and carbon cloth GDL materials currently used in PEM fuel cells. It is evident from this analysis that a lower value of GDL thermal conductivity (e.g. between 0.5 and 1.0 W/mK), while maintaining a high electronic conductivity to avoid voltage loss due to electron transport resistance, would dramatically benefit fuel cell performance. This emphasizes the need for the development of GDL with novel microstructure and morphology, which will essentially act as a thermal insulator in the limiting sense and an electronic conductor.

The abundant use of carbon-based materials inherent to the conventional fuel cell design forms the base for exploiting current techniques for the micro and nanofabrication of carbon, like Carbon MEMS/NEMS (Carbon Micro-/Nano-ElectroMechanical Systems), in the development of component and system architecture for miniature PEM fuel cell systems.

Madou and co-workers [296, 329] have recently pioneered the C-MEMS technique for fabricating bipolar plates and demonstrated their integration in a miniature PEM fuel cell. The bipolar plate fabrication method and its integration into a single cell architecture is detailed in [296] and summarized below:

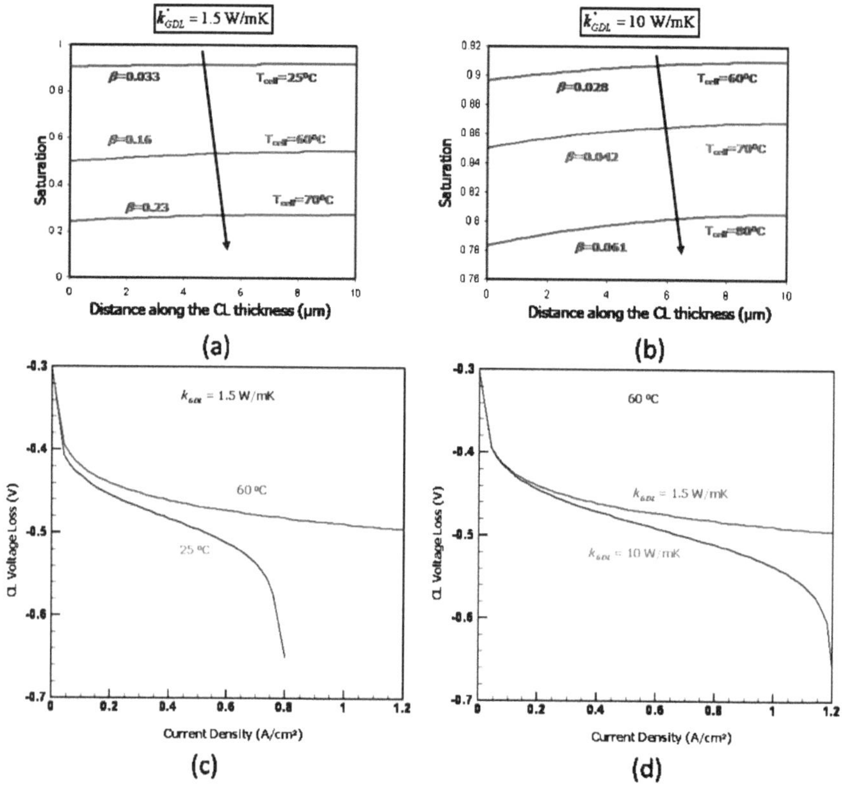

Fig. 5.34 (**a**) Liquid water saturation distributions along the CL thickness for different cell operating temperatures and with GDL thermal conductivity of 1.5 W/mK; (**b**) Liquid water saturation distributions along the CL thickness for different cell operating temperatures and with GDL thermal conductivity of 10 W/mK; (**c**) Effect of the cell operating temperature on the CL voltage loss prediction; (**d**) Effect of the GDL thermal conductivity on the CL voltage loss prediction

(1) Fluidic channel walls and separators are machined from thick polymer sheets and bonded together to create fluidic plates.
(2) The structures are converted into carbon by pyrolysis. A physical binder acts also as an electrical binder.
(3) Commercial fuel cell electrodes are combined with an activated Nafion® membrane to create a membrane electrode assembly (MEA).
(4) The MEA, fluidic plates, and gas inlet/outlets are brought and sealed together by epoxy.

Figure 5.35 shows the fluidic plate structures before carbonization and a three-layer carbon bipolar plate structure made by first bonding Cirlex® fluidic channel walls to a 5 mm Kapton® sheet and then converting the entire structure into carbon. The design and integration of a miniature PEM fuel cell stack using the C-MEMS

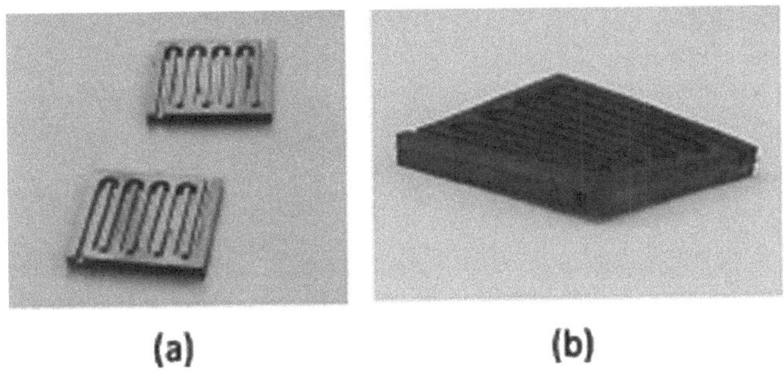

Fig. 5.35 (**a**) Fluidic plate structures before carbonization; (**b**) A three-layer carbon bipolar plate structure made by first bonding Cirlex® fluidic channel walls to a 5 mil Kapton® sheet then converting the entire structure into carbon. Reprinted from [296] with permission from Elsevier.

fabrication technique has also been successfully demonstrated by Madou and co-workers [329]. Figure 5.36 shows a three-cell PEM fuel cell stack configuration. The performance in terms of polarization (I-V curve) and power density characteristics of this microfabricated PEM fuel cell stack are shown in Fig. 5.37. The C-MEMS technique has also been successfully deployed for the development of a direct photosynthetic/metabolic micro bi-fuel cell with improved power capability by Moriuchi and co-workers [315] and is currently being used in the development of an enzyme-based bio fuel cell (detailed below). These recent developments in micro fuel cell component fabrication and system integration truly highlight the enormous

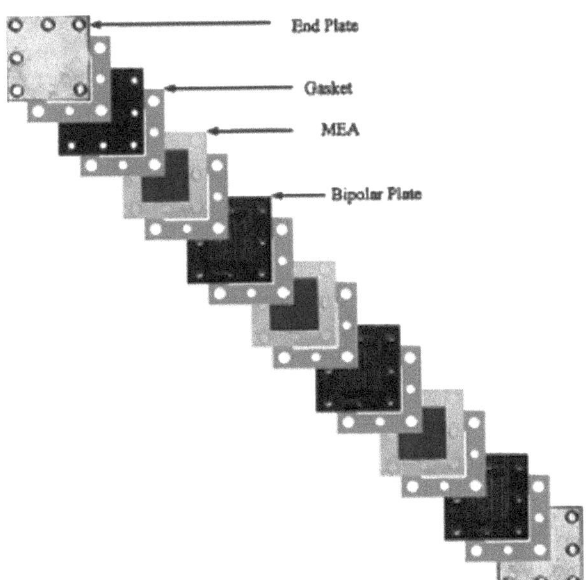

Fig. 5.36 Configuration of a three-cell fuel cell stack. Reprinted from [329] with permission from Elsevier.

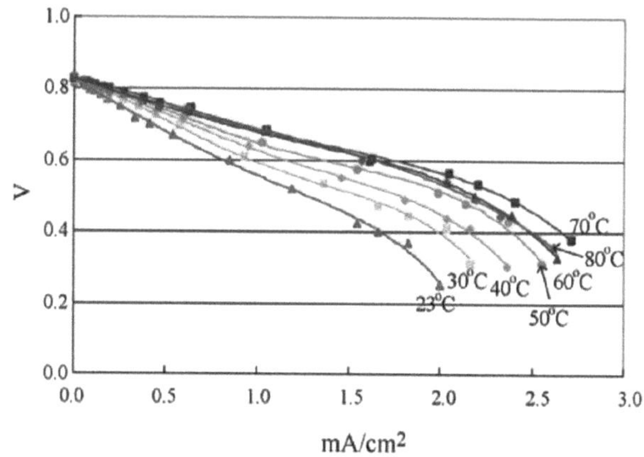

Fig. 5.37 Effect of operating temperature on the fuel cell polarization curve. Reprinted from [329] with permission from Elsevier

promise of the C-MEMS/NEMS techniques in the development of future portable power devices for clean electricity generation.

Concurrent to the success of C-MEMS/NEMS-based devices, it is worth discussing the promise and recent developments of carbon nanotube (CNT) based miniature PEM fuel cell component fabrication and MEMS integration toward improved power output. Kuriyama and co-workers have recently demonstrated the use of multi-walled carbon nanotubes (MWCNT) as a promising carbon material in micro/nano fuel cells. MWCNTs exhibit good electrical conductivity, gas permeability, catalyst support properties and most importantly, suitability for their integration into existent MEMS processes. The MWCNTs were employed for both the gas diffusion and catalyst layers, which were integrated as a single layer with intimate electromechanical contact. The key concept is the integrated anode/cathode configuration, which electromechanically integrates all the components including the flow channel, current collector, gas diffusion layer and catalyst layer. These are deposited and grown on the silicon wafer substrate that serves as the flow channel structure. In a conventional fuel cell, electromechanical contact between all components is controlled by applying external mechanical pressure. However, uniform pressure control in miniature fuel cells requires more complicated assembly and results in higher costs. In contrast, in this configuration all components are deposited and grown layer-by-layer with good electromechanical contact, which eliminates the need for pressurizing the components such as gaskets, fasteners and end plates. It also decreases cell resistance. The three-component micro fuel cell configuration with the polymeric membrane sandwiched between the integrated anode/cathode assembly is shown schematically in Fig. 5.38. The unique aspect of the fabrication process is that the silicon dioxide layer, which is serving as the foundation of MWCNTs growth, is also employed as a sacrificial layer to make the MWCNTs suspend by themselves over the circular holes of a MoSi current collector. Representative I-V and power density characteristics, as a measure for the

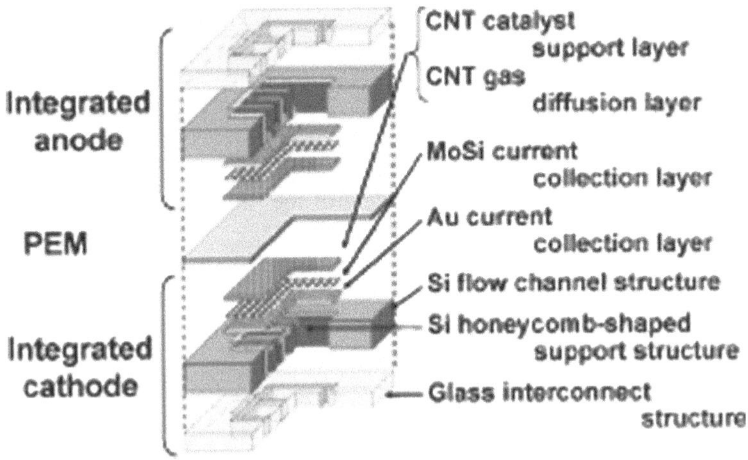

Fig. 5.38 Schematic design of MEMS-based three-component fuel cell. PEM is sandwiched by the integrated anode and cathode. The three components are assembled using a bonding technique without any fasteners. Reprinted from [294] with permission from Elsevier

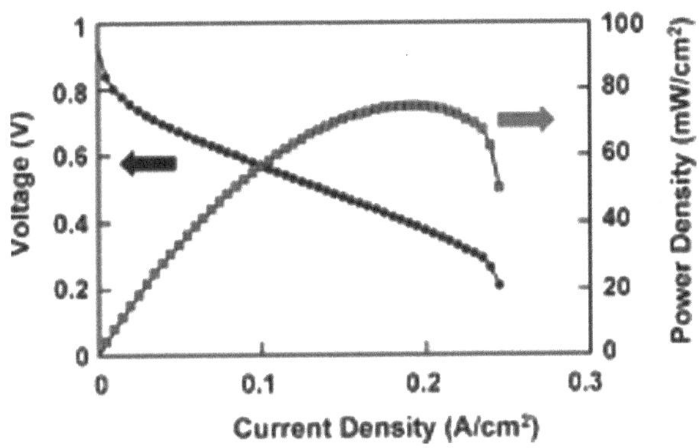

Fig. 5.39 *I–V* and power density characteristic curves of the prototype (active cell area: 1 cm^2; electrolyte: GORE-SELECT®, 30 μm thick; H2: 8.0 sccm; air: 8.0 sccm; no humidification and back-pressure) [294]. Reprinted from [294] with permission from Elsevier

fuel cell prototype performance, are shown in Fig. 5.39. The details of the fabrication process, fuel cell prototype integration, MWCNT characterization and cell performance are reported in [294].

Other recent developments in CNT based 3D architectures show promise in modifying the thermal and electronic properties. Prasher et al. [332] has recently demonstrated that a 3D fibrous bed with CNT arrays behaves as a thermal insulator with effective thermal conductivity as low as 0.2 W/mK while maintaining significantly high electronic conductivity. This further underscores the tremendous potential of C-MEMS/NEMS techniques [294, 296, 327] along with layer-by-layer 3D assembly of CNT based structures [333] in the fabrication of components and system integration toward dramatic improvement of micro fuel cells performance for powering emergent engineering devices.

5.6.3.2 Enzyme-Based Bio Fuel Cells

Basic enzymatic biofuel cells contain many of the same components as a hydrogen/oxygen fuel cell. However, rather than employing metallic electrocatalysts at the anode and the cathode, the electrocatalyst used are oxidoreductase enzymes. This is a class of enzymes that can catalyze oxidation–reduction reactions. Since these enzymes are selective electrocatalysts, the separator could be an electrolyte solution, gel, or polymer, or not be present at all.

The miniaturization of bio-fuel cells is mostly driven by its use in portable electronics and biomedical devices. These applications require continued energy supply and small dimensions, and both can be achieved using enzymatic bio-fuel cells. Even when the substrates used for the immobilization of enzymes are generally metals (mostly gold and platinum), carbon offers an excellent alternative given its favorable electrochemical properties. Furthermore, Carbon-MEMS technology represents a good option for the miniaturization of enzymatic bio-fuel cells because it is a relatively easy way to miniaturize carbon electrodes.

In order to miniaturize fuel cell components (mainly, the anode and the cathode), the main challenge is to reduce the size of the electrodes while keeping large current densities. Single carbon fiber fuel cells have been already demonstrated by Heller et al. [334]. This approach might be enough for applications that need very small amounts of power (sensors, transmitters, etc.), but in applications that need larger power and energy densities (body implants, power sources for electronic devices, etc.), the amount of enzyme immobilized in one single carbon fiber is not enough to power the entire device.

The Carbon-MEMS process brings several advantages when used as substrate material for the immobilization of enzymes in bio-fuel cell applications because:

1. It is possible to fabricate carbon electrodes with high surface area, which offers more immobilization sites per footprint. This increases the specific current substantially, due to the high local enzyme concentration.
2. Its surface presents a micro porous structures, with pore dimension small enough to trap the enzyme (size of the enzymes is about 80 Å) and big enough to let the electrolyte pass through. This also may lead to longer lifetime of the bio-fuel cell [335].
3. It is possible to fabricate bio-fuel cells with both cathode and anode on the same surface.
4. The integration of carbon nanotubes and carbon nanofibers with Carbon-MEMS could allow achieving direct electrical connection between enzyme and carbon electrode.
5. Fractal structures can be fabricated and integrated with Carbon-MEMS.

As pointed out above, one of the main advantages that Carbon-MEMS brings to the miniaturization of bio-fuel cells is the possibility for enhancing the total surface area of the electrodes by combining carbon nano-tubes and carbon nano-fibers with pre-existing three-dimensional microstructures. As it has been demonstrated by Calabrese-Barton et al. the addition of carbon nano-fibers to carbon cloth improves the total current density of the bio-fuel cell by one order of magnitude [336]. This combined top-down (lithography) and bottom-up (chemical vapor deposition) approach is bringing the possibility of creating enzyme-based bio-fuel cells into reality (see the section on fractal electrodes in the discussion about glucose sensors)

Even though the use of Carbon-MEMS represents a big advantage over current technologies, there are other big challenges to be overcome in order to make this approach a success. The most relevant one is to be able to covalently attach the enzyme to the Carbon-MEMS surface, respecting both angle and distance of the redox center from the carbon electrode. By achieving this, the obtained current will be higher than in the case of a multilayer of enzymes connected through a conducting wire (molecular wire) to the electrode surface. Although such achievement might not make much of a difference in a sensor configuration, it would represent a significant 50% improvement for an energy conversion device.

It is important to note that an extensive future research effort is warranted to develop suitable C-MEMS/NEMS based techniques in order to infuse transformational breakthroughs imperative to the success of the emerging electrochemical energy conversion systems which will ultimately pave the way toward securing a sustainable energy future.

References

1. Ferchault de Reaumur R (1722) L'art de convertir le fer forge en acier, et l'art d'adoucir le fer fondu, ou de faire des ouvrages de fer fondu aussi finis que le fer forge. Available from http://www.books.google.com.
2. Lavoisier A (1789) Traite Elementaire de Chimie. Available from http://gallica.bnf.fr/.
3. Shevlin PB (1972) Formation of atomic carbon in the decomposition of 5-tetrazolyldiazonium chloride. J. Am. Chem. Soc. 94:1379–1380.
4. Dewar MJS, Nelson DJ, Shevlin PB, and Biesiada KA (1981) Experimental and theoretical investigation of the mechanism of deoxygenation of carbonyl compounds by atomic carbon. J. Am. Chem. Soc. 103:2802–2807.
5. Sladkov AM, Kasatochkin VI, Kudryavstev YP, and Korshak VV (1968) Synthesis and properties of valuable polymers of carbon. Russ. Chem. Bull. 17:2560–2566.
6. Sergushin IP, Kudryavstev YP, Elizen VM et al. (1977) X-ray Electron and X-ray spectral study of Carbyne. J. Struct. Chem. 18:698–700.
7. Whittaker AG (1978) Carbon: A new view of its high-temperature behavior. Science. 200:763–764.
8. Smith PPK and Buseck PR (1982) Carbyne forms of carbon: Do they exist? Science 216:984–986.
9. Whittaker AG (1985) Carbyne forms of carbon: Evidence for their existence. Science 229:485–486.
10. Lagow RJ, Kampa JJ, Wei H et al. (1995) Synthesis of linear acetylenic carbon: The "sp" carbon allotrope. Science 267:362–367.

11. Baughman RH (2006) Chemistry: Dangerously seeking linear carbon. Science. 312:1009–1010.
12. Frondel C and Martin UB (1967) Lonsdaleite, a Hexagonal polymorph of diamond. Nature. 214:587–589.
13. Bundy FP and Kasper JS (1967) Hexagonal diamond – a new form of carbon. J. Chem. Phys. 46:3437–3446.
14. Kroto HW, Heath JR, O'Brien SC, Curl RF, and Smalley RE (1985) C60: Buckminsterfullerene. Nature. 318:162–163.
15. Novoselov KS, Geim AK, Morozov SV et al. (2004) Electric field effect in atomically thin carbon films. Science. 306:666–669.
16. Meyer JC, Geim AK, Katsnelson MI, Novoselov KS, Booth TJ, and Roth S (2007) The structrue of suspended graphene sheets. Nature. 446:60–63.
17. Geim AK and Kim P (2008) Carbon wonderland. Scientific American. 298:68–75.
18. Rode AV, Gamalay EG, and Luther-Davies B (2000) Formation of cluster-assembled carbon nano-foam by high repetition-rate laser ablation. Appl. Phys. A. 70: 135–144.
19. Rode AV, Gamalay E, and Luther-Davies B (2002) Electronic and Magnetic properties of carbon nanofoam produced by high-repetition-rate laser ablation. Appl. Surf. Sci. 197:644–649.
20. Rode AV, Gamalay E, Christy AG et al. (2004) Unconventional magnetism in all-carbon nanofoam. Phys. Rev. B. 70:054407-1-054407-9.
21. Rode AV, Gamalay E, Christy AG et al. (2005) Strong paramagnetism and possible ferromagnetis in pure carbon nanofoam produced by laser ablation. J. Magn. Magn. Mater. 290–291:298–301.
22. Sullivan JP, Friedmann TA, and Hjort K (2001) Diamond and amorphous carbon MEMS. MRS Bulletin. April:309–311.
23. Robertson J (2002) Diamond-like amorphous carbon. Mat. Sci. Eng. R. 37:129–281.
24. Hauser JJ (1977) Electrical, structural and optical properties of amorphous carbon. J. Non-Cryst. Solids. 23:21–41.
25. Andersson L (1981) A review of recent work on hard i-C films. Thin Solid Films. 86:193–200.
26. Robertson J (1986) Amorphous carbon. Adv. Phys. 35:317–375.
27. Robertson J (1991) Hard amorphous (diamond-like) carbons. Prog. Solid State Chem. 21:199–333.
28. Fitzer E, Kochling KH, Boehm HP, and Marsh H (1995) Recommended terminology for the description of carbon as a solid. Pure Appl. Chem. 67:473–506.
29. Anderson PW (1958) Abscence of diffusion in certain random lattices. Phys. Rev. 109:1492–1505.
30. Dobbs HS (1974) Vitreous Carbon and the brittle fracture of amorphous solids. J. Phys. D: Appl. Phys. 7:24–34.
31. Bragg RH and Hammond ML (1965) X-ray study of pyrolytic graphites and glassy carbons. Carbon. 3:340–340.
32. Rothwell WS (1968) Small-angle x-ray scattering from glassy carbon. J. Appl. Phys. 39:1840–1845.
33. McFeely SP, Kowalczyk SP, Ley L, Cavell RG, Pollak RA, and Shirley DA (1974) X-ray photoemission studies of diamond, graphite, and glassy carbon valence bands. Phys. Rev. B. 9:5268–5278.
34. Nishikawa K, Fukuyama K, and Nishizawa T (1998) Structure change of glass-like carbon with heat treatment, studied by small angle x-ray scattering: i. glass-like carbon prepared from phenolic resin. Jpn. J. Appl. Phys. 37:6486–6491.
35. Fukuyama K, Nishizawa T, and Nishikawa K (2001) Investigation of the pore structure in glass-like carbon prepared from furan resin. Carbon. 39:2017–2021.

36. Iwashita N, Park CR, Fujimoto H, Shiraishi M, and Inagaki M (2004) Specification for a standard procedure of X-ray diffraction measurements on carbon materials. Carbon. 42: 701–714.
37. Dekanski A, Stevanovic J, Stevanovic R, Nikolic BZ, and Jovanovic VM (2001) Glassy carbon electrodes: I. Characterization and electrochemical activation. Carbon 39:1195–1205.
38. Oberlin A and Oberlin M (1983) Graphitizability of carbonaceous materials as studied by TEM and X-ray diffraction. J. Microsc. 132:353–363.
39. Yamada S and Sato H (1962) Some physical properties of glassy carbon. Nature 193:261–262.
40. Lewis JC, Redfern B, and Cowlard FC (1963) Vitreous carbon as a crucible material for semiconductors. Sol. State Electron. 6:251–254.
41. Cowlard FC and Lewis JC (1967) Vitreous carbon – a new form of carbon. J. Mater. Sci. 2:507–512.
42. Cahn W and Harris B (1969) Newer forms of carbon and their uses. Nature. 221:132–141.
43. Halpin MK and Jenkins GM (1969) Interaction of glassy carbon with alkali metal vapours. Proc. R. Soc. Lond. A. 313:421–431.
44. Jenkins GM and Kawamura K (1976) Polymeric Carbon – Carbon Fibre, Glass and Char. Cambridge University Press, London, New York.
45. Wang C, Jia G, Taherabadi LH, and Madou MJ (2005) A novel method for the fabrication of high-aspect ratio C-MEMS structures. J. MEMS. 14:348–358.
46. Jenkins GM, Kawamura K, and Ban LL (1972) Formation and structure of polymeric carbons. Proc. R. Soc. Lond. A. 327:501–517.
47. Kakinoki J (1965) A model for the structure of "glassy carbon". Acta. Cryst. 18:578.
48. Jenkins GM and Kawamura K (1971) Structure of glassy carbon. Nature. 231:175–176.
49. Pesin LA (2002) Structure and Properties of glass-like carbon. J. Mater. Sci. 37:1–28.
50. Shiraishi M (1984) Kaitei Tansozairyo Nyumon (Introduction to Carbon Materials). Tanso Zairyo Gakkai 29–33.
51. Fedorov VB, Shorshorov MK, and Khakimova DK (1978) Carbon and its interaction with metals [in Russian]. Metallurgiya, Moscow:119–129.
52. Pesin LA and Baitinger EM (2002) A new structural model of glass-like carbon. Carbon. 40:295–306.
53. Yoshida A, Kaburagi Y, and Hishiyama Y (1991) Microtexture and magnetoresistance of glass-like carbons. Carbon. 29:1107–1111.
54. Fitzer E, Schaefer W, and Yamada S (1969) The formation of glasslike carbon by pyrolysis of polyfurfuryl alcohol and phenolic resin. Carbon. 7:643.
55. Vohler O, Reiser P, Martina R, and Ovrehoff D (1970) New forms of Carbon. Angew. Chem. Int. Ed. Engl. 9:414–425.
56. Kawamura K and Kimura S (1983) Glass-like carbon made from epoxy resin cured with 2, 4, 6-Trinitrophenol. Bull. Chem. Soc. Jpn. 56:2499–2503.
57. Aggarwal RK, Bhatia G, Bahl OP, and Malik M (1988) Development of glass-like carbon from phenol formaldehyde resins employing monohydric and dihydric phenols. J. Mater. Sci. 23:1677–1684.
58. Callstrom MR, Neenan TX, McCreery RL, and Alsmeyer DC (1990) Doped glassy carbon materials (DGC): low-temperature synthesis, structure, and catalytic behavior. J. Am. Chem. Soc. 112:4954–4956.
59. Neenan TX, Callstrom MR, Bachman BJ, McCreery RL, and Alsmeyer DC (1990) Doped glassy carbon materials (DGC): Their synthesis from polymeric precursors and investigation of their properties. Br. Polym. J. 23:171–177.
60. Hutton HD, Huang W, Alsmeyer DC et al. (1993) Synthesis, characterization, and electrochemical activity of halogen-doped glassy carbon. Chem. Mater. 5:1110–1117.
61. Schueller OJA, Brittain ST, and Whitesides GM (1997) Fabrication of glassy carbon microstructures by pyrolysis of microfabricated polymeric precursors. Adv. Mater. 9: 477–480.

62. Shah HV, Brittain ST, Huang Q, Hwu S-, Whitesides GM, and Smith DW (1999) Bis-o-diynylarene (BODA) Derived polynaphthalenes as precursors to glassy carbon microstructures. Chem. Mater. 11:2623–2625.
63. Hishiyama Y, Igarashi K, Kanaoka I et al. (1997) Graphitization behavior of Kapton-derived carbon film related to structure, microtexture and transport properties. Carbon. 35: 657–668.
64. Kim J, Song X, Kinoshita K, Madou M, and White R (1998) Electrochemical studies of carbon films from pyrolyzed photoresist. J. Electrochem. Soc. 145:2314–2319.
65. Ranganathan S, McCreery RL, Majji SM, and Madou M (2000) Photoresist-derived carbon for microelectromechanical systems and electrochemical applications. J. Electrochem. Soc. 147:277–282.
66. Singh A, Jayaram J, Madou M, and Akbar S (2002) Pyrolysis of negative photoresists to fabricate carbon structures for microelectromechanical systems and electrochemical applications. J. Electrochem. Soc. 149:E78–E83.
67. Malladi K, Wang C, and Madou M (2006) Fabrication of suspended carbon microstructures by e-beam writer and pyrolysis. Carbon. 44:2602–2607.
68. Lee L (1965) Mechanisms of thermal degradation of phenolic condensation polymers. II. Thermal stability and degradation schemes of epoxy resins. J. Polym. Sci: Part A. 3: 859–882.
69. Fitzer E and Schafer W (1970) The effect of crosslinking on the formation of glasslike carbons from thermosetting resins. Carbon. 8:353–364.
70. Gac NA, Spokes GN, and Benson SW (1970) Thermal degradation of nadic methyl anhydride-cured epoxy novolac. J Polym Sci A. 8:593–608.
71. Jenkins GM, Kawamura K, and Ban LL (1972) Formation and structure of polymeric carbons. Proc. R. Soc. Lon. A. 327:501–517.
72. Lyons AM, Wilkins CW, and Robbins M (1983) Thin pinhole-free Carbon films. Thin Solid Films. 103:333–341.
73. Nakagawa H and Tsuge S (1987) Studies on thermal degradation of epoxy resins by high-resolution pyrolysis-gas chromatography. J. Anal. Appl. Pyrolysis. 12:97–113.
74. Chen KS and Yeh RZ (1996) Pyrolysis kinetics of epoxy resin in a nitrogen atmosphere. J. Hazard. Mater. 49:105–113.
75. Beyler CL and Hirschler MM (2002) Thermal decomposition of polymers. In: SFPE Handbook of Fire Protection Engineering.
76. Ma CM, Chen C, Kuan H, and Chang W (2004) Processability, Thermal, Mechanical, and Morphological Properties of Novolac Type-Epoxy Resin-Based Carbon-Carbon Composite. J. Comp. Mater. 38:311–322.
77. Mehrotra BN, Bragg RH, and Rao AS (1983) Effect of heat treatment temperature (HTT) on density, weight and volume of glass-like carbon (GC). J. Mater. Sci. 18:2671–2678.
78. Spain IL (1981) The electronic transport properties of graphite, carbons, and related materials. In: Walker Jr. PL and Thrower PA (eds) Chemistry and Physics of Carbon, Marcel Dekker, Inc., New York.
79. Pocard NL, Alsmeyer DC, McCreery RL, Neenan TX, and Callstrom MR (1992) Doped glassy carbon: A new material for electrocatalysis. J. Mater. Chem. 2:771–784.
80. Weintraub E and Miller LB (1915) Microphone. 1156509.
81. Davidson HW (1962) The properties of G. E. C. impermeable carbon. Nucl. Eng. 7:159–161.
82. Redfern B and Greens N (1963) Bodies and Shapes of Carbonaceous Materials and Processes for their Production. US 3109712.
83. Kotlensky WV and Martens HE (1965) Tensile properties of glassy carbon to 2,900 C. Nature. 206:1246–1247.
84. Zittel HE and Miller FJ (1965) A Glassy-carbon electrode for voltammetry. Anal. Chem. 37:200–203.
85. Lewis JB, Murdoch R, and Moul AN (1969) Heat of combustion of vitreous carbon. Nature. 221:1137–1138.

86. Benson J (1969) Carbon offers advantages as implant material in human body. NASA Technical Brief.
87. Benson J (1971) Elemental carbon as a biomaterial. J. Biomed. Mater. Res. 5:41–47.
88. Von Fraunhofer JA, L′estrange PR, and Mack AO (1971) Materials science in dental implantation and a promising new material: vitreous carbon. Biomed. Eng. 6:114–118.
89. Hucke EE, Fuys RA, and Craig RG (1973) Glassy Carbon: a potential Dental Implant Material. J. Biomed. Mater. Res. 7:263–274.
90. Stanitski CL and Mooney V (1973) Osseus attachment to vitreous carbons. J. Biomed. Mater. Res. 7:97–108.
91. Hobkirk JA (1976) Tissue reactions to implanted vitreous carbon and high purity sintered alumina. J. Oral. Rehabil. 4:355–368.
92. Haubold A (1977) Carbon in prosthetics. Annals of the New York Academy of Sciences. 283:383–395.
93. Maropis PS, Molinari JA, Appel BN, and Baumhammers A (1977) Comparative study of vitreous carbon, pyrolytic carbon, pyrolytic graphite/silicon-carbide, and titanium implants in rabbit mandibules. Oral Surg. Oral Med. Oral Pathol. 43:506–512.
94. Sherman AJ (1978) Bone reaction to orthodontic forces on vitreous carbon dental implants. Am. J. Orthod. 74:79–87.
95. Jenkins GM and Grigson CJ (1979) The fabrication of artifacts out of glassy carbon and carbon-fiber-reinforced carbon for biomedical applications. J. Biomed. Mater. Res. 13: 371–394.
96. Williams MW (1972) Optical properties of glassy carbon from 0 to 82 eV. J. Appl. Phys. 43:3460–3463.
97. Taylor RJ and Humffray AA (1973) Electrochemical studies on glassy carbon electrodes I. Electron-transfer kinetics. J. Electroanal. Chem. 42:347–354.
98. Nathan MI, Smith JE, and Tu KN (1974) Raman spectra of glassy carbon. J. Appl. Phys. 45:2370–2370.
99. Blaedel WJ and Jenkins RA (1974) Steady-state voltammetry at glassy carbon electrodes. Anal. Chem. 46:1952–1955.
100. Taylor RJ and Humffray AA (1975) Electrochemical studies on glassy carbon electrodes II. Oxygen reduction in solutions of high pH. J. Electroanal. Chem. 64:63–84.
101. Norvell VE and Mamantov G (1977) Optically transparent vitreous carbon electrode. Anal. Chem. 49:1470–1472.
102. Shigemitsu T and Matsumoto G (1979) Electrical Properties of glassy-carbon Electrodes. Med. Biol. Eng. Comput. 17:470.
103. Bourdillon C, Bourgeois JP, and Thomas D (1980) Covalent Linkage of glucose oxidase on modified glassy carbon electrodes. Kinetic phenomena. J. Am. Chem. Soc. 102:4231–4235.
104. Van der Linden, WE and Dieker JW (1980) Glassy carbon as electrode material for electro analytical chemistry. Anal. Chim. Acta. 119:1–24.
105. Hebert NE, Snyder B, McCreery RL, Kuhr WG, and Brazil SA (2003) Performance of pyrolyzed photoresist carbon films in a microchip capillary electrophoresis device with sinusoidal voltammetric detection. Anal. Chem. 75:4265–4271.
106. McCreery RL (1991) Carbon electrodes: structural effects on electron transfer kinetics. In: Bard AJ (ed) Electroanalytical Chemistry, Marcel Dekker, New York.
107. Kinoshita K (1988) Carbon, Electrochemical and Physicochemical Properties. Wiley-Interscience, New York.
108. Fitzer E, Mueller K, and Schaefer W (1971) The chemistry of the pyrolytic conversion of organic compounds to carbon. In: Walker Jr. PL (ed) Chemistry and Physics of Carbon, Marcel Dekker, Inc., New York.
109. Oberlin A (1989) High-Resolution TEM studies of carbonization and graphitization. In: Thrower PA (ed) Chemistry and Physics of Carbon, Marcel Dekker, Inc., New York.
110. Wang J (1981) Reticulated vitreous carbon – a new versatile electrode material. Electrochim. Acta. 26:1721–1726.

111. Friedrich JM, Ponce-de-Leon C, Reade GW, and Walsh FC (2004) Reticulated vitreous carbon as an electrode material. J. Electroanal. Chem. 561:203–217.
112. Schueller OJA, Brittain ST, Marzolin C, and Whitesides GM (1997) Fabrication and Characterization of Glassy Carbon MEMS. Chem. Mater. 9:1399–1406.
113. Schueller OJA, Brittain ST, and Whitesides GM (1999) Fabrication of glassy carbon microstructures by soft lithography. Sens. Actuators A: Phys. 72:125–139.
114. Schueller OJA, Brittain S, and Whitesides GM (2000) Fabrication of Carbon Microstructures.6143412.
115. Zhao X, Xia Y, Schueller OJA, Qin D, and Whitesides GM (1998) Fabrication of microstructures using shrinkable polystyrene films. Sens. Actuators A: Phys. 65:209–217.
116. Kostecki R, Song X, and Kinoshita K (2000) Influence of geometry on the electrochemical response of carbon interdigitated microelectrodes. J. Electrochem. Soc. 147:1878–1881.
117. Ranganathan S and McCreery RL (2001) Electroanalytical performance of carbon films with near-atomics flatness. Anal. Chem. 73:893–900.
118. Lorenz H, Despont M, Fahrni M, LaBianca N, Vettiger P, and Renaud P (1997) SU-8: a low-cost negative resist for MEMS. J. Micromech. Microeng. 7:121.
119. Shaw JM, Gelorme JD, LaBianca NC, Conley WE, and Holmes SJ (1997) Negative photoresists for optical lithography. IBM J. Res. Dev. 41:81–94.
120. Wang C and Madou M (2005) From MEMS to NEMS with carbon. Biosens. Bioelectron. 20:2181–2187.
121. Park BY, Taherabadi L, Wang C, Zoval J, and Madou MJ (2005) Electrical properties and shrinkage of carbonized photoresist films and the implications for carbon microelectromechanical systems devices in conductive media. J. Electrochem. Soc. 152:J136–J143.
122. Wang C, Taherabadi L, Jia G, Madou M, Yeh Y, and Dunn B (2004) C-MEMS for the manufacture of 3D microbatteries. Electrochem. Solid-State Lett. 7:A435–A438.
123. Galobardes F, Wang C, and Madou M (2006) Investigation on the solid electrolyte interface formed on pyrolyzed photoresist carbon anodes for C-MEMS lithium-ion batteries. Diam. Relat. Mater. 15:1930–1934.
124. Min H, Park BY, Taherabadi L et al. (2008) Fabrication and properties of a carbon/polypyrrole three-dimensional microbattery. J. Power Sources. 178:795–800.
125. Teixidor GT, Zaouk RB, Park BY, and Madou MJ (2008) Fabrication and characterization of three-dimensional carbon electrodes for lithium-ion batteries. J. Power Sources. 183: 730–740.
126. Park BY and Madou MJ (2006) Design, fabrication, and initial testing of a miniature PEM fuel cell with micro-scale pyrolyzed carbon fluidic plates. J. Power Sources. 162:369–379.
127. Lin P, Park BY, and Madou MJ (2008) Development and characterization of a miniature PEM fuel cell stack with carbon bipolar plates. J. Power Sources. 176:207–214.
128. Xu H, Malladi K, Wang C, Kulinsky L, Song M, and Madou M (2008) Carbon post-microarrays for glucose sensors. Biosens. Bioelectron. 23:1637–1644.
129. Turon Teixidor G, Gorkin III RA, Tripathi PP et al. (2008) Carbon microelectromechanical systems as a substratum for cell growth. Biomed. Mater. 3:034116.
130. Park BY and Madou MJ (2005) 3-D electrode designs for flow-through dielectrophoretic systems. Electrophoresis. 26:3745–3757.
131. Martinez-Duarte R, Rouabah HA, Green NG, Madou M, and Morgan H (2007) Higher Efficiency and Throughput in Particle Separation with 3D Dielectrophoresis with C-MEMS. Proceedings of uTAS 2007: Paris, France, October 7–11 1:826–828.
132. Martinez-Duarte R, Andrade-Roman J, Martinez SO, and Madou MJ (2008) A High Throughput Multi-stage, Multi-frequency Filter and Separation Device based on Carbon Dielectrophoresis. Proceedings of NSTI Nanotech 2008: Boston, MA, June 1–5 3: 316–319.
133. Martinez-Duarte R, Cito S, Collado-Arredondo E, Martinez SO, and Madou M (2008) Fluido-dynamic and Electromagnetic Characterization of 3D Carbon Dielectrophoresis with Finite Element Analysis. Proceedings of NSTI Nanotech 2008: Boston, MA, 3:265–268.

134. Martinez-Duarte R, Cito S, Collado-Arredondo E, Martinez SO, and Madou MJ (2008) Fluido-dynamic and electromagnetic characterization of 3D carbon dielectrophoresis with finite element analysis. Sens. Transd. J. 3:25–36.
135. Martinez-Duarte R, Gorkin R, Abi-Samra K, and Madou MJ (2009) The integration of 3D carbon dielectrophoresis on a rotating platform. Proc. Transd. 2009: Denver, CO, June 21–25: 2147–2150.
136. Madou MJ, Park BY, and Paradiso AC (2006) Method and Apparatus for Dielectrophoretic Separation.US 2006/0260944A1.
137. Park BY, Zaouk R, and Madou M (2004) Validation of lithography based on the controlled movement of light-emitting particles. Proc. Soc. Photo Opt. Instrum. Eng. 5374:566–578.
138. Martinez-Duarte R, Madou MJ, Kumar G, and Schroers J (2009) A Novel Method for Amorphous Metal Micromolding using Carbon MEMS. Proc. Transd. 2009.: June 21–25: 188–191.
139. Park BY, Zaouk R, Wang C, and Madou MJ (2007) A case for fractal electrodes in electrochemical applications. J. Electrochem. Soc. 154:P1–P5.
140. Sharma CS, Kulkarni MM, Sharma A, and Madou M (2009) Synthesis of carbon xerogel particles and fractal-like structures. Chem. Eng. Sci. 64:1536–1543.
141. Jeong OC and Konishi S (2008) Three-dimensionally combined carbonized polymer sensor and heater. Sens. Actuators. A. 143:97–105.
142 oleman JN, Khan U, Blau WJ, and Gun'ko YK (2006) Small but strong: A review of the mechanical properties of carbon nanotube–polymer composites. Carbon. 44: 1624–1652.
143. Hafner JH, Bronikowski MJ, Azamian BR et al. (1998) Catalytic growth of single-wall carbon nanotubes from metal particles. Chem. Phys. Lett. 296:195–202.
144. Madou MJ (2009) Manufacturing Techniques for Microfabrication and Nanotechnology. CRC Press, Boca Raton, FL.
145. Metz TE, Savage RN, and Simmons HO (1992) In Situ Control of Photoresist Coating Processes. Semicond. Int. 15:68–69.
146. Thompson LF (1994) An introduction to lithography. In: Thompson LF, Willson CG, and Bowden MJ (eds) Introduction to Microlithography, 2nd edn. American Chemical Society, Washington, DC.
147. Webster RJ (1998) Thin film polymer dielectrics for high-voltage applications under severe environments. Master of Science in Electrical Engineering, Virginia Polytechnic Institute and State University.
148. Chilton JA and Goosey MT (1995) Special Polymers for Electronics and Optoelectronics, Chapman and Hall, London, 351.
149. Madou M and Florkey J (2000) From Batch to Continuous Manufacturing of Microbiomedical Devices. Chem. Rev. 100:2679–2691.
150. Reichmanis R, Nalamasu O, Houlihan FM, and Novembre AE (1999) Radiation chemistry of polymeric materials. Polym. Int. 48:1053–1059.
151. Shaw JM, Gelorme JD, LaBianca NC, Conley WE, and Holmes SJ (1996) Negative photoresists for optical lithography. IBM J. Res. Dev. 41:81–94.
152. Ito H (1996) Chemical amplification resists: History and development within IBM. IBM J. Res. Dev. 41:69–80.
153. Gelorme JD, Cox RJ, and Gutierrez SAR (1989) Photoresist Composition and Printed Circuit Boards and Packages made therewith. 4882245.
154. Angelo R, Gelorme J, Kucynski J, Lawrence W, Pappas S, and Simpson L (1992) Photocurable epoxy composition with sulfonium salt photoinitiator.5102772.
155. Harris TW (1976) Chemical Milling. Clarendon Press, Oxford.
156. Reznikova EF, Mohr J, and Hein H (2005) Deep Photo-Lithography Characterization of SU-8 Resist Layers. Microsys. Tech. 11:282–291.
157. Lee SJ, Shi W, Maciel P, and Cha SW (2003) IEEE University/Government/Industry Microelectronics Symposium. Boise, ID, 389–390.

158. Bertsch A, Lorenz H, and Renaud P (1999) 3D Microfabrication by Combining Microstereolithography and Thick Resist UV Lithography. Sens. Act. A. 73:14–23.
159. Yang R and Wang W (2005) A numerical and experimental study on gap compensation and wavelength selection in UV-Lithography of Ultra-High Aspect Ratio SU-8 Microstructures. Sens. Act B. 110:279–288.
160. del Campo A and Greiner C (2007) SU:8 a photoresist for high-aspect-ratio and 3D submicron lithography. J. Micromech. Microeng. 17:R81–R95.
161. Wolf S and Tauber RN (2000) Silicon Processing for the VLSI Era. Lattice Press, Sunset Beach.
162. Arnone C (1992) The laser-plotter: A versatile lithographic tool for integrated optics and microelectronics. Microelectron. Eng. 17:483–486.
163. Duffy DC, McDonald JC, Schueller OJA, and Whitesides GM (1998) Rapid prototyping of microfluidic systems in poly(dimethylsiloxane). Anal. Chem. 70:4974–4984.
164. Sure A, Dillon T, Murakowski J, Lin C, Pustai D, and Prather DW (2003) Fabrication and characterization of three-dimensional silicon tapers. Optics Express. 11:3555–3561.
165. Windt DL and Cirelli RA (1999) Amorphous carbon films for use as both variable-transmission apertures and attenuated phase shift masks for deep ultraviolet lithography. J. Vac. Sci. Technol. B17:930–932.
166. Stauffer JM, Oppliger Y, Regnault P, Baraldi L, and Gale MT (1992) Electron beam writing of continuous resist profiles for optical applications. J. Vac. Sci. Technol. B10:2526.
167. Le Barny P (1987) In: Stroeve P and Franses E (eds) Molecular Engineering of Ultrathin Polymeric Films, Elsevier, New York.
168. Moreau WM (1987) Semiconductor Lithography: Principles, Practices and Materials. Springer, New York.
169. Cataldo A (1999) IBM Uses X-Ray Lithography to Build Prototypes. Semiconductor Business News.
170. Ehrfeld W and Schmidt A (1998) Recent Developments in Deep X-Ray Lithography. J. Vac. Sci. Technol. B16:3526–3534.
171. Rosolen GC (1999) Automatically aligned electron beam lithography on the nanometer scale. Appl. Surf. Sci. 144–145:467–471.
172. Skjolding LHD, Teixidor GT, Emnéus J, and Montelius L Negative UV–NIL (NUV–NIL) – A mix-and-match NIL and UV strategy for realisation of nano- and micrometre structures. Microelectronic Engineering. 86:654–656.
173. Editorial (1991) Ion Beam Focused to 8-nm Width. Res. Dev. September:23.
174. Sanchez JL, van Kan JA, Osipowicz T, Springham SV, and Watt F (1998) A high resolution beam scanning system for deep ion beam lithography. Nucl. Instr. Meth. Phys. Res. B136–B38:385–389.
175. van Kan JA, Sanchez JL, Xu B, Osipowicz T, and Watt F (1999) Micromachining using focused high energy ion beams: Deep ion beam lithography. Nucl. Instr. Meth. Phys. Res. B148:1085–1098.
176. Seliger RL, Ward JW, Wang V, and Kubena RL (1979) A high intensity scanning ion probe with submicrometer spot size. Appl. Phys. Lett. 34:310–312.
177. Brodie I and Muray JJ (1982) The Physics of Microfabrication. Plenum Press, New York.
178. Youn SW, Takahashi M, Goro H, and Maeda R (2006) A study on focused ion beam milling of glassy carbon molds for the thermal imprinting of quartz and borosilicate glasses. J. Micromech. Microeng. 16:2576–2584.
179. Youn SW, Takahashi M, Goto H, Kobayashi T, and Maeda R (2006) The effect of heat-treatment conditions on mechanical and morphological properties of a FIB-milled glassy carbon mold with micro patterns. J. Micromech. Microeng. 16:1277–1284.
180. Youn SW, Takahashi M, Goto H, Maeda R, and Kobayashi N (2006) AFM, SEM and Nano/micro-indentation studies of the FIB-milled glassy carbon surface heat-treated at different conditions. DTIP 2006: April 26–28, 2006.

181. Youn SW, Takahashi M, Goto H, and Maeda R (2006) Microstructuring of glassy carbon mold for glass embossing – Comparison of focused ion beam, nano/femtosecond-pulsed laser and mechanical machining. Microelectronic Eng. 83:2482–2492.
182. Youn SW, Takahashi M, Goto H, and Maeda R (2007) Fabrication of micro-mold for glass embossing using focused ion beam, femto-second laser, eximer laser and dicing techniques. J. Mater. Process. Technol. 187–188:326–330.
183. Chou SY, Krauss PR, and Renstrom PJ (1995) Imprint of sub-25 nm vias and trenches in polymers. Appl. Phys. Lett. 67:3114–3116.
184. Chou SY, Krauss PR, and Renstrom PJ (1996) Imprint Lithography with 25-Nanometer Resolution. Science 272:85–87.
185. Colburn M, Johnson S, Stewart M et al. (1999) Step and flash imprint lithography: a new approach to high-resolution patterning. Proc. SPIE. 3676:379–389.
186. Reyes DR, Iossifidis D, Auroux P, and Manz A (2002) Micro Total Analysis Systems. 1. Introduction, Theory and Technology. Anal. Chem. 74:2623–2636.
187. Auroux P, Iossifidis D, Reyes DR, and Manz A (2002) Micro total analysis systems. 2. Analytical standard operations and applications. Anal. Chem. 74:2637–2652.
188. Dittrich PS, Tachikawa K, and Manz A (2006) Micro total analysis systems. Latest advancements and trends. Anal. Chem. 78:3887–3907.
189. Janasek D, Franzke J, and Manz A (2006) Scaling and the design of miniaturized chemical-analysis systems. Nature. 442:374–380.
190. Rheea M and Burns MA (2008) Microfluidic assembly blocks. Lab Chip. 8:1365–1373.
191. Andersson H and van den Berg A (2003) Microfluidic devices for cellomics: a review. Sens. Actuators B: Chem. 92:315–325.
192. Chung TD and Kim HC (2007) Recent Advances in miniaturized microfluidic flow cytometry for clinical use. Electrophoresis. 28:4511–4520.
193. Duffy DC, McDonald JC, Schueller OJA, and Whitesides GM (1998) Rapid prototyping of microfluidic systems in poly(dimethylsiloxane). Anal. Chem. 70:4974–4984.
194. Xia Y and Whitesides GM (1998) Soft lithography. Ann. Rev. Mat. Sci. 28:153–184.
195. Zhao X, Xia Y, and Whitesides GM (1997) Soft lithographic methods for nano-fabrication. J. Mater. Chem. 7:1069–1074.
196. Anderson JR, Chiu DT, Jackman RJ et al. (2000) Fabrication of topologically complex three-dimensional microfluidic systems in PDMS by rapid prototyping. Anal. Chem. 72: 3158–3164.
197. Becker H and Gartner C (2000) Polymer microfabrication methods for microfluidic analytical applications. Electrophoresis. 21:12–26.
198. Becker H and Locascio LE (2002) Polymer microfluidic devices. Talanta. 56:267–287.
199. del Campo A and Arzt E (2008) Fabrication approaches for generating complex micro- and nanopatterns on polymeric surfaces. Chem. Rev. 108:911–945.
200. Izumi Y, Katoh M, Ohte T, Ohtani S, Kojima A, and Saitoh N (1996) Heating effects on modifying carbon surface by reactive plasma. Appl. Surf. Sci. 100–101:179–183.
201. Pohl HA (1978) Dielectrophoresis. Cambridge University Press, Cambridge.
202. Pohl HA (1958) Some effects of nonuniform fields on dielectrics. Appl. Phys. 29: 1182–1188.
203. Sanchis A, Brown AP, Sancho M et al. (2007) Dielectric characterization of bacterial cells using dielectrophoresis. Bioelectromagnetics. 28:393–401.
204. Minerick AR, Zhou R, Takhistov P, and Chang H-C (2003) Manipulation and characterization of red blood cells with alternating current fields in microdevices. Electrophoresis. 24:3703–3717.
205. Hughes MP, Morgan H, and Rixon FJ (2001) Dielectrophoretic manipulation and characterization of herpes simplex virus-1 capsids. Eur. Biophys. J. Biophys. Lett. 30:268–272.
206. Markx GH and Davey CL (1999) The dielectric properties of biological cells at radiofrequencies: applications in biotechnology. Enzyme Microb. Technol. 25:161–171.

207. Crane JS and Pohl HA (1968) A study of living and dead yeast cells using dielectrophoresis. J. Electrochem. Soc. 115:584–586.
208. Morgan H and Green NG (2003) AC Electrokinetics: Colloids and Nanoparticles. Research Studies Press LTD, Hertfordshire, England.
209. Ramos A, Morgan H, Green NG, and Castellanos A (1998) Ac electrokinetics: a review of forces in microelectrode structures. J. Phys. D.: Appl. Phys. 31:2338–2353.
210. Castellanos A, Ramos A, González A, Green NG, and Morgan H (2003) Electrohydrodynamics and dielectrophoresis in microsystems: scaling laws. J. Phys. D. 36:2584–2597.
211. Rousselet J, Markx GH, and Pethig R (1998) Separation of erythrocytes and latex beads by dielectrophoretic levitation and hyperlayer field-flow fractionation. Colloids and Surfaces A: Physiochem. Eng. Aspects. 140:209–216.
212. Hughes MP, Morgan H, Rixon FJ, Burt JPH, and Pethig R (1998) Manipulation of herpes simplex virus type 1 by dielectrophoresis. Biochimica et Biophysica Acta-General Subjects. 1425:119–126.
213. Becker FF, Wang X-, Huang Y, Pethig R, Vykoukal J, and Gascoyne PRC (1995) Separation of human breast cancer cells from blood by differential dielectric affinity. Proc. Natl. Acad. Sci. USA 92:860–864.
214. Flanagan LA, Lu J, Wang L et al. (2008) Unique dielectric properties distinguishing stem cells and their differentiated progeny. Stem Cells. 26:656–665.
215. Lapizco-Encinas BH, Simmons BA, Cummings EB, and Fintschenko Y (2004) Insulator-based dielectrophoresis for the selective concentration and separation of live bacteria in water. Electrophoresis 25:1695–1704.
216. Srivastava SK, Pullen SA, and Minerick AR (2008) Dielectrophoretic characterization of erythrocytes: Positive ABO blood types. Electrophoresis 29:5033–5046.
217. Cummings EB and Singh AK (2003) Dielectrophoresis in Microchips Containing Arrays of Insulating Posts: Theoretical and Experimental Results. Anal. Chem. 75:4724–4731.
218. Cummings EB and Singh AK (2000) Dielectrophoretic trapping without embedded electrodes.; September 18–19, 2000 Copyright 2003 SciSearch Plus. 4177:164–173.
219. Lapizco-Encinas BH, Ozuna-Chacón S, and Rito-Palomares M (2008) Protein manipulation with insulator-based dielectrophoresis and DC electric fields. J. Chromatogr. A 1206: 45–51.
220. Zheng LF, Brody JP, and Burke PJ (2004) Electronic manipulation of DNA, proteins, and nanoparticles for potential circuit assembly. Biosens. Bioelectron. 20:606–619.
221. Tracy NI and Ivory CF (2008) Protein separation using preparative-scale dynamic field gradient focusing. Electrophoresis 29:2820–2827.
222. Betts WB (1995) The potential of dielectrophoresis for the real-time detection of microorganisms in foods. Trends Food Sci. Technol. 6:51–58.
223. Klodzinska E and Buszewski B (2009) Electrokinetic detection and characterization of intact microorganisms. Anal. Chem. 81:8–15.
224. Markx GH, Huang Y, Zhou XF, and Pethig R (1994) Dielectrophoretic characterization and separation of microorganisms. Microbiology. 140:585–591.
225. Suzuki M, Yasukawa T, Shiku H, and Matsue T (2005) Separation of live and dead microorganisms in a micro-fluidic device by dielectrophoresis. Bunseki Kagaku. 54:1189–1195.
226. Castellarnau M, Errachid A, Madrid C, Juarez A, and Samitier J (2006) Dielectrophoresis as a tool to characterize and differentiate isogenic mutants of Escherichia coli. Biophys. J. 91:3937–3945.
227. Demierre N, Braschler T, Muller R, and Renaud P (2008) Focusing and continuous separation of cells in a microfluidic device using lateral dielectrophoresis. Sens. Actuators B: Chem. 132:388–396.
228. Vahey MD and Voldman J (2008) An equilibrium method for continuous-flow cell sorting using dielectrophoresis. Anal. Chem. 80:3135–3143.

229. Fatoyinbo HO, Hoettges KF, and Hughes MP (2008) Rapid-on-chip determination of dielectric properties of biological cells using imaging techniques in a dielectrophoresis dot microsystem. Electrophoresis. 29:3–10.
230. Gascoyne P and Vykoukal J (2004) Dielectrophoresis-based sample handling in general-purpose programmable diagnostic instruments. Proc. IEEE. 92:22–42.
231. Wang XB, Yang J, Huang Y, Vykoukal J, Becker FF, and Gascoyne PRC (2000) Cell separation by dielectrophoretic field-flow-fractionation. Anal. Chem. 72:832–839.
232. Gascoyne PRC, Wang XB, Huang Y, and Becker FF (1997) Dielectrophoretic separation of cancer cells from blood. IEEE Trans. Ind. Appl. 33:670–678.
233. Müller T, Fiedler S, Schnelle T, Ludwig K, Jung H, and Fuhr G (1996) High frequency electric fields for trapping of viruses. Biotechnol. Tech. 10:221–226.
234. Ermolina I, Milner J, and Morgan H (2006) Dielectrophoretic investigation of plant virus particles: Cow pea mosaic virus and tobacco mosaic virus. Electrophoresis. 27:3939–3948.
235. Grom F, Kentsch J, Müller T, Schnelle T, and Stelzle M (2006) Accumulation and trapping of hepatitis A virus particles by electrohydrodynamic flow and dielectrophoresis. Electrophoresis 27:1386–1393.
236. Berman D, Rohr ME, and Safferman RS (1980) Concentration of polio virus in water by molecular filtration. Appl. Environ. Microbiol. 40:426–428.
237. Hughes MP (2002) Strategies for dielectrophoretic separation in laboratory-on-a-chip systems. Electrophoresis. 23:2569–2582.
238. Hughes MP (2002) Nanoelectromechanics in Engineering and Biology. CRC Press, Boca Raton, FL.
239. Simmons BA, McGraw GJ, Davalos RV, Fiechtner GJ, Fintschenko Y, and Cummings EB (2006) The development of polymeric devices as dielectrophoretic separators and concentrators. MRS Bulletin 31:120–124.
240. Cummings EB (2003) Streaming dielectrophoresis for continuous-flow microfluidic devices. IEEE Eng. Med. Biol. Mag. 22:75–84.
241. Wang L, Flanagan L, and Lee AP (2007) Side-wall vertical electrodes for lateral field microfluidic applications. J. MEMS. 16:454–461.
242. Voldman J, Gray M, Toner M, and Schmidt M (2002) A Microfabrication-based Dynamic Array Cytometer. Anal. Chem. 74:3984–3990.
243. Illiescu C, Xu GL, Samper V, and Tay FEH (2005) Fabrication of a dielectrophoretic chip with 3D silicon electrodes. J. Micromech. Microeng. 15:494–500.
244. Fatoyinbo H, Kamchis D, Whattingham R, Ogin SL, and Hughes M (2005) A high-throughput 3-D composite dielectrophoretic separator. IEEE Trans. Biomed. Eng. 52: 1347–1349.
245. de Paula CC, Garcia Ramos A, da Silva AC, Cocchieri Botelho E, and Rezende MC (2002) Fabrication of glassy carbon spools for utilization in fiber optic gyroscopes. Carbon. 40: 787–788.
246. Iacono ST, Perpall MW, Wapner PG, Hoffman WP, and Smith Jr. DW (2007) Carbonization and thermal expansion of glassy carbon derived from bis-ortho-diynylarenes. Carbon 45:931–935.
247. Hull R (1999) Properties of Crystalline Silicon. The Institution of Engineering and Technology, London.
248. Anne A, Blanc B, Moiroux J, and Saveant JM (1998) Facile derivatization of glassy carbon surfaces by N-Hydroxysuccinimide esters in view of attaching biomolecules. Langmuir. 14:2368–2371.
249. Nowall WB, Wipf DO, and Kuhr WG (1998) Localized avidin/biotin derivatization of glassy carbon electrodes using SECM. Anal. Chem. 70:2601–2606.
250. Fleischmann M, Pons S, Rolison DR, and Schmidt PP (1987) Ultramicroelectrodes. Datatech Systems Inc. Morganton, NC.
251. Hadi M, Rouhollah A, Yousefi M, Taidy F, and Malekfar R (2006) Electrochemical Characterization of a pyrolytic carbon film electrode and the effect of anodization. Electroanalysis 18:787–792.

252. McCreery RL (2008) Advanced carbon electrode materials for molecular electrochemistry. Chem. Rev. 108:2646–2687.
253. Lawrence NS, Beckett EL, Davis J, and Compton RG (2002) Advances in the voltammetric analysis of small biologically relevant compounds. Anal. Biochem. 303:1–16.
254. Aoki K and Tanaka M (1989) Time-dependence of diffusion-controlled currents of a soluble redox couple at interdigitated microarray electrodes. J. Electroanal. Chem. 266:11–20.
255. Niwa O and Tabei H (1994) Voltammetric measurements of reversible and quasi-reversible redox species using carbon film based interdigitated array microelectrodes. Anal. Chem. 66:285–289.
256. Dam VAT, Olthuis W, and van den Berg A (2007) Redox cycling with facing interdigitated array electrodes as a method for selective detection of redox species. Analyst. 132: 365–370.
257. Bard AJ, Crayston JA, Kittlesen GP, Varco Shea T, and Wrighton MS (1986) Digital simulation of the measured electrochemical response of reversible redox couples at microelectrode arrays: consequences arising from closely spaced ultramicroelectrodes. Anal. Chem. 58:2321–2331.
258. Yang X and Zhang G (2005) Diffusion-controlled redox cycling at nanoscale interdigitated electrodes. Proceedings of the COMSOL Multiphysics Conference, Boston MA.
259. Iwasaki Y and Morita M (1995) Electrochemical Measurements with Interdigitated Array Microelectrodes Current Separations. Curr. Sep. 14:2–8.
260. Spegel C, Heiskanen A, Pedersen S, Emneus J, Ruzgas T, and Taboryski R (2008) Fully automated microchip system for the detection of quantal exocytosis from single and small ensembles of cells. Lab Chip. 8:323–329.
261. Spegel C, Heiskanen A, Skjolding LHD, and Emneus J (2008) Chip based electroanalytical systems for cell analysis. Electroanalysis 20:680–702.
262. Tabei H, Takahashi M, Hoshino S, Niwa O, and Horiuchi T (1994) Subfemtomole detection of catecholamine with interdigitated array carbon microelectrodes in HPLC. Anal. Chem. 66:3500–3502.
263. Fruchter L, Crepy G, and Le Mehaute A (1986) Batteries, identified fractal objects. J. Power Sources. 18:51–62.
264. Pajkossy T (1991) Electrochemistry at fractal surfaces. J. Electroanal. Chem. 300:1–11.
265. Whitesides GM and Grzybowski B (2002) Self-assembly at all scales. Science; Science. 295:2418–2421.
266. Maddison DS (1993) Fractal analysis of polypyrrole deposition. Synth. Met. 57:3544–3549.
267. Wang C, Zaouk R, and Madou M (2006) Local chemical vapor deposition of carbon nanofibers from photoresist. Carbon 44:3073–3077.
268. Khan GF and Wernet W (1997) Design of enzyme electrodes for extended use and storage life. Anal. Chem. 69:2682–2687.
269. de la Guardia M (1995) Biochemical sensors: The state of the art. Microchim. Acta. 120:243–255.
270. Gorton L (1995) Carbon paste electrodes modified with enzymes, tissues, and cells. Electroanalysis 7:23–45.
271. Heller A (1990) Electrical wiring of redox enzymes. Acc. Chem. Res. 23:128–134.
272. Mao F, Mano N, and Heller A (2003) Long tethers binding redox centers to polymer backbones enhance electron transport in enzyme "wiring" hydrogels. J. Am. Chem. Soc. 125:4951–4957.
273. Wang J and Angnes L (1992) Miniaturized glucose sensors based on electrochemical codeposition of rhodium and glucose oxidase onto carbon-fiber electrodes. Anal. Chem. 64:456–459.
274. Ciriello R, Cataldi T, Centonze D, and Guerrieri A (2000) Permselective behavior of an electrosynthesized, nonconducting thin film of poly(2-naphthol) and its application to enzyme immobilization. Electroanalysis. 12:825–830.
275. Vielstich W, Lamm A, and Gasteiger H (2003) Handbook of Fuel Cells: Fundamentals, Technology, Applications. John Wiley & Sons, Ltd. Chichester.

276. Tamaki T and Yamaguchi T (2006) High-surface-area three-dimensional biofuel cell electrode using redox-polymer-grafted carbon. Ind. Eng. Chem. Res. 45:3050–3058.
277. Arechederra RL, Treu BL, and Minteer SD (2007) Development of glycerol/O2 biofuel cell. J. Power Sources. 173:156–161.
278. Brunel L, Denele J, Servat K et al. (2007) Oxygen transport through laccase biocathodes for a membrane-less glucose/O2 biofuel cell. Electrochem. Commun. 9:331–336.
279. Lim KG and Palmore GTR (2007) Microfluidic biofuel cells: The influence of electrode diffusion layer on performance. Biosens. Bioelectron. 22:941–947.
280. Report on Basic Research Needs to Assure a Secure Energy Future (2003) U. S. Department of Energy. Available from http://www.science.doe.gov/bes/reports/list.html.
281. Thomas EV, Bloom I, Christophersen JP, and Battaglia VS (2008) Statistical methodology for predicting the life of lithium-ion cells via accelerated degradation testing. J. Power Sources. 184:312–317.
282. Sakaue E (2005) Micromachining/nanotechnology in direct methanol fuel cell. Digest IEEE Conference on MEMS: Miami Beach, FL, Jan 30–Feb 3 1:600–605.
283. Jankowski AF, Hayes JP, Graff RT, and Morse JD (2002) Micro-fabricated thin film fuel cells for portable power requirements. Materials Research Society Proceedings. San Francisco, CA, April 1–5 730:93–98.
284. Hayase M, Kawase T, and Hatsuzawa T (2004) Miniature 250 μm thick fuel cell with monolithically fabricated silicon eletrodes. Electrochem. Solid-State Lett. 7:A231–A234.
285. Modroukas D, Modi V, and Frechette LG (2005) Micromachined silicon structures for free-convection PEM fuel cells. J. Micromech. Microeng. 15:S193–S201.
286. Min KB, Tanaka S, and Esashi M (2002) MEMS-based polymer electrolyte fuel cell. Electrochemistry 70:924–927.
287. Kravitz SH, Apblett CA, Schmidt CF, Beggans MH, and Hecht AM (2002) A silicon micro fuel cell with a porous silicon nitride membrane electrode assembly. Abstracts of the Electrochemical Society Meeting: Salt Lake City, UT, October 20–25 1:708.
288. Kelley SC, Deluga GA, and Smyrl WH (2000) A miniature methanol/air polymer electrolyte fuel cell. Electrochem. Solid-State Lett. 3:407–409.
289. Lee SJ, Chang-Chien A, Cha SW et al. (2002) Design and fabrication of a micro fuel cell array with "flip-flop" interconnection. J. Power Sources. 112:410–418.
290. Motokawa S, Mohamedi M, Momma T, Shoji S, and Osaka T (2004) MEMS-based design and fabrication of a new concept micro direct methanol fuel cell (u-DMFC). Electrochem. Commun. 6:562–565.
291. D′Arrigo G, Spinella C, Arena G, and Lorenti S (2003) Fabrication of miniaturized Si-based electrocatalytic membranes. Mater. Sci. Eng. C. 23:13–18.
292. Erdler G, Frank M, Lehmann M, Reinecke H, and Muller C (2006) Chip integrated fuel cell. Sensor Actuat. A-Phys. 132:331–336.
293. Meyers JP and Maynard HL (2002) Design considerations for miniaturized PEM fuel cells. J. Power Sources 109:76–88.
294. Kuriyama N, Kubota T, Okamura D, Suzuki T, and Sasahara J (2008) Design and fabrication of MEMS-based monolithic fuel cells. Sensor Actuat. A-Phys. 145–146:354–362.
295. Dyer CK (2002) Fuel cells for portable applications. J. Power Sources 106:31–34.
296. Park BY and Madou MJ (2006) Design, fabricaion, and initial testing of a miniature PEM fuel cell with micro-scale pyrolyzed carbon fluidic plates. J. Power Sources 162:369–379.
297. Gottesfeld S and Wilson MS (2000) Polymer electrolyte fuel cells as potential power sources for portable electronic devices. In: Osaka T and Datta M (eds) Energy Storage Systems for Electronics Devices, Gordon and Breach Science Publishers, Singapore.
298. Arico AS, Srinivasan S, and Antonucci V (2001) DMFCs: from fundamental aspects to technology development. Fuel Cells 1:133–161.
299. Müller J, Frank G, Colbow K, and Wilkinson D (2003) Transport/kinetic limitations and efficiency losses. In: Vielstich W, Gasteiger HA, and Lamm A (eds) Handbook of Fuel Cells, John Wiley and Sons Ltd., England.

300. Neergat M, Friedrich KA, and Stimming U (2003) New material for DMFC MEAs. In: Vielstich W, Gasteiger HA, and Lamm A (eds) Handbook of Fuel Cells, John Wiley and Sons Ltd., England.
301. Narayanan SR, Valdez TI, and Rohatgi N (2003) DMFC system design for portable applications. In: Vielstich W, Gasteiger HA, and Lamm A (eds) Handbook of Fuel Cells, John Wiley and Sons Ltd., England.
302. Lu G and Wang CY (2004) Two-phase microfluidics, heat and mass transport in direct methanol fuel cells. In: Sunden B and Faghri M (eds) Transport Phenomena in Fuel Cells, WIT Press, Boston, MA.
303. Gottesfeld S (2007) Polymer electrolyte and direct methanol fuel cells. In: Bard AJ and Stratman M (eds) Encyclopedia of Electrochemistry, John Wiley and Sons Ltd., Weinheim.
304. Shaffer CE and Wang CY (2009) Performance modeling and cell design for high concentration methanol fuel cells. In: Vielstich W, Yokokawa H, and Gasteiger HA (eds) Advances in Electrocatalysis, Materials, Diagnostics and Durability, John Wiley and Sons Ltd., England.
305. Choban ER, Markoski LJ, Wieckowski A, and Kenis PJA (2004) Microfluidic fuel cell based on laminar flow. J. Power Sources. 128:54–60.
306. Kjeang E, McKechnie J, Sinton D, and Djilali N (2007) Planar and three-dimensional microfluidic fuel cell architectures based on graphite rod electrodes. J. Power Sources. 168:379–390.
307. Kjeang E, Djilali N, and Sinton D (2009) Microfluidic fuel cells: A review. J. Power Sources. 186:353–369.
308. Li A, Chan SH, and Nguyen NT (2007) A laser-micromachined polymeric membraneless fuel cell. J. Micromech. Microeng. 17:1107–1113.
309. Litster S and Djilali N (2008) Theoretical performance analysis of microstructured air-breathing fuel cells. Electrochem. Solid-State Lett. 11:B1–B5.
310. Jiang Y, Wang X, Zhong L, and Liu L (2006) Design, fabrication and testing of a silicon-based air-breathing micro direct methanol fuel cell. J. Micromech. Microeng. 16:S233–S239.
311. Cao J, Zou Z, Huang Q et al. (2008) Planar air-breathing micro-direct methanol fuel cell stacks based on micro-electronic-mechanical-system technology. J. Power Sources. 185:433–438.
312. Chiao M, Lam K, and Lin L (2006) Micromachined microbial and photosynthethic fuel cells. J. Micromech. Microeng. 16:2547–2553.
313. Barton SC, Gallaway J, and Atanassov P (2004) Enzymatic biofuel cells for implantable and microscale devices. Chem. Rev. 104:4867–4886.
314. Kerzenmacher S, Ducree J, Zengerle R, and von Stetten F (2008) Energy harvesting by implantable abiotically catalyzed glucose fuel cells. J. Power Sources. 182:1–17.
315. Moriuchi T, Morishima K, and Furukawa Y (2008) Improved power capability with pyrolyzed carbon electrodes in micro direct photosynthetic/metabolic bio-fuel cell. Int. J. Precision Eng. Manuf. 9:23–27.
316. Topcagic S and Minteer SD (2006) Development of a membraneless ethanol/oxygen biofuel cell. Electrochim. Acta. 51:2168–2172.
317. O'Hayre R, Braithwaite D, Hermann W et al. (2003) Development of portable fuel cell arrays with printed-circuit technology. J. Power Sources. 124:459–472.
318. More KL and Reeves KS (2005) Microstructural characterization of PEM fuel cell MEAs. DOE Hydrogen Program Annual Merit Review Proceedings: Arlington, VA, May 23–26 1.
319. Gottesfeld S and Zawodzinski TA (1997) Polymer electrolyte fuel cells. In: Tobias C (ed) Advances in Electrochemical Science and Engineering, Wiley and Sons, New York.
320. Eikerling M, Kornyshev AA, and Kulikovsky AA (2007) Physical modeling of cell components, cells and stacks. In: Macdonald DD and Schmuki P (eds) Encyclopedia of Electrochemistry, VCH-Wiley, Weinheim.
321. Eikerling M, Kornyshev AA, and Kulikovsky AA (2006) Water in polymer electrolyte fuel cells: Friend or foe? Phys. Today. 59:38–44.

322. Debe MK (2003) Novel catalysts, calatyst support and catalysts coated membrane methods. In: Vielstich W, Lamm A, and Gasteiger HA (eds) Handbook of Fuel Cells – Fundamentals, Technology and Applications, John Wiley and Sons Ltd., England.
323. Debe MK, Schmoeckel AK, Vernstrom GD, and Atanasoski R (2006) High voltage stability of nanostructured thin film catalysts for PEM fuel cells. J. Power Sources. 161:1002–1011.
324. Mukherjee PP, Sinha PK, and Wang CY (2007) Impact of gas diffusion layer structure and wettability on water management in polymer electrolyte fuel cells. J. Mater. Chem. 17:3053–3089.
325. Mathias MF, Roth J, Fleming J, and Lehnert W (2003) Diffusion media materials and characterization. In: Lietsich W, Lamm A, and Gasteiger HA (eds) Handbook of Fuel Cells – Fundamentals, Technology and Applications, John Wiley and Sons Ltd., Chicester.
326. Weber AZ and Darling RM (2007) Understanding porous water transport plates in polymer-electrolyte fuel cells. J. Power Sources. 168:191–199.
327. Wang CY (2004) Fundamental models for fuel cell engineering. Chem. Rev. 104:4727–4766.
328. Wang CY (2003) Two-phase flow and transport. In: Anonymous (ed) Handbook of Fuel Cells – Fundamentals, Technology and Applications, John Wiley and Sons Ltd., Chicester.
329. Lin P, Park BY, and Madou MJ (2008) Development and characterization of a miniature PEM fuel cell stack with carbon bipolar plates. J. Power Sources. 176:207–214.
330. Mukherjee PP (2007) Pore-Scale Modeling and Analysis of the Polymer Electrolyte Fuel Cell Catalyst Layer. PhD Dissertation, The Pennsylvania State University.
331. Mukherjee PP and Wang CY (2008) A Catalyst layer Flooding Model for Polymer Electrolyte Fuel Cells. Proceedings of ASME Fuel Cell 2008: Denver, CO, June 16–18 1.
332. Prasher RS, Hu XJ, Chalopin Y et al. (2009) Turning carbon nanotubes from exceptional heat conductors into insulators. Phys. Rev. Lett. 102:105901–105904.
333. Lee SW, Kim B, Chen S, Shao-Horn Y, and Hammond PT (2009) Layer-by-Layer assembly of all carbon nanotube ultrathin films for electrochemical applications. J. Am. Chem. Soc. 131:671–679.
334. Mano N, Mao F, and Heller A (2003) Characteristics of a miniature compartment-less Glucose–O2 biofuel cell and its operation in a living plant. J. Am. Chem. Soc. 125:6588–6594.
335. Klotzbach T, Watt M, Ansari Y, and Minteer SD (2006) Effects of hydrophobic modification of chitosan and Nafion on transport properties, ion-exchange capacities, and enzyme immobilization. J. Membr. Sci. 282:276–283.
336. Barton SC, Sun Y, Chandra B, White S, and Hone J (2007) Mediated enzyme electrodes with combined micro- and nanoscale supports. Electrochem. Solid-State Lett. 10:B96–B100.

Chapter 6
Mechanical Micromanufacturing: An Overview

P.K. Mishra, S.B. Patil, S.S. Pardeshi, and S.R. Kajale

Abstract "The most startling advances have their origin at the boundaries of the specialties, where the techniques developed in one field are applied with fertile effect, to the subject matter of another... If this cross-fertilization dwindles, the rate of scientific advance will almost surely dwindle as well, and so any thing that encourages cross-fertilization is all to the good. . ." said the great Philosopher, Isaac Asimov.

Keywords Electroforming · Laser processing · Laser sintering · Micro Grooving · Micro Electrodischarge machining (micro EDM) · Micro milling · Micro welding · Micro WEDM · LIGA · SLIGA · Surface modification · Surface integrity · Selective heat treatment · Thermal processing · Ultrasonic processing

6.1 Introduction

In aeronautics, reactors, automobiles, or electronics industries *manufacturing* is the main activity to create wealth for any nation, industrialized or developing. It is well established that in the present or future needs the miniaturization of any components, which play the dominant role to speed up the action and efficiency (for lower inertial components) of the systems. This also enables the systems to operate in a holistic pattern with mechatronics approach with appropriate technological developments and innovations.

The most contribution for the dramatic improvements in standards of living (if not the quality of life) of people, enjoyed over the last forty years is primarily due to the new products of technology, e.g.:

P.K. Mishra (✉)
Microsystems Engineering Laboratory, Department of Mechanical Engineering, College of Engineering, Pune, Maharashta, India-411005
e-mail: pk.pkmishra@gmail.com

S. Chakraborty (ed.), *Microfluidics and Microfabrication*,
DOI 10.1007/978-1-4419-1543-6_6, © Springer Science+Business Media, LLC 2010

- Domestic appliances such as refrigerators, microwave ovens, automatic washing machine, television, VCR/DVD, CD, CCD cameras, convenience foods etc.
- Computers, telecommunication, satellite TV, fax and copier systems etc.
- Automobiles, high-speed trains and aircrafts.
- Medical and biomedical appliances.

The development in the methods of producing these products due to the advances in materials, design, manufacturing technology and their judicious management, have brought and would bring in such dramatic effects on our culture, in terms of the improved living standards.

6.2 The Problems

The problems to catering to the needs are mostly for the follows reasons.

- Population growth.
- Continuous draining out of natural resources and the needs for the development of newer materials to meet the diversified functionality (in the multiscale design approach) of the components.
- The products now possesses feature with more functions and needs fewer discrete components (in Mechatronics approach).
- The component parts would be of more complexes in shape, utilizing improved materials (with higher strength to weight ratio) and processes for higher precision and quality.
- The manufacturing systems would then be more complex and are to be simplified to adopt more variables to control, with better flow and integration of information for higher accuracy, surface integrity and ecological conditions [1].
- The processes need to be efficiently designed to meet the above requirements removing the drawbacks of the conventional practices.

6.3 The Solutions

To cater to the industrial needs, one must bring in the innovative changes to the existing (traditional or conventional) by nontraditional (unconventional) practices, the schemes shown in Figs. 6.1 and 6.2 are self-explanatory, and

- Sustain productivity with increased strength of work material (to bring in compactness, environmental fitness and higher strength to weight ratio, working in any Inertial systems) e.g. stainless steel, titanium, nimonics and similar HSTR alloys, fiber-reinforced-composite, stellites (cobalt based alloys) ceramics, silicon and other difficult to process materials;
- Maintain productivity with components with desired shape, accuracy and surface integrity requirements (for better quality [2] and integration); and

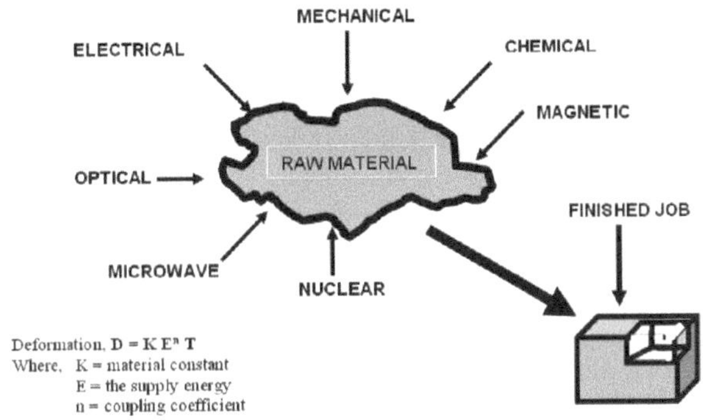

Fig. 6.1 Nontraditional energy approach to manufacturing [24]

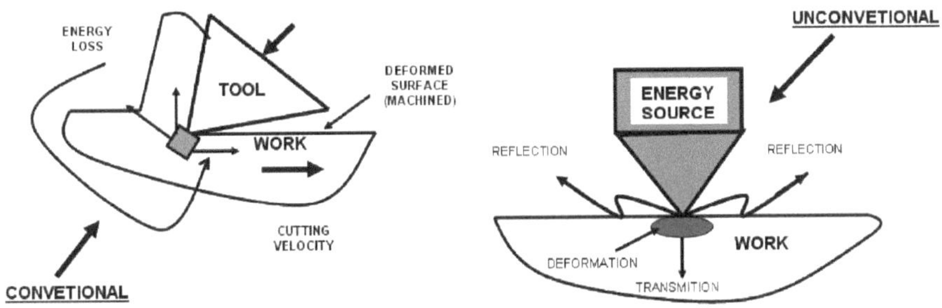

Fig. 6.2 Conventional and unconventional energy approach to manufacturing [24]

- Improve the capability of integration, automations and decreasing their sophistication (decreasing the investment cost) requirements, e.g. converting 3D control to 1D control of work-tool movements (higher performance).

While manufacturing, in achieving higher precision and accuracy [3] in components, one must also think of

- Low-stress condition (clamping, deformation etc.) or complete elimination on the job,
- Vibration free operation, and
- Elimination/minimization of errors in tool path motion for the generation mechanism/s with the principles of zero play between the perfect kinetic references and kinematic pairs.

To meet these challenges, one must change his/her attitude towards manufacturing processes without having any prejudices over the mundane conventional practices, using mechanical means. The adoption of the energy to which the (work) material shows weakness, seemed to be a good proposition to deform, as the processing technology would be beneficial.

The simplicity in concentrating the form of the energy to which the material is weaker and directing it in a desired contoured path on a low-inertial system, brings in the concept of manufacturability (subtractive or additive) of components having diversified material structure, shape, size, accuracy and finish with a tailored surface integrity.

If the processing principles are analyzed critically, it must be observed that the non-contact and practically stress free processing are not jeopardized by material strength unlike traditional processes. A spectrum of non-traditional processes, mostly used is shown in Fig. 6.5 and many more are in the horizon (research stage). Moreover, it is to be clear that the magnitude of the control volume of deformation would be decided on the basis of energy requirement, i.e. the smaller is the energy input the smaller is the deformation in manufacturing, whether cutting, welding, alloying, heat treatment, etc. It is not explicitly dependent on the size on the energy level provided by the machine and its accuracy. This is evident from the figure below (Figs. 6.3 and 6.4):

Since, the mechanism of each process involves different energy forms, the surface integrity assigned to the work would differ from each other and the component

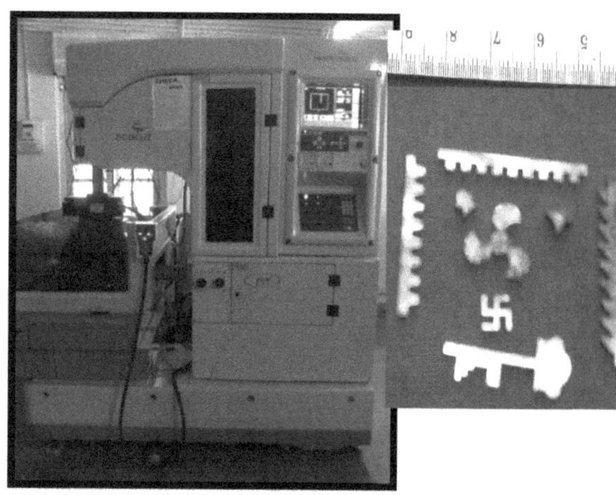

Fig. 6.3 The 4-axis CNC available at Microsystems Lab in Mechanical Engineering Department of COEP is capable of processing microcomponents

Fig. 6.4 Micro-welding is possible with the 3-axis CNC Laser System available at Microsystems Lab in Mechanical Engineering Department of COEP

performance characteristics would also change. Or in the other words, two or more processes can be combined (hybridized) to take the advantage of the worthiness of the constituent processes.

So when components requiring surface different properties to improve their performance characteristics (as evident from stress free and crack free, non-directional surface finish, surface hardness etc.) can be processed easily by hybridized nontraditional processes even in conjunction with conventional ones. This gives a way to the idea of flexible machines in processing a job component even in one station!

Though nonconventional or nontraditional machining methods are always talked upon, but it would be wiser to survey upon the manufacturing spectrum covering machining, heat treatment, welding, alloying, cladding etc using the technique or techniques when hybridized.

Moreover, most of these nontraditional processes can be used for macro to micromanufacturing and even some of them can be extended to nano level.

Manufacturing to higher precision is also gathering momentum as well [4], where,

- Precision engineering
- Microengineering
- Nanotechnologies

play a very dominant role in the future, miniaturization of components and systems are detrimental. Moreover, miniaturization improves the strength to weight ratio, sensitivity to thermal response (heating and cooling for higher surface to volume ratio) etc., however requires a newer materials (e.g. silicon) for increased stress level.

One more thing detrimental for higher precision/accuracy is the minimization of number kinematic linkages and mechanical forces. The fundamental requirements of machines to achieve higher precision are through: *a.* precision of perfect kinematic reference; *b.* precision of perfect kinematic pair; *c.* construction to prevent noise, either internal or external; and finally, *d.* the accuracy in movements and its assessment.

When one talks of manufacturing for miniaturization, it is pertinent to discuss some the processes, which would sustain the productivity under complex surface configurations (mono-block design to avoid loss of precision in assembly), stringent material properties etc.

Micromanufacturing or microfabrication techniques involve processes where concentrated deformation energy at selective path or areas is applied to bring in the required changes. They may be classified as in the following table (Fig. 6.5).

In the development of flexible machines, to achieve precision irrespective of job specifications, it is prudent to hybridized different techniques at one station.

While processing, it may be convenient also to start at a point source on the job and then to extend it to any contour (a generation principle) to the requirement would be beneficial as well.

For an example in thermal processing, a single laser beam may be used to process a material for heat treatment, alloying and machining on a surface simultaneously with different process conditions at a speed faster than the conventional techniques.

MICROMANUFACTURING			
FORMING		**MACHINING (Subtractive)**	**GENERATIVE (Addition or Creation)**
CASTING [5]	FORMING [7]	PHOTOLITHOGRAPHY [11]	Laser Assisted CVD [18]
INJECTION [6]	STAMPING [7]	LIGA & SLIGA [11]	Micro Stereolithography [19]
VACUUM CASTING [6]	LASER FORMING [8]	DIMOND Milling [12]	Electro Deposition [20]
	ELECTROFORMING [9]	MicroAJM [13]	Micro Plasma Spraying [21]
	Electrohydro FORMING [10]	MicroUSM [14]	EB Deposition [22]
		MicroECM [14]	FDM using ECD [23]
		MicroEDM/EDG [15]	
		LASER Micromachining [16]	
		PLASMA Micromachining [17]	
		EB Micromachining [17]	
		DRY ETCHING [17]	

Fig. 6.5. Micromanufacturing Laser/EB/ED Micro-heat-treatment/alloying/cladding/welding [24]

Micro-patterns and casting systems, e.g. Solidscape from Solidscape, Inc., USA is capable of producing patterns with minimum feature size of 0.25 mm and 0.1% dimensional error [5] and PPC 3200s from Schultheiss GmbH, West Germany is a high temperature casting unit for casting metal components of materials with melting temperature up to 2100°C [3]. Minimum feature size of about 0.5 mm in materials: steel, nickel-chromium alloys, gold, silver or Ti-alloys is possible.

The Battlefield Microsystem50 [6] from UK MIG has the accurate dosing and high injection rates required for micromoulding combined with features such as an integrated clean room and modular assembly which makes it an appealing choice for manufacturers of medical and MEMS devices. Fabrication of 3-D controlled surface microgeometry by the technology of plasticity is being done at Materials Fabrication Lab., RIKEN, Japan.

The University of Bradford/Rondol Micromoulding In-Line Compounder [5] has developed machines with small shot masses associated with the micromoulding process. Residence times of polymer melt can be quite high even in the small extrusion screws found in these machines. This can be a problem when processing temperature sensitive materials. The MIC machine was a result of a design originally conceptualized two years ago to address this kind of problem. The machine consists of a 16 mm twin screw extruder feeding a metering and injection piston system, allowing compounding of novel materials such as nanocomposites directly followed by injection. This process subjects the material to a single heating–cooling cycle and causes lower thermal degradation than two separate compounding/moulding processes. The screw is starve fed which reduces the amount of material in the barrel when compared with a single screw machine and it is capable of efficiently feeding even very low molecular weight materials. However, IIT Kharagpur has taken an initiative in developing micro powder feeder with single 4 mm screw with ultrasonic assisted levitation.

From the same company, the Rondol High-Force Micro-Injection Moulding Machine uses a single electric motor with a system of mechanical toggles to clamp the mould and inject the shot in one process. The machine is very efficient, takes up very little space and can mould shots ranging from 15 g to a fraction of a gram.

Hot Embossing [7] is in great demand for low-cost methods for high volume production of micro-components and-systems, requiring a low-cost substrate material in one hand and an easy way of microstructuring in the other.

Micro Electroforming [9] is again extremely powerful technique to produce micro metallic parts as a stand-alone process or by hybridization (as in LIGA and SLIGA). Precision electroforming is the process whereby a metal object is built-up,atom-by-atom, onto a contoured surface called a mandrel. Nickel or another desired metal is deposited onto a contoured mandrel to a precise tolerance and thickness in order to manufacture objects with a unique shape or surface detail. Often the contoured original or mandrel is a "one-of-a-kind" object that has been specifically created for its use as the original. Some examples are CD and DVD masters, and special lens molds. Often the surfaces and contours of electroforming mandrels are made to tolerances exceeding $\pm$ 0.05 μm. Electroforming can faithfully reproduce these tolerances.

Through manufacturing of components of pure materials is successful, yet process technology for alloys are in search. Metallic microstructures with relatively large thickness (i.e. 10–1000 μm) in the field of MEMS can be manufactured. For example, metallic microstructures with higher thickness for higher structural rigidity and/or higher actuation force are required, greatly needed in the actuation system. Metallic micronozzles and microchannels, as the essential fluidic parts in inkjet heads, are in need with the growth of printer market. Metallic microstructures with small feature and high aspect ratio are demanded as precise hot-embossing masters for plastic micromachining.

Micro Electrohydro forming of polymers/metals thin sheet components are possible with micro spark discharge energy from condensers [10].

Micro Laser forming [8] can be used for micro forming of tubes and thin sheets of materials. The laser-induced thermal distortion is used to the advantage bending of materials, here a tube.

Many more, e.g. spray forming, plasma forming, spark deposition forming etc. are yet to come into practice. Higher accuracy and precision are achieved by micro machining processes that are in vogue.

The most prevalent method in MEMS/MS technology is Photolithography [11], discussed by many. In the process (as in the scheme, Fig. 6.6) a suitable mask is created on a substrate and its thickness is controlled by single or multi layer spay or spinning, and fixed. After generation of the pattern either by contact/projection

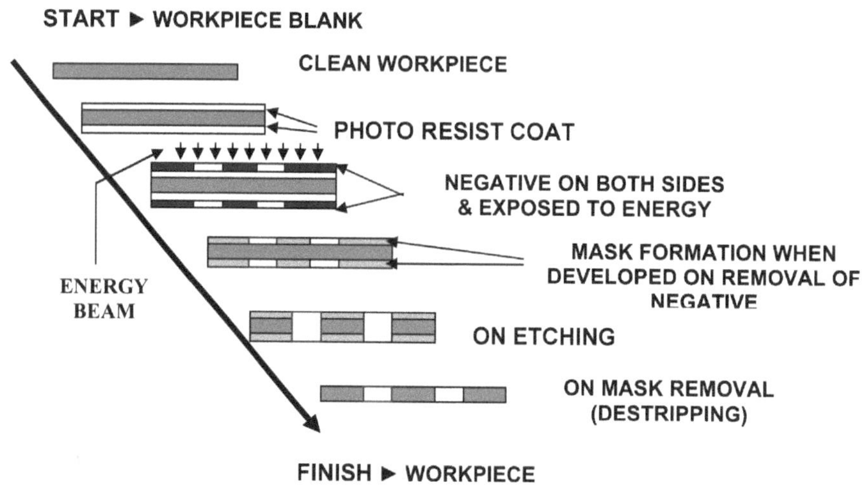

Fig. 6.6 Basic scheme of photolithography [24]

printing or pattern-generator, it is metal coated to fill the voids, inspected and repaired. Then the resist is stripped to leave the chromium pattern on the surface before etching (micromachining). However, the resolution of the pattern depends on the illuminating radiation wavelength, mask thickness, depth of focus etc. as shown in the equations below.

For Proximity and Contact Printing:

$$resolution = \sqrt{\lambda(gap\ between\ mask\ and\ resist\ +\ resist\ thickness)}$$

where λ is the wave length of illuminating radiation.

For Projection Printing:

$$resolution = (K_a\lambda)/(NA);\ Depth\ of\ focus = (K_b\lambda)/NA^2$$

where K_a is a process constant ~ 0.75,

K_b is a process constant ~ 0.5 and

NA is the numerical aperture of the optical system.

To increase resolution, ultraviolet (UV) and deep ultraviolet (DUV) sources, include excimer lasers, which operate at wavelengths of 248 nm, 193 nm, and less are used; however wavelength 193 nm and below are associated with problems in its absorption. Much shorter [11] wavelength of the electrons is one reason for increased resolution using E-beam lithography. X-ray lithography uses collimated rays as the exposing energy. Being much shorter in wavelength than the previous, x-ray provides higher lateral resolution. For its penetrating power, x-ray deep into the photoresist, micromanufacturing of microstructures with "great height" and "high aspect ratio" are possible. Examples of micro gear and even inclined slot machining (with inclined beam projection) has been achieved using LIGA process by Friedrich Craig [14].

LIGA (Lithographie, Galvanoformung, Abformung): Lithography, electroplating, and molding are hybridized process to manufacture micro components. X-ray lithography is used to create a mold with PMMA resist, placed on a thin metallic base. The PMMA resist is ablated to the plating base and, when developed, reveals the metallic (plating) base. The plating base is used as the cathode as in an electroforming process to fabricate a metallic microstructure within the mold. The metallic negative of the PMMA mold is used as a mold insert for subsequent injection molding or hot embossing. These create duplicates of the PMMA mold fabricated previously by the x-ray. This helps replicate the components for mass production of microsystems. Some examples of microcomponents and microsystems fabricated by x-ray lithography are shown [14]. The spherical surfaces is obtained by softening on controlled heating (can also achieved by any energy beam).

Mechanical machining (micro milling) is still used widely to manufacture (finish) component using top-down technology. It is the most flexible method for creating 3D surfaces from variety of engineering materials with features that range from tens of micrometers to a few millimeters in size. Traditionally, ultra-precision machining uses diamond tool-cutting operations, but because of high demand for machining ferrous materials, a micro-milling operation using carbide tools is considered the material, geometry of micro-tools and precision of machine tool are important factors in this operation [25]. Miniaturized machine tools, referred to as mesoscale machine tools have been proposed as a way to manufacture, micro/mesoscale mechanical components. However, a through study of dynamic behaviour of these machine tools is required for the successful development of its machine structure [26].

The runout of the tool tip even within the microns greatly affects the accuracy of micro-milling as opposed to the conventional milling (Bao and Tansel, 2000b). Micro-milling is associated with sudden tool failure due to highly unpredictable cutting action (Bao and Tansel, 2000c). The tool deflection in the micro-milling in the micro-milling greatly affects the chip formation and accuracy on the desired surface as compared to conventional milling (Dow et al., 2004). The tool edge radius (typically between 1 and 5 μm) and its uniformity along cutting edge is highly important as the chip thickness becomes a comparable in size to the edge radius (Lucca, 1993; Melkote and Endres and Seo, 1998). Micro-milling may result in surface generation with burrs and increased roughness due to the ploughing-dominated cutting and the side flow of the deformed material when the cutting edge becomes worn and blunter (Lee and Dornfeld, 2002) [27].

The crystalline texture of the materials resulting from its processing could lead to variation of the chip thickness. In addition, such variations could be caused by changes in shear angle from grain to grain due to varying material properties such as elastic modulus (E). However, it should be said that anisotropic cutting conditions resulting from these effects may be attenuated or eliminated by refining the grain structure or strain hardening the material before machining.

The defects in the crystalline structure influence strongly the material properties and affect directly the metal cutting conditions. In the micro-milling, during the cutting process micro-cracks along the grain boundaries develop and also dislocation slip occur in metal's crystal structure. The specific processing energy is required to initiate the chip formation depends directly on the ability of metals to produce dislocation slips. By enhancing mechanical strength of metals the mobility of the dislocations is reduced and higher cutting forces will be required to move sharp tool through the material. Also, during the cutting, the dislocation density increases due to the formation of new dislocations and dislocation multiplication. It is considered that refinement of the grain structure could lead to more "favorable" conditions during cutter-material interactions and thus result in better machining response, especially at micro-scale.

One problematic area of the micro-milling is the low stiffness of the tools. The section modulus of the cylindrical bars decreases cubically with decreasing diameter. Therefore, end mills with small diameters have a low ability to compensate forces and torques without failure. Process forces lead to tool deflection and, as a consequence, to a dimensional deviation from the desired contour. In unfavorable conditions this may lead to tool breakage. The other problems are high tool wear and its assessment, and development of technique of microfabrication of the end milling cutter.

For the above reasons non-traditional/non-conventional processes with proper hybridizations are essential.

Mechanical micromaching is also used to micro mill of components with features down to 50–100 μm, with aspect ratio 3–10 for non-ferrous materials, and 1–2 for ferrous materials with Kern HSPC 2216 [11]. The cutting tools, double or multi flute are manufactured by EB or Ion Beam machining [10]. A lot of work is being done at Northwestern University [28] and MEL, NIST Japan on the development of micro machining (turning) with micro work-handling systems.

Marcel Achtsnick [13] from Delft University of Technology reports of using micro Abrasive-Air-Jetting to machine minimum dimensions reachable to about 30 μm in width and down to 1 μm in depth. It uses 3–50 μm size in a working area of about $600 \times 600 \times 300$ mm is carefully closed and exhausted.

National Research Laboratory of Korea [15] uses micro USM to machine holes as small as 70 μm in ceramics. Masuzawa Laboratory of University of Tokyo has developed micro USM to achive small features precisely. One of most delighting development is micro Water Jet which not only can be used for achieving micro features but also aggressively used in surgery and biomedical applications.

Micro ECM [15] is the method to produce micro tools in large quantities by electro-chemical etching method. Micro ECM can produce sharp tips used in various fields such as electrochemistry, cell biology, elecrodischarge machining, field-ion, electron microscopy, nano electronics, and field emitters. Moreover, it can be used in surface finishing of machined feature. With the use masking technique, it is now possible to use micro Electrochemical machining to etch electrically conductive components with micro features, is under development at IIT Kharagpur.

However, the replenishing of the electrolyte into the working zone would have to be improvised with Ultrasonic assistance.

However, each process has its limitations but the most advantageous that the processes are non-contact type, giving the chances for hybridization. On suitable mixing of the processes in conjunction with conventional one would improve the surface integrity and production speed and precision, giving a chance for future developments.

In the domain of manufacturing, besides machining, other processing techniques, e.g. heat-treatment, welding, alloying etc. are also necessary for micromanufacturing. These being in the domain of thermal processing, the existing nontraditional methods like EDM; Laser, Electron and Ion-beam are used successfully to tailor the surfaces (surface rather than bulk manufacturing technology). These methods may be considered as futuristic technology for the development of flexible machine. It would be clear from the basics of the following Fig. 6.7, on the application of the very localized heat source at the point "A" on the surface, a desired change (micro-structural change, melting or evaporation) is observed, while at a sub-surface point "B" an undesired (but sometimes advantageous) change at lower temperature occurs. This undesired change might be referred as thermal damage, distortion or accuracy etc. in the processing operation. If one looks at the problem, finds that the application of high energy-rate results in faster rise of temperature at "A" as compared to that at "B" (conduction heat transfer). Gradually this temperature difference increases with increase energy dumping rate! So it is possible to achieve evaporation on the surface with or without any change at the subsurface point, "B". These phenomena can easily be predicted with a proper thermal analysis to tailor the surface. It is obvious that multiple processing operations is also possible by a single source of thermal energy with variable energy density and interaction time (Table 6.1) bringing in the concept of flexible machine/s. However, there is no direct form thermal energy input (excepting Plasma Processing) but the energy absorption generates heat within a few micron depths below the surface (adsorption).

The widely used thermal processing is by Electrical Discharge and Laser. Micro EDM is of course the best for micromachined metallic components [16] as shown in Fig. 6.8.

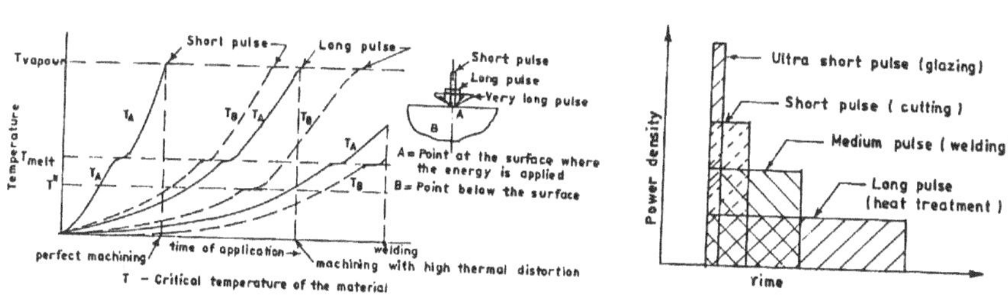

(a) Temperature rise with interaction time

(b) Pulse definition

Fig. 6.7 A basic concept of thermal processing [24]

Table 6.1 Working domain for material processing

Processing	Specific Energy (J/cm^2)	Power Density (W/cm^2)	Interaction Time (Sec)
Shock Hardening	$10–10^2$	$10^8–10^9$	$10^{-8}–10^{-6}$
Glazing	$1–10$	$10^6–10^7$	$10^{-6}–10^{-4}$
Machining	$10^3–10^4$	$10^5–10^6$	$10^{-5}–10^{-3}$
Welding	$10^3–10^4$	$10^5–10^6$	$10^{-2}–10^{-4}$
Transformation Hardening	$10^4–10^5$	$10^4–10^5$	$10^{-4}–10$

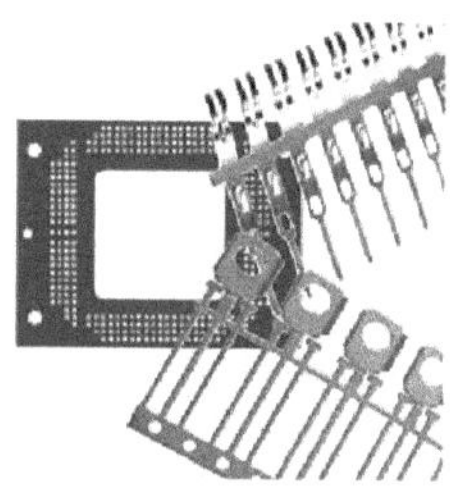

Microelectronics Parts[16]

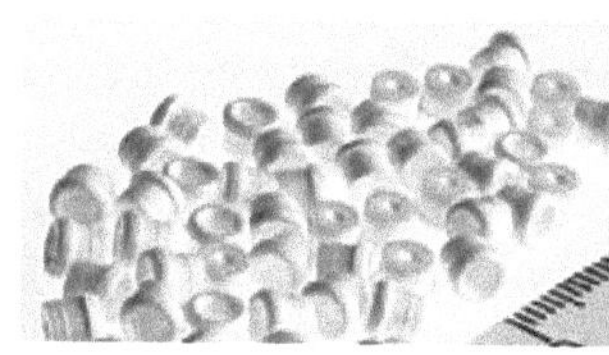

Micro Molds: Aerosol spray nozzle with high quality finishing on precise micro holes [16]

Fig. 6.8 EDMed micro-components [16]

6.4 Futuristic Manufacturing (Laser Based)

Micro manufacturing activity if can be maintained at one station would be the best solution, where the Lasers have many role to play for ideal hybridization. An attempt to this illustrates the feasibility.

The figure below (Fig. 6.9) shows the multiple processing operations by a single source (in this case, a Laser system).

While mentioning about Laser microwelding, trial (Figs. 6.10, 6.11 and 6.12) has been attempted [29] besides the many others, globally. Microwelding with Electrochemical discharge [30] is also successful.

When one comes to Generative manufacturing processes, Electrochemical Discharge Fused deposition finds application micro-FDM [30]. The new microstereolithography (μSTL) machine in the CMF [19] is a system for creating 3 dimensional shapes from an AutoCAD design and a photo reactor with liquid resin. It has Lateral and Vertical resolution: 10 μm, Maximum field size: 10.24 mm × 7.68 mm and Structural height: up to 5 mm. 3-D printing is also commercially viable systems, useful for use of Rapid manufacturing of micro to nano level manufacturing.

There are many more innovations are in pipeline all over the world to produce micro components from any material with high precision and accuracy. However to mention here, a method of integration of different laser systems (Excimer and CO_2) have been integrated with a 4-axis CNC manipulation to process micro to macro level manufacturing at IIT Kharagpur (Fig. 6.13). Few micro machined samples are shown in Fig. 6.14.

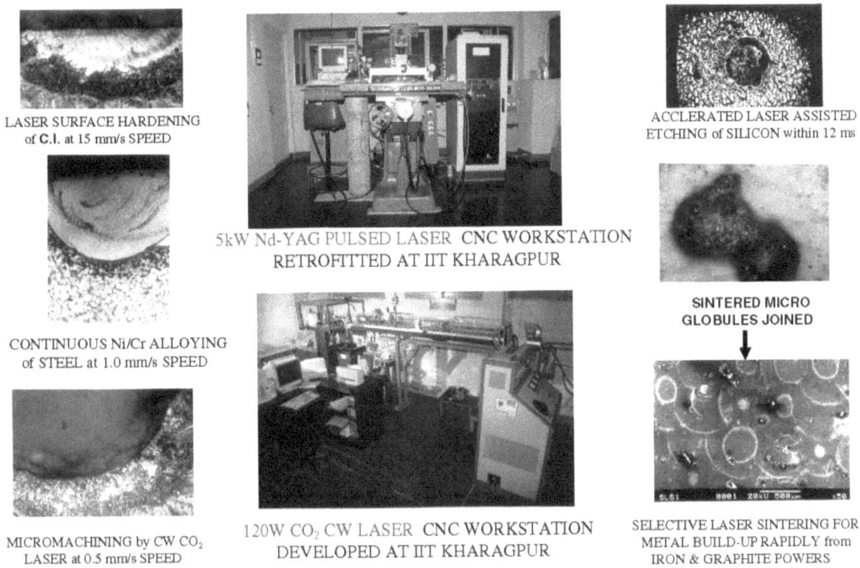

Fig. 6.9 Laser processing as flexible manufacturing tool [31]

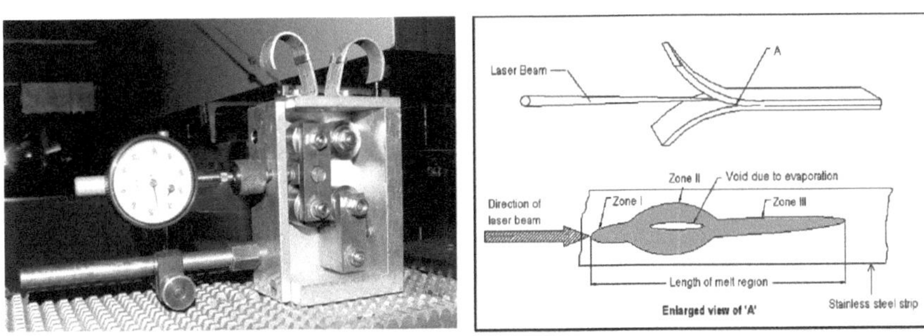

Fig. 6.10 Photograph of set up and welding procedure

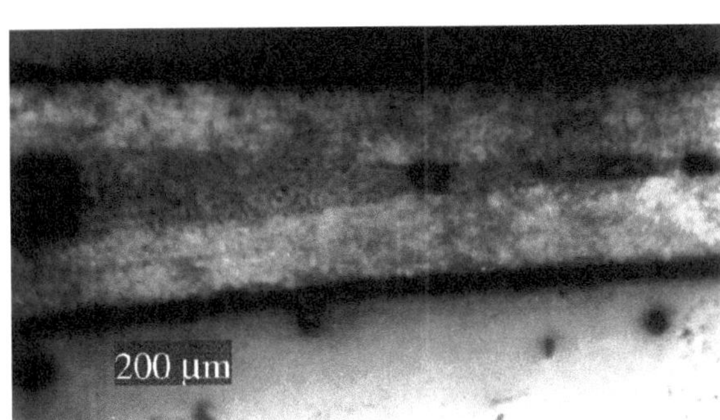

Fig. 6.11 Optical micrograph showing a good welded zone

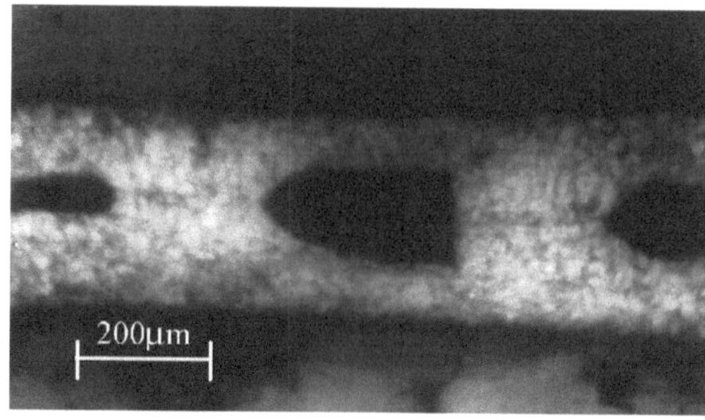

Fig. 6.12 Optical micrograph showing void in the welded zone

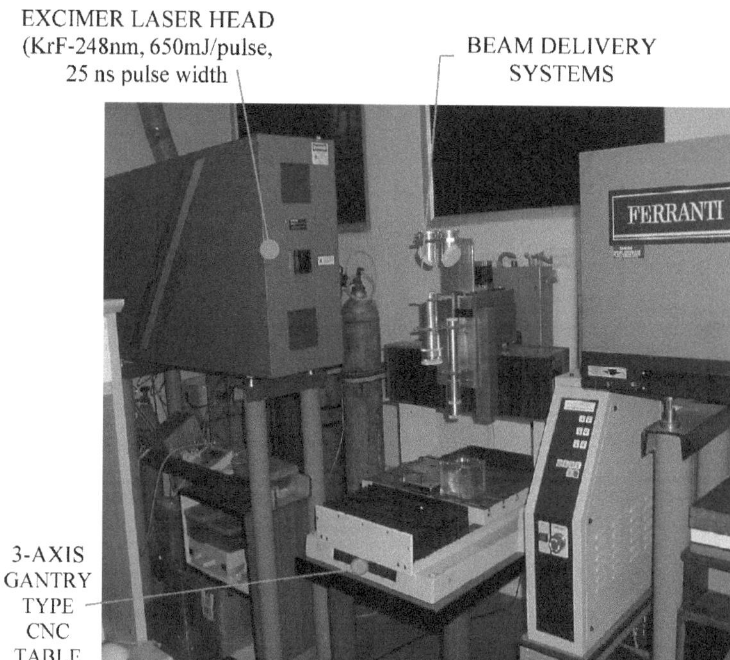

Fig. 6.13 A flexible manufacturing system using an excimer and a CO_2 Laser System

The days are not far, when the manufacturing engineers can manufacture components in micro to macro level in one machine and would be able to deliver the components directly from the design communicated through Internet to the machine, including the material development inside the fabricating machine! However, proper modeling of the processes is to be developed with micro energy transfer and micro mechanics principles.

Few Innovative ideas

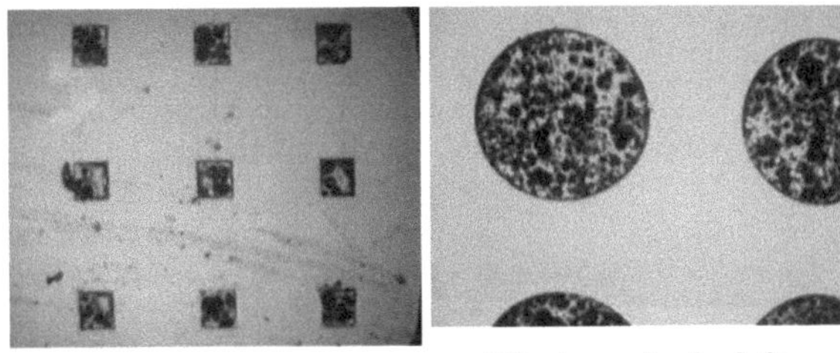

70-micron square hole   100-micron circular hole

Fig. 6.14 Micromachined PMMA with Excimer laser

Earth worm principles (vibrating wire); Surface wave techniques (acoustic wave, laser); Wire sawing (with impregnated diamond wire or diamond paste), Hot wire technique.

Acknowledgment We sincerely convey my gratitude to Dr. Partha Saha who has given his untiring assistance to buildup the systems available in EPP and Laser lab and experimentations at IIT Kharagpur.

We sincerely acknowledge the constant encouragement and every support extended by Dr. A D Sahasrabudhe in the establishment of Microsystems Engineering Lab at Mechanical Engineering Department of CoEP. We also acknowledgement the concerned faculty members, Ms. S S Bhavikatti, Mrs. P J Muley, Mr. V K Haribhakta and Mr. S V Wagh for their untiring efforts in the Lab development.

Further Readings

1. Altshuller G (2005) The Innovation Algorithm, Technical Innovation Center Inc, Worcester, MA.
2. Benidict GF (1987) Nontraditional Manufacturing Processes, Marcel Dekker, Inc. NY.
3. Bhattacharyya A (1973) New Technology, The Institute of Engineers, India, Calcutta.
4. Campbell SA (1996) The Science and Engineering of Microelectronic Fabrication, Oxford University Press .
5. Ehelich DJ and Tsao JY (1989) Laser Microfabrication: Thin Film Processing and Lithography, Academic Press, New York.
6. Elwenspoek M and Jansen H (1998) Silicon Micromachining, Cambridge University Press, Cambridge.
7. Eric Drexler K (1992) Nanosystems: Molecular Machinery, Manufacturing and Computation, Wiley-Interscience Pb., USA.
8. Fatkow S and Remobold U (1997) Microsystem Technology and Microrobotics, Springer, New York.
9. Fukuda T and Menz W (1998) Micro Mechanical Systems: Principle and Technology, Elsevier, Amsterdam.
10. Gibson I Ed. (2005), Advanced Manufacturing Technology for Medical Applications, John Wiley and Sons Ltd, New York.
11. Helvajan H Ed. (1999) Microengineering of Aerospace Systems, The Aerospace Press, California.
12. Hsu TR (2002) MEMS and Microsystems Design and Manufacture, Tata McGraw-Hill India.
13. Jain Vijay K (2002) Advanced Machining Processes, Allied Publishers. New Delhi.

14. Kotake S and Tien CL Ed. (1994) Molecular and Microscale Heat Transfer, Begell House, Inc.
15. Madou Marc J and Boca Raton (2002) Fundamentals of Microfabrication: the science of miniaturization, 2nd ed., CRC Press, New York.
16. McGeough JA (2002) Micromachining of Engineering Materials, Marcel Dekker, Inc.
17. McGeough JA (1988) Advanced Machining Methods, Chapman and Hall .
18. Metev SM and Veiko VP (1998) Laser Assisted Micro-Technology, Springer, New York.
19. Nakejawa, H. (1994), Principles of Precision Engineering, Oxford University Press .
20. Ratner M, and Ratner D (2003) Nanotechnology (A Gentle Introduction to the Next Big Ideas, Pearson Education, Inc.
21. Taniguchi N (1989) Energy Beam Processing of Materials, Clarendon Press, Oxford.
22. Tien CL, Majumder A, and Gerner FM, Ed. (1997), Microscale Energy Transport, Taylor & Francis, London.
23. Zheng C (2005) Micro- Nano Fabrication: Technology and Applications, Springer, New York.

References

1. Snoeys R, Stallens F, and Dakeyser W (1986) Current trends in Non-conventional material removal processes. Annals CIRP, 35(2):467–480.
2. Nakazawa H (1994) Principles of Precision Engineering, Oxford University Press.
3. Taniguchi N (1983) Current status and future trends of ultra precision machining and ultra fine materials processing. Annals of CIRP 32(2):1–8.
4. Taniguchi N (1994) The state of Art of Nanotechnologies for processing of ultra precision and ultrafine products, Precision Eng. 16(1):6–24.
5. A. Rankin (Dec. 2002) Micromoulding, Medical Device Technology Magazine, www.solid-scape.com/pm.html, http://www.ukmig.com/ukmig/MIC.html.
6. www.ukmig.com/ukmig; www.riken.go.jp/lab; www/mat-fab/personal/H.Ike/ SMG.html.
7. Geiger M et al. Microforming, Annals of CIRP, keynote paper, V 50/2/2001, 445–459.
8. Li W and Lawrence Y (Nov. 2001) Laser Bending of Tubes: Mechanism, Analysis, and Prediction, Tr of ASME, V-123, 676–681. www.fzk.de/pmt/, http://www.imm-mainz.de.
9. http://rf-serviceplace.com/, www.commercialisation.strath.ac.uk/ Research Expertise/ Manufacturing Engineering.html.
10. Spiro P, Electroforming: A comprehensive survey of theory, practice and commercial applications, ISBN:085218039X, http://www.strictlynano.com/, Guttmann M and Moritz H, Improved Handling of Wafers for Micro-Electroforming, Letzte Änderung Dienstag, 27 Jan. 2004, Thin-Film Forming of Cluster Diamond-Dispersed Aluminum Composite by Dynamic Compaction, 0-87849-857-5.
11. Danny Blank, Introduction to Microengineering: MEMS Micromachines MST http://www.dbanks.demon.co.uk/ueng/.
12. www.kern-microtechnic.com.
13. Achtsnick M, Micro-Abrasive-Air-Jetting: Micromachining of Brittle Materials, PTO, TU Delft.
14. Friedrich C, Precision Manufacturing Processes: applied to miniaturization technology, Michigan Technological University.
15. Masuzawa T and Egashira K, Microultrasonic machining by the application of workpiece vibration, Annals of CIRP, V 48/1/1999, http://prema.snu.ac.kr/.
16. Yu ZY, Masuzawa T, Fujino M, Micro-EDM for three-dimensional cavities – Development of uniform wear Method, Ann. CIRP, V 47/1/1998, www.sarix.com, www.panasonicfa.com, www.unl.edu/nmrc/microEDM, Masuzawa T, Tönshoff HK, Three- dimensional micromachining machine tools, Keynote Paper, Annals of CIRP, V 46/2/1997, 621–628.

17. McAuley SA et al. (2001) Silicon micromachining using a high-density plasma source, J. Phys. D: Appl. Phys. **34**(21 September 2001): 2769–2774, Elwenspoek M, Jansen H, Silicon Micromachining, Cambridge University Pr., 1998.
18. http://www.cmf.rl.ac.uk
19. http://www.cmf.rl.ac.uk
20. Goetz J (2001) Electro-Spark deposition coating for replacement electroplating, Pollution Prevention, pp-1147, agoetz@pica.army.mil
21. http://www.ecn.nl/_files/tsc/ECN-B–03-006_Micro_PlasmaSpraying.pdf
22. http://dmtwww.epfl.ch/ioa/research/reports/Hoffmann/Hoffmann98-2.pdf
23. Sengupta A, Choudhury A, Mishra PK and Ghosh A (2001) Development of layered manufacturing system using FDM-ECD, Proc. of 5th India-Japan Joint Seminar, Manufacturing of Advanced Composite, 9–10.
24. Mishra PK (1997) Nonconventional Machining, Narosa Publication House, New Delhi.
25. Chae J, Park SS and Freiheit T (2006) Investigation of micro-cutting operations. Int. J. Machine Tools Manufacture, 46:313–332.
26. Sang Won Lee, Rhet Mayor & Jun Ni (February 2006), "Dynamic analysis of a mesoscale machine tool", Transaction of the ASME, 128:194–203.
27. Dhanorkar A and Ozel T (2008) Meso/micro scale milling for micro-manufacturing Int. J. Mechatronics Manufacturing Syst. Vol. 1, No. 1.
28. www.mech.northwestern.edu
29. Ganesh J, Misra D, Saha P and Mishra PK (2002) A feasibility study of microwelding of stainless steel thin sheets using pulsed Nd-YAG laser. Proc. AIMTDR conference 337–342.
30. Sengupta A, Choudhury A, Mishra PK and Ghosh A (2002) Development of micro welding using Electro-chemical discharge, Proc. 20th AIMTDR Conference, 363–368.
31. Mishra PK, and Saha P (1992–2002) Compiled from works done in Laser Laboratory, Mechanical Engineering Department, IIT Kharagpur, India.

Chapter 7
Bio-Inspired Adhesion and Adhesives: Controlling Adhesion by Micro-Nano Structuring of Soft Surfaces

Abhijit Majumder, Ashutosh Sharma, and Animangsu Ghatak

Abstract Although the man made synthetic adhesives have quite high adhesion because of their viscoelasticity or irreversible chemical bonding, they are not reusable and are often prone to particulate contamination and cohesive failure. On the other hand, attachment pads found at the feet of different insects and climbing animals like geckos show high adhesion, self-cleaning and reusability. Decades of research have confirmed that the patterns and structures present at the surface of or buried inside the natural adhesive pads have rendered them these amazing qualities. These observation inspired scientists and researchers to mimic the structures to fabricate soft, synthetic reusable adhesives. This chapter will present a brief review on those efforts with a focus on structure and mechanism of patterned bio-adhesives and synthetic pressure sensitive adhesives.

Keywords Adhesion enhancement · bio-mimetic · gecko adhesive · hairy adhesive · patterned surface · fibrillar interface · surface structuring/pattering · micro-fabricated adhesive · sub-surface structures · microfluidic adhesive · reusable adhesive · carbon noano-tube.

7.1 Introduction

Almost all of the insects and many vertebrates like lizards can attach themselves to a stationary object, can climb on a vertical surface and can even walk upside down [1, 2]. This amazing climbing ability has been intriguing scientists and common folks alike for ages. As the Indian legend goes, Tanaji Malusare, the military leader of Maratha warrior-king Shivaji, used lizards as gripping device to climb on a rock cliff and mount a surprise attack on his opponent in the battle of Sinhagad, in 1670 [3]. There has been intensive activity to understand the mechanism of insect and gecko adhesion and in trying to mimic them. Among other things, a climbing

A. Sharma and A. Ghatak (✉)
Indian Institute of Technology Kanpur, Kanpur-208016, India
e-mail: ashutos@iitk.ac.in; aghatak@iitk.ac.in

S. Chakraborty (ed.), *Microfluidics and Microfabrication*,
DOI 10.1007/978-1-4419-1543-6_7, © Springer Science+Business Media, LLC 2010

animal uses its claws and/or spines to negotiate a rough surface [4]. However, when the surface roughness becomes comparable to the claw tip diameter, this mechanism fails [5]. In that situation, attachment is achieved by using the adhesive/attachment pads present at their feet and other body parts. To have a reliable adhesion, the pad must be compliant enough so that it can come to a close contact with the adhering surface, which in general has microscale roughness. In nature, the compliance of the adhesive pads is achieved in two different ways namely by decreasing material modulus and/or by surface structuring. For many insects with comparatively low body weight, the pads are made of soft material and are relatively smooth. In second type, which is found in species with higher body mass like geckos, the pads are highly patterned/hairy [6]. There may or may not be any body secretion involved in their adhesion process. When there is no body secretion involved, the adhesive is called "dry adhesives". The natural adhesives, irrespective of their type, are reusable, self cleaning, easy to detach and free from cohesive failure [7]. All these qualities are essential for effortless and rapid locomotion of the species. Such an adhesive are in great demand in MEMS and microelectronics industry for integrating micro and nano devices into complete device architecture [8, 9]. Other than that they can be very useful in automobile, robotics and biomedical applications too [10–14]. However none of the manmade adhesive possesses all of these qualities at the same time, be it our daily use adhesive tape or acrylic based reactive glues. Because of this, geckos and insect seems to be the source of inspiration for fabricating an adhesive which will be reusable yet strong. This chapter will discuss the present day knowledge about dry adhesive pads and will review the recent efforts in mimicking them.

7.2 Synthetic Adhesives: Strong but not Reusable

Before we enter into our discussion on bio and bio-inspired adhesion, let us first understand how the synthetic adhesives work and where and why they fall short. So, what is the source of adhesion in all these different adhesives? The first obvious component which contributes to the overall adhesive energy is the surface energy. Bringing two surfaces in contact creates one new interface in place of two free surfaces. Energy associated with this process is $\Delta E = \gamma_{12} - (\gamma_1 - \gamma_2)$. When the intermediate medium between these two surfaces is vacuum or air, the energy associated ΔE is always negative, indicating that the adhesion is a preferred situation. For other intermediate media, possibility of adhesion greatly depends on the interaction between the medium and the individual surfaces. This energy of adhesion is commonly referred as thermodynamic work/energy of adhesion which falls in 0.01–0.1 J/m^2 range [15]. However, for any good adhesive, adhesive strength is found to be not less than 1 J/m^2 which is at least one order of magnitude higher than the thermodynamic work of adhesion. The other contributors to the overall adhesion energy account of commonly used adhesives are permanent (plastic) deformation, viscous flow in the bulk of the material, micro-interlocking and interfacial friction [16].

The adhesives that we use everyday must possess both liquid and solid properties. When applied on a surface, it should be "liquid" enough to flow and wet the surface nicely. Unless it comes in intimate contact with the surface, the adhesive strength remains low. However, when a load is applied, the adhesive should not behave like a liquid any more. It should have the strength to bear that load so that it does not separate simply because of cohesive failure [16, 17]. Chemical glues achieve this by solvent evaporation (e.g. polystyrene), chemical reactions (e.g. cyanoacrylate, two-part resin) or by a change in temperature (e.g. hot melts) [18]. There is another class of adhesives called pressure sensitive adhesives (PSA) which are used in adhesive tapes. These adhesives are thin films of polymers with low moduli. The low modulus is required for an intimate initial contact with the adhering surface. The upper limit of tensile elastic modulus at 1 Hz for a good PSA is given as 100 kPa [19]. With a little applied pressure, these polymeric films behave like a liquid and wet the surface. During debonding process, the thin film of such a soft polymer undergoes different instability driven morphologies like fingering, cavitation and fibrillation [20, 21]. The same morphological development can be seen for elastic adhesives as well [22–26]. These morphologies allow the adhesive to stretch a lot, resulting in a large separation distance of the surfaces which may at times extend to a few millimeters before a complete separation [15]. The final separation may take place either by snapping of the fibers, i.e., cohesive failure or by detachment of the fibers from their base substrate, i.e., by adhesive failure. In this whole process, a large amount of energy gets dissipated due to plastic deformation and bulk viscous flow. The energies thus spent during the process of debonding add up to give the final fracture toughness of an adhesive tape [27].

The adhesive energy achieved by the some of the above mentioned adhesives like chemical glues or hot melts can be as large as 1000 J/m^2 which is quite satisfactory for all practical purposes [28]. For a soft adhesive like PSAs, the adhesive energy typically falls in the range of 100 J/m^2 [29, 30]. However, they fall short in terms of reusability. After sticking two surfaces with these adhesives, if one of the surfaces is removed, it becomes difficult to stick it again with the same strength. The reasons of this non-reusability are permanent change of state and composition in case of reactive glues and permanent change of shape, cohesive fracture and particulate contamination in case of PSAs [20, 31].

However, as we have already mentioned, natural adhesive pads are free from these shortcomings. The mystery of their super adhesive power thus lies in the micro-nano structures of their adhesive pads, rather than in chemical bonding or visco-plastic dissipation.

7.3 Structures of Bio-Adhesives

7.3.1 Surface Patterns

Surface of natural adhesive pads found in the nature is often highly patterned and contains multiscale structures. The length scales of these patterns vary from millimeter to a few nanometers. The structure of the pad becomes finer and more intricate with increase in the body mass of the species [32]. The scanning electron microscopy images of some of the natural adhesive pads shown in

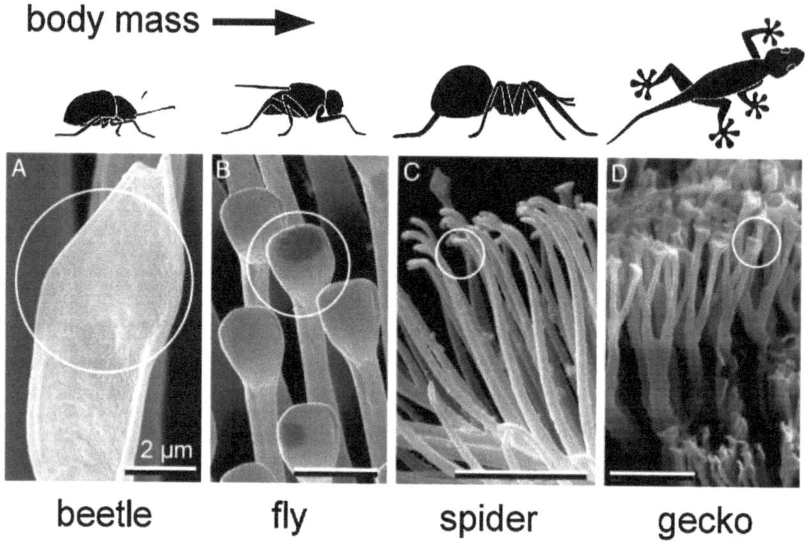

Fig. 7.1 Surface structures present on the adhesive pads of different animals. As the body mass of the species increases, the structure becomes finer. Reproduced from reference [32] © (2003) National Academy of Sciences, U.S.A

Fig. 7.1 bring out this fact. Among all of the climbing animal species studied, geckos have the highest body mass and the most complex pad structure. The adhesive pads found in geckos and in many other species are hierarchical in nature. For example, thousands of hair-like structures called setae are arranged in groups in the adhesive pad of Tokay gecko [33–35]. Each of these groups is called lamellae which are placed over blood sinuses that work as a sort of hydraulic suspension. A foot of Tokay gecko has nearly 6.5 million keratinous setae (~14,000 setae/mm^2), each branching out into 100–1000 finer triangular projections called spatula [36–38]. A spatula is 200–500 nm in length 5 nm in thickness and is connected to the setae by the apex of the triangle [34, 38]. The shape of the spatula varies from species to species. Using lamellae to spatula, because of their continuously decreasing length scale, the gecko pad can adjust to millimeter to nanometer scale of surface roughness and can hold 20.1 N of force parallel to the surface with 227 mm^2 pad area [38]. However, adhesion force measurement of individual setae shows that with 6.5 million setae on the toes, one gecko should be able to hold a shear load of 133 kg if all of the setae are attached maximally [37, 39]. Interestingly, even with such a high adhesion, a gecko can detach its feet in just 15 ms without any measurable detachment force! [40].

7.3.2 Sub-Surface Patterns

Other than these surface patterns, the natural adhesive pads have subsurface structures too. Figure 7.2A shows the adhesive pad of the insect *Rhodinus prolixus* [41]. In this picture, a large sac filled with fluid can easily be seen.

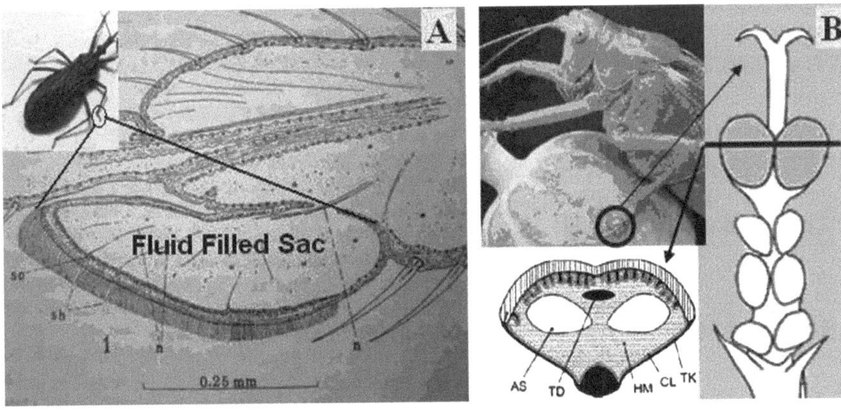

Fig. 7.2 Examples of sub-surface structures filled with different fluids found in insect (**A**). *Rhodnius prolixus* (© Adapted with kind permission from Royal Society, ref [41]) and (**B**). *Tettigonia viridissima* (bush cricket). AS: Air sack CL: Epidermal cell layer HM: Haemolymph, TD: Tendon of the claw flexor mussle, TK:Tanned cuticle. (© Adapted from ref [42] with kind permission from Springer Science+Business Media)

Another example of sub-surface structure can be found in insect *Tettigonia viridissima*, commonly known as bush cricket [42–45]. The attachment pads of this species are not hairy but have micron sized hexagonal structures at their surface [44, 45]. However, underneath the outer layer of the adhesive pad, fibers of diameter $\sim$1 μm branch into nanometer size finer fibers in a hierarchical fiber structure. The space between the fibers is filled with liquids [42, 43]. Other than that, they have two large air sacks floating in a pool of haemolymph in each of their adhesive pads (Fig. 7.2B) [43]. Although, the scientists and entomologists have known the presence of this subsurface fluid carrying structure for long, their possible role in adhesion mechanism has been beginning to be appreciated only very recently [46–48]. In different studies related to natural adhesive pads, these fluid filled structures have been guessed to work as a hydraulic suspension [9] and/or a mechanism to provide additional flexibility of the pad [43].

7.4 Physics of Adhesion

The mechanism of adhesion in the dry adhesive pads has been a matter of debate continuing well over a century. In this period, many different mechanisms have been proposed from time to time including the concept of suction cups, capillary adhesion, friction, electrostatic attraction, micro-inter locking and a near perfect use of van der Waal's force [1, 34, 38, 39, 49–60]. By 1969, other than capillary adhesion and van der Waals force, all the other hypotheses were discarded. The current opinion, with some caution, favors the intermolecular van der Waals force being the principal underlying force responsible for the dry bio-adhesions [57]. Although the van der Waals interaction is most universal, it is also the weakest of all intermolecular forces and its strength decays rapidly with the intersurface separation distance. This indicates that the surface geometry should have as much or even greater role than the specific surface chemistry in gecko/insect adhesion. Other mechanisms like capillary adhesion mediated by a thin film of adsorbed water or by body secretions may be present in some of the specific situations and play critical role to enhance adhesion [59].

7.5 Role of the Structure in Adhesion

As already mentioned in the Section 7.2.2 and shown in the Fig. 7.1, the dry adhesive pads are mostly hairy, covered with millions of slender fibers. As the body mass of the animal increases the number density of the fibers as well as their branching increases. Therefore, the question that arises is: how fibrils and other structures in naturally occurring adhesive pads help in enhancing adhesion? In the next section, we address some aspects of this question.

7.5.1 Roughness Compatibility: How Surface Structures Engender Better Adhesion to Real Surfaces

Intuitively, it may seem that the hairy/patterned pads should have less adhesion because of its less area of contact compared to a similar smooth adhesive! Such a view assumes that the adhesive force is proportional to the contact area and a smooth surface seems to give the maximum contact area. However, for two solid surfaces, microscopic surface roughness brings the actual contact down to a small fraction of the surface area. To achieve a good contact, the adhesive pad or a good adhesive must deform locally to conform to the topology of the adhering surface irrespective of its roughness. The thermodynamic work of adhesion is expressed as:$\Delta\gamma A$ where, $\Delta\gamma = \gamma_{12} - (\gamma_1 + \gamma_2)$, is the energy of adhesion per unit area and A *is* the actual atomic contact area. The actual work of adhesion also includes the work done to deform the adhesive layer to bring it in greater conformal contact, thus increasing A. However, the stored elastic energy in this process tends to decrease the work of debonding [61, 62], because the elastic stresses generated favor separation of the interfaces. Thus, in order to have higher strength of adhesion, the stored elastic energy contribution needs to be minimized [62]. One way to reduce the stored elastic energy is by relaxation of stresses by viscous flow or plastic deformations so that the elastic stresses induced at the time of achieving a good conformal adhesion can eventually relax, making the separation difficult. Further, the ability of a body to deform without storing much elastic energy is called compliance which can be enhanced either by decreasing the elastic modulus or by making the structure more slender by reducing its lateral dimension. Both of these latter mechanisms have been used in animal kingdom leading to two different classes of adhesive pads: the first kind are relatively smooth and soft pads where capillary adhesion mediated by liquid secretion may also be present. The second kind of adhesive pads are covered with millions of slender micro-nano scale hairs. Although the material of the hair may have very high modulus, the effective modulus of the pad as a whole is low because of its hierarchical and hairy structure [29]. This can be explained by considering a dense array of thin curved fibers of length L, radius R, elastic modulus E and aerial density N. For an applied load F, if this array experiences a normal deformation u, then the normal stress thus generated will be $\sigma = NF = Nku$, where k is the spring constant of each of the fiber. Now, let us consider an elastic slab of thickness L and effective modulus E^* such that for same deformation u, it will give the same stress σ. So, we can write, $Nku = E^* u/L$ giving rise to $E^* = NkL$. The spring constant now can be replaced by CER^4/L^3, where, C is a dimensionless number of typically order 10, which depends on the shape of the fiber. So, by plugging the typical values of $R/L \sim 0.02$ and $N = 10^4$ setae per mm^2 for a setal array of gecko into the expression for E^*, we obtain $E^* \approx 10^{-4}E$ which signifies the importance of fibrillar surface in terms of decreasing effective modulus [63]. Experimental results also show that the modulus of the hair found in the pad of *T. gecko* are made of β keratin having modulus as high as 3 GPa but the effective modulus of the gecko setal array is only 100 kPa, interestingly similar to that of PSAs [64]. As a result, the gecko pad can follow the roughness of the surface intimately without storing much elastic energy which is essential to fully deploy the short- range van der Waals attractive force that are most effective at contact [29, 61–63]. As the structures have different length scales varying from millimeter to nanometers, the adhesive pads can fully attach to rough and smooth surfaces alike.

7.5.2 Fracture Mechanics Aspects of Adhesion and Debonding

The above mentioned argument to explain the performance of a hairy adhesive holds good for a rough adhering surface, but it does not explain the observation that the hairy pads perform rather well on perfectly smooth surfaces too! Moreover, pads of many insects do not have the hairs but comparatively large scale patterns. For example, adhesive pads of the bush cricket have micron size hexagonal patterns [42–45]. These observations indicate that the argument based on maximal contact area alone is not sufficient to explain the better performance of a patterned adhesive over a smooth one. To address this inadequacy, different theories and hypothesis based on fracture mechanics have been proposed which can be clubbed into three major groups. These three major groups are contact splitting theory, uniform stress distribution and crack arrest mechanism.

In *contact splitting*, calculation based on the JKR type contact model [65] shows that if the contact area is broken down into "n" number of finer sub-contacts, adhesion increases by a factor of n^r. Here, r is known as splitting efficiency value of which depends on the shape of the tip [33, 38, 66–70]. For example, for a purely elastic hemispherical fiber tip of radius R when in contact with a flat and rigid substrate, JKR analysis predicts the finite pull-off force as $F_C = \frac{3}{2}\pi R\gamma$, where γ is the thermodynamic work of adhesion per unit area. Now, if the fiber-tip branches into n sub-contacts, each with radius $R/\sqrt{n}$ for self similar scaling, or with radius R for curvature invariance, the total pull off force increases by $\sqrt{n}$ and n times, respectively [32]. Here it has been implicitly assumed that all the fibers of a pad across the contact area are uniformly stressed. This is not true when the pad detaches from the surface by peeling because in that case, the stress is concentrated near the edge of the pad.

Uniform stress distribution theory assumes that when two surfaces are separated from contact by an external pull-off force, a uniformly distributed tensile stress over the contact area gives maximum adhesive strength [71–73]. However, for large soft and deformable surfaces, stresses do not distribute evenly over the entire contact area, but are rather concentrated near the edge of the joint [65]. Further, the area of contact does not remain uniform during the pull-off, but breaks into an array of bridges interdispersed by cavities that enlarge from their edges as the separation increases [22–26]. For a soft elastic adhesive layer, the surface instability creates a mean spacing of about 3–4 h between the adhesive bridges and the areas of local adhesive failures or cavities, where h is the film thickness [22–26]. As the load or intersurface separation increases, the intensity of stress concentration at the edge of the cavities reaches a critical value at which the crack starts to propagate from their edges and progressively decreases the contact area, eventually breaking the complete joint [25, 26]. Under this situation, the adhesion strength of the whole contact is not really fully utilized, but limited to only a small fraction of the members which are in contact are highly stressed and failure occurs by incremental crack propagation. The strength of the joint is thus greatly decreased compared to the theoretical estimate assuming a uniform separation without the formation of cavities and isolated points of contact [23, 25, 26]. In contrast, if the contacting bodies are so designed that at the time of pull-off, the two bodies could be uniformly separated over the entire contact region, then the tensile stress will also be uniformly distributed, assuming the molecular interaction to be determined only by the distance between two bodies. Gao and Yao [71] showed that for a fibrillar adhesive, this situation can be realized by reducing the diameter of the fiber below a critical size. In that situation, the contact becomes shape insensitive i.e the adhesive strength becomes independent of the local geometry of the crack tip and the critical stress for crack propagation cannot be reached before it attains the theoretical adhesion strength. A similar concept is put forward by Hui et al. [72] where they have shown that when the diameter of the fiber is brought down below a critical value, the failure does not occur by crack propagation, but the entire interface fails at once. They have termed this condition as “flaw insensitive region”. In this situation, it is possible to attain the critical stress to be equal to the interfacial strength because the failure is no more controlled by the edge singularity [73]. The critical radius of the fiber is found to depend on the modulus of the fiber and its surface energy.

Even for a large contacting member, crack propagation can be arrested by surface patterning using *crack arrest* mechanism [74–76]. Experiments with surface patterned model adhesive have shown that a moving crack gets arrested when it meets a material discontinuity (Fig. 7.3). To reinitiate its propagation, further increase of

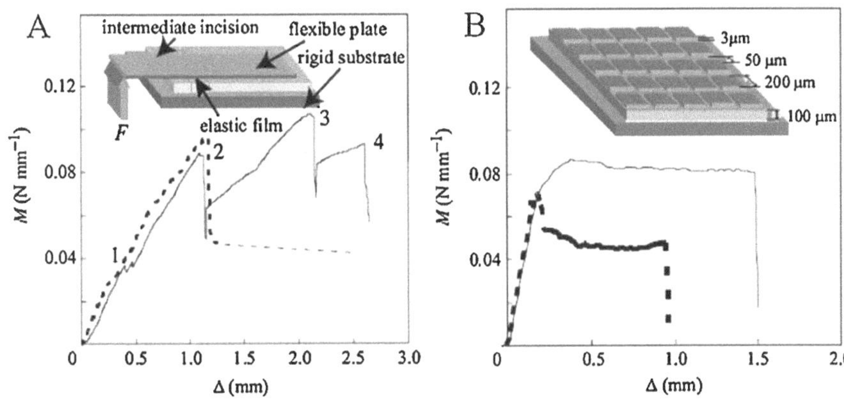

Fig. 7.3 Peeling over a patterned adhesive: (**A**). Adhesive with incisions shows that a patterned interface will fail by multiple crack initiation rather than single crack propagation. Each peak in the plot of peeling torque *M* represents crack arrest and crack initiation near to an incision. (**B**). If the patterns are closely packed, instead of multiple rise and drop in peeling torque, sustained high adhesion strength is observed. (© Reproduced with kind permission from Royal Society, ref [74])

stress is required which manifests itself as enhanced fracture toughness for a surface patterned adhesive. The reason behind this observation is that the interface starts to fail only when a critical cohesive tensile stress is reached and the work required to stretch the polymer is much higher than the surface energy needed to create two new surfaces [67, 77]. However, as long as the crack propagates over a smooth surface, the stress level is maintained at the crack tip because the strain energy released can be transferred to its immediately adjacent load bearing portion of the structure. However, when crack meets a discontinuity/cut/incision in its journey, its further progress is arrested. While the segment behind the discontinuity relaxes at zero load dissipating the stored elastic energy, the segment ahead of it has to be stressed to the critical level again to reinitiate the crack propagation process again. Thus, for a film with close-spaced patterns, this crack re-initiation energy needs to be supplied continuously resulting in increased toughness of the interface. Similar argument has been put forward for fibrillar surface too where in case of the failure of a single fiber, the major part of the stored strain energy in that fiber gets dissipated because of materials damping [29, 61, 72–76, 78]. As this energy is much higher than the interfacial work of adhesion, a fibrillar adhesive surface is tougher than a smooth one. A similar idea was presented by Lake and Thomas [79] for cross-linked rubbery network where the fracture toughness of elastomers is about two orders of magnitude higher than the energy required to break a unit area of covalent bonds. They rationalized this observation with the hypothesis that the breaking of a single bond in a chain releases all the stored energy in the chain between cross-links and this energy is much greater than the energy required to break a single bond.

Although these theories do not always converge to the same understanding, all of them point out that dividing a single contact area into many smaller sub-contacts enhances adhesion. They also suggest that the adhesive strength can be optimized by bringing the pattern dimension below some critical value which depends on the material properties and crack geometry. However, increment of adhesion by decreasing the pattern size is even theoretically limited by fracture strength of the material, theoretical contact strength and self adhesion of fibers and practically by available fabrication technology [66]. Taking clue from these observations and understandings, many bio-inspired adhesives with different micro-fabricated patterns have been tried with varying degrees of success.

7.6 Micro-Fabricated Bio-Mimicked Adhesives

7.6.1 Adhesive with Surface Patterns

In 2002, Autumn et al. [57] demonstrated that van der Waals interactions play the principle role in gecko adhesion. The millions of spatula present in the adhesive pads of gecko and other insects facilitate close contact, an essential condition to utilize short range interactions. As the van der Waals interaction depends more on geometry than on surface chemistry, performance of these natural dry adhesives primarily depends on their structure. This observation and the knowledge that the surface of the adhesive pad of gecko and many insects are highly patterned encouraged both theoretical [29, 61–63, 71–73, 80–82] and experimental works on patterned/hairy adhesives.

The first synthetic nano-hair that resembles hairs at gecko pad was synthesized by Autumn et al. in 2002 by micromoulding PDMS and polyester in a master mould [57]. Figure 7.4 [83] describes the fabrication process schematically. It shows that the master mould was prepared by indenting a wax surface with an AFM tip. The conical nano-hairs thus prepared had a tip diameter between 230 and 440 nm. With a flat AFM probe, the perpendicular adhesive force was measured to be 181 nN and 294 nN for PDMS and polyester respectively [57] which is comparable to that of a biological hair having adhesive force in the range of 50–300 nN [56]. For synthetic hairs, 47–63% of their adhesive force can be well accounted considering van der Waals interaction while the rest could be due to polar interactions, other adhesive effects and surface roughness effect [57, 83]. Calculation based on JKR mechanics [65] further shows that adhesive strength of a synthetic adhesive can be increased by decreasing the spatula dimension, thus increasing its number density [57]. The authors finally suggested that a microfabricated bio-inspired adhesive

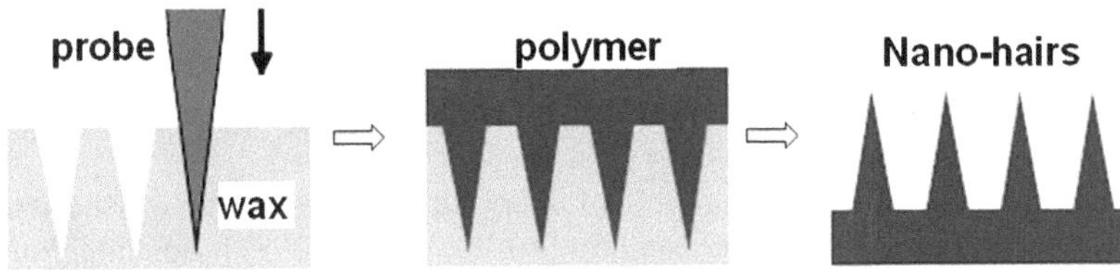

Fig. 7.4 The first synthetic gecko inspired nano-hair prepared by indenting a wax with AFM tip and then molding polymer into that wax mould. Reproduced with permission from Ref [83] © 2002 IEEE

need not necessarily mimic the complex structure of gecko pad but adopting the essential design principles should be sufficient. In the same year, Sitti and Fearing [83] prepared adhesive by casting a polymer inside the pores of a nanoporous membrane. According to them, the basic design features of a hairy adhesive should be high aspect ratio of the fibers with micrometer to nanometer scale diameter, maximum possible hair density, maximum stiffness, low surface energy and high tensile strength. In a later work by one of the authors, angle made by the fiber with the base was also considered as another design parameter [84]. While high aspect ratio and high hair density enhance the adhesion, high material stiffness and low surface energy prevent the slender hair from bunching together [29].

Following the understanding that patterning a surface can enhance adhesion, Geim et al. in 2003 [85] for the first time prepared a bio-inspired adhesive patch which they called "Gecko tape". It was an array of micro-pillars prepared by micropatterning a thin polyimide film of area 1 cm^2. The micro-pillars were 2 μm long, with a diameter ~500 nm and a periodicity of 1.6 μm, as shown in Fig. 7.5A. They showed conclusively that micropatterning a surface can enhance its adhesive strength. While the tape with a soft backing shows 3 N of adhesive strength, a similar

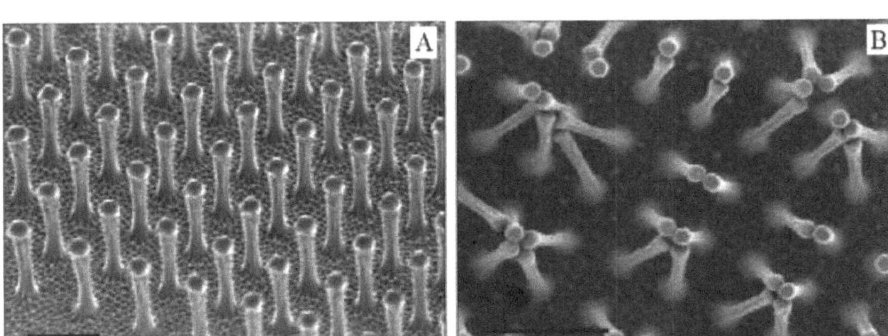

Fig. 7.5 Gecko-inspired adhesive tape. (**A**). An array of micropillars made by micropatterning polyimide surface. (**B**). Bunching/mating/condensation of microfibers due to self-adhesion causes significant reduction in adhesive strength. Reprinted by permission from Macmillan Publishers Ltd: [Nature Material] (Ref. [85]), copyright (2003)

but smooth polyimide film exhibit only a small adhesion ($<10^{-3}$ N). However, durability of the tape still remains an issue as after several detachment-attachment cycles, its adhesive property degraded because of breaking of hairs and lateral bunching (Fig. 7.5B).

The problem of bunching/clumping of fibers still remains pertinent issue and has been addressed by many of the researchers [10, 29, 71, 72, 82–87]. Bunching/mating of hairs together can significantly reduce the adhesive strength because it prevents an individual hair to come in full contact with the substrate by restricting its movement. Hui et al. [86] gave the design criteria to prevent bunching as $\frac{L}{2a}\left(\frac{2\gamma_s}{3E^*a}\right)^{1/4} < \sqrt{w/a}$, by equating elastic energy penalty for bending to the adhesion energy gain when two fibers bend and adhere laterally. Here, L,a,γ_s,w and E^* stand for length and radius of the fiber, surface energy of the fiber material, inter-fiber spacing and effective modulus of the pad respectively. Other researchers too arrived to the similar expression for non-mating condition [84]. This equation shows that clumping can be prevented by increasing the hair spacing and stiffness. However, increasing hair spacing reduces the hair density resulting in lower adhesion. Similarly, higher stiffness of an individual hair decreases the adhesive strength by reducing overall roughness compatibility. Because of these opposing design criteria, prevention of clumping has always been a challenge. Nature has overcome this challenge by utilizing three different criteria together. First, the hairs in gecko pad are made of stiff keratin protein with elastic modulus ~1–3 GP [63–64], second, the hair material is hydrophobic that is low energy surface [39] and third the pad is hierarchical in structure. Due to the high stiffness of the fibers, bending which is a prerequisite for clumping is an energy intensive process. This energy penalty could be overcome, if the surface energy gain due to bunching was high. However, as the pad is a low energy hydrophobic material [39], the energy reward in clumping is low, thus preventing the hairs from bunching. These two conditions together could greatly decrease the compliance of the pad and affect its adhesive property. However, the hierarchical structure of the pad from setae to spatula stalk to spatula maintains its high compliance. While the stiffer base structures stop the hairs from bending and bunching, the final attachment element spatulas are soft enough to comply with rough surfaces because of their finer structure. Taking clue from this understanding, several innovative designs have been suggested to prevent self-mating without sacrificing the adhesive quality. For example Sitti and Fearing [83] has suggested a two-level hierarchical structure of a longer, stiffer base (setal structure) with fine terminal hairs (spatula). Similarly, Northen and Turner [8, 9] fabricated a very interesting four-level hierarchical structure (Fig. 7.6a, b and c). In their design, nanorods of 50–200 nm diameter and 2 μm length were arranged over a photolithographically designed 2 μm thick silicone dioxide platform. The platform had a square central base and four horizontally extended "S" shaped hands of length 100–150 μm . In the next level of hierarchy, each of such platforms was supported by a single pillar of diameter 1 μm and height ~50 μm . They termed this whole assembly as "multi-scale integrated compliant structure (MICS)". Finally 2,500 of such MICS were arranged in a square

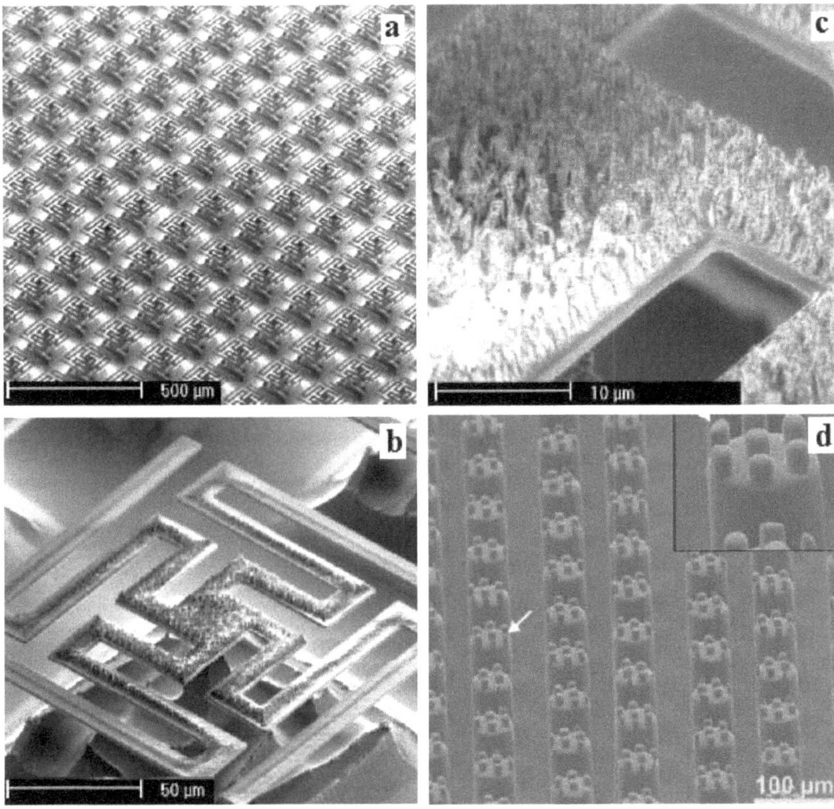

Fig. 7.6 Hierarchical design of synthetic gecko-inspired adhesive. (**A**). An array of multi-scale integrated compliant structures (MICS). (**B**). Image of an individual MICS which is a silicon dioxide platform having four "S" shaped "hands" and supported by a single-crystal silicon pillar. (**C**). The platform is covered with polymeric nanorods (Reproduced with permission from ref [9] © 2005 IEEE). (**D**). Array of two-level hierarchical PDMS pillars (ref [88], © Wiley-VCH Verlag GmbH & Co. KGaA. Reproduced with permission)

packing arrangement to give to final adhesive patch. They found that while a similar smooth surface showed negligible adhesion strength of 0.1 Pa, patterning the surface with nanorods enhanced the adhesion up to 6.5 Pa. Further, integrating the nanorods with platform and pillars increased the adhesion strength by more than a factor of 3 up to 22 Pa made possible by using the hierarchical assembly as described above. Interestingly, adding a hydrophobic coating of fluorocarbon to the nanorods improved adhesion further, probably by preventing self mating of the rods. This design clearly demonstrated the superiority of multi-scale hierarchical design over a single scale structure. However, some of the later efforts actually showed reverse trend where a two-level hierarchical structure (Fig. 7.6d) had lower adhesive strength than a single-level structure [88, 89]. This observation goes against the theoretical prediction of beneficial effect of hierarchical structure [90–93]. This anomaly was attributed to the decreased packing density of the micro-structures for a hierarchical adhesive than that of a single-level structure [88].

To understand the underlying physics in patterned adhesive and to explore the effect of different geometric parameters, different surface patterns have been studied in last 5 years. Those patterns/structures include multiple incisions and chocolate bar pattern [74, 75], low aspect ratio columns or "pancakes" [94], micro-fibrils and micro-columns [10–14, 83, 85, 87, 95–97], fibril with extended end [68, 98] or mushroom shaped columns [99, 100], hierarchical structures [8, 9, 88, 89], fibrils terminated with a thin film [101, 102], a complex CNT based fibrillar structure [103–107], etc. It was shown that in spite of the suggestion by earlier researchers for high aspect ratio structures, a polymer surface decorated with low-aspect ratio posts (pancakes) [94] gives adhesion enhancement by about 4 times over a similar smooth film. Ghatak et al. [74] and Chung and Chaudhury [75] showed that simple incisions or cuts at the surface can also increase the fracture energy by arresting crack propagation. Every time the propagating crack meets a cut or discontinuity of material, it stops there and further energy is required to be supplied to re-initiate the crack (see Section 7.4.2). For a peeling geometry,they identified a length scale called "stress decay length" along the direction of crack propagation as $k^{-1} = \left(Dh^3/12\mu\right)^{1/6}$, where D, h and μ are rigidity of the adherent plate, thickness and modulus of the film respectively. If the pattern dimension is comparable to, or less than, the stress decay length, the fracture toughness increases considerably. Similar results and similar explanation has been put forward by other researchers also where they tested the adhesive performance of a thin PVB sheet with fibers of millimeter scale dimensions [87]. Although in this experiment, the fibers are much larger in dimension compare to the gecko and other insects' foot hair, the experiment is capable of capturing the inherent physics of adhesion enhancement by crack arrest mechanism. The experimental result shows that even with the large fibers, the adhesive energy increases by about an order of magnitude over a similar smooth surface. In all these cases, the enhanced adhesion is explained in the light of dissipation of stored elastic energy as described in Section 7.4.2. A fiber needs to be stretched to a critical extent before it fails. The elastic energy thus stored in each fiber during stretching gets dissipated when the fiber fails. This loss in elastic energy is hypothesized as the cause of enhanced adhesion and this hypothesis is supported by other researchers as well [29, 61, 72–76].

Other than the geometry of the over-all structure, local geometry of the contact tip is also an important parameter that controls adhesion in patterned adhesives. After comparing the performance of a fibriallar adhesive with that of its negative replica with holes, it was found that irrespective of the shape of the fibers (circular, rectangular or elliptical) or the holes in replica, the pull-off force scales linearly with contact perimeter [67]. This observation can be explained well in the light of the contact splitting argument (ref Section 7.5.2) which suggests that the pull-off force for a patterned adhesive is governed by the real contact perimeter and not by the contact area. Other than flat tip end, many other tip geometries like mushroom shaped, spherical tip, flat tip with rounded edge, spatula tip and concave tip have also been tried (Fig. 7.7) [68, 97]. It has been found that while mushroom shaped

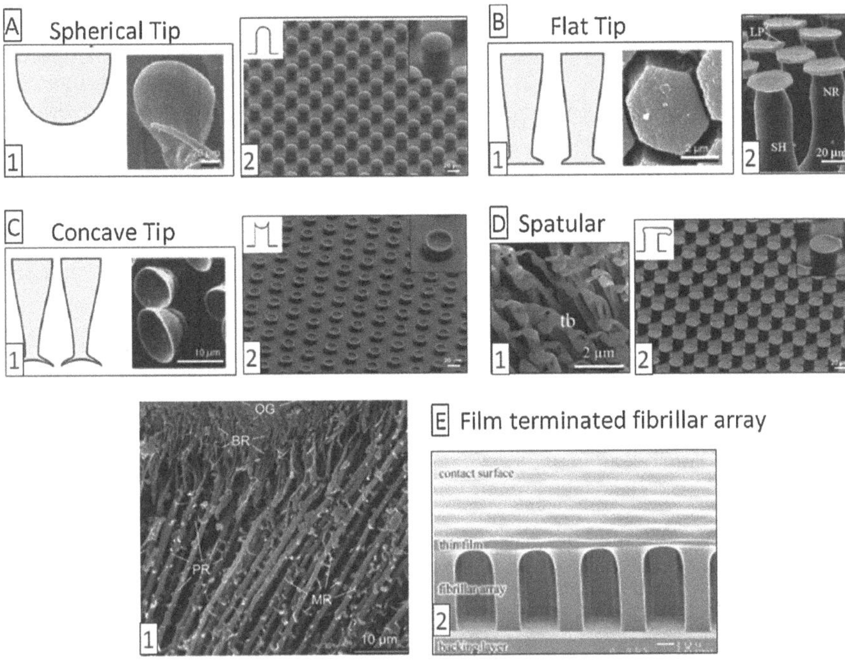

Fig. 7.7 A representative but not exhaustive graphical presentation of various tip geometries found in nature (all images labeled as 1) and in their corresponding bio-inspired synthetic adhesives (all images labeled as 2). For natural pads (all images as 1), the species are (**A1**). Bug *Pyrrhocoris apterus,* (**B1**). Beetle *Chrysolina fastuosa,* (**C1**). Male beetle *Dytiscus marginatus* (**D1**). *Tokay gecko, Gecko gecko* and (**E1**). Internal structure of insect *Tettigonia viridissima.* Image reproduced from: (**A1**), (**B1**) and (**C1**): (© Adapted in part with kind permission from Royal Society, ref [69]), Image (**D1**) (Reprinted with permission from ref [63], Copyright [2003], American Institute of Physics), Image (**E1**). (© Reproduced from ref [42] with kind permission from Springer Science+Business Media). Images (**A2**), (**C2**), (**D2**). (Adapted in part with permission from Ref [68]. Copyright 2005 American Chemical Society). Image (**B2**). (© Adapted in part with kind permission from Royal Society, ref [99]), Image (**E2**). (Reproduced from reference [101] © (2007) National Academy of Sciences, U.S.A)

and spatula shaped tip gives much higher adhesion than a flat punch pillar, the other geometries give either similar or lower adhesion. In particular, the pull-off values for concave tip are significantly lower than that for other tip geometries. For all the different tips, pull-off force increases with applied pre-load at the beginning finally reaching a plateau. The adhesion of mushroom shaped tip has also been tried under water to find a promising performance [90]. However, other than the mushroom shaped tip, the pull-off force has not been scaled with real contact perimeter for any other tip geometry. So, it cannot be concluded whether change noticed in adhesion is a result of changed contact perimeter, change in actual contact area or increased compliance because of the thin terminal structure. It can only be assumed that the effect observed is most probably a combined effect of all these three factors. However, it can be concluded that addition of a thin terminal film to a fibrillar array enhances adhesion because thin terminal member enhances contact area, increases compliance and prevents fibers from self-bunching (Fig. 7.7E2) [101, 102].

Other than using polymers as fabrication material, arrays of multi-walled carbon nanotubes (MWCNT) have also been tried in some recent research efforts in mimicking gecko adhesive pad. MWCNT is a very promising candidate in fabrication of fibrillar adhesives because of its high tensile and flexural strength, high elastic modulus and ease of growing into high aspect ratio structure [108]. In 2005, Yurdumakan et al., used array of vertically aligned MWCNT of diameter 10–20 nm and length $\sim$65 μm as synthetic gecko tape [103]. The MWCNT was prepared by chemical vapor deposition on quartz or silicon substrate and then was partially embedded into matrix of polymetylmethacrylate (PMMA). Adhesion test of this fibrillar adhesive was performed by the tip of a scanning probe microscope in which the probe was brought in contact to the carbon brush, pushed into and then retracted back. The maximum negative force in the force-displacement curve of this process during retraction gives an estimation of adhesive strength. The calculated adhesive strength of this pad ($1.6 \pm 0.5 \times 10^{-2}$ nN/nm^{-2}) was 200 times higher than the estimated adhesive force of a gecko's setae (10^{-4} nN/nm^{-2}) [56]. However, this impressive adhesion will not be observed for a macroscopic contact. In microscopic measurement, the probe penetrates into the nanotube brush resulting into side contacts between the probe tip and nanotubes. This multiple side contacts drastically enhance the observed adhesion. However, a macroscopic object, instead of penetrating into the fiber brush, will largely lie on the tip of the fibers resulting into predominantly point contact. As a result, observed adhesion will be significantly low. Side contacts also cause a counterintuitive effect of bunching of nanotubes on adhesion. While bunching of the fibers is reported to be detrimental, MWCNT shows the reverse trend. The reason behind this anomaly is the fact that the bunching of MWCNTs allows the microscopic probe to penetrate inside the fibre brush resulting into considerable side contacts unlike vertically aligned densely packed or horizontally aligned nanotubes which show almost negligible adhesion [109]. Other than bunching, an array of CNT also shows entanglement near the surface due to its growth in random direction at the initial stage [103]. When shear force is applied, the near-surface entanglements get horizontally aligned leading to predominant line contact instead of point contacts. However, if a normal load is applied, the nanotubes come off the surface by point-by point detachment. As a result, CNT based adhesives exhibit high adhesion in shear ($\sim$100 N/cm^2 so far) but easy normal lift-off ($\sim$10 N/cm^2) [106]. This directionality of the adhesive is very much desirable in applications like wall climbing robots. Moreover, they show good self-cleaning property too which is another desirable quality for a reusable adhesive [107]. The adhesive has been found to performs equally well on variety of adhering surfaces like glass mica and Teflon irrespective of their hydrophobicity. A more recent study has found that a multiscale structure of MWCNT where nanometer scale CNTs are arranged to form micrometer scale pattern gives much higher adhesive force. A 1 cm^2 area of CNT pattern can support a shear force of 36 N, which is nearly 4 times higher than a gecko foot, 10 times higher than polymer pillars and 4 times higher than unpatterned carbon nanotube patches on silicon [105].

7.6.2 Adhesion with Buried Sub-Surface Patterns: Tuning Adhesion at Smooth Elastic Surfaces

In Section 7.2.2, it is already mentioned that other than these surface patterns, the natural dry adhesive pads have subsurface structures role of which in adhesion is not yet clearly known. In a quest to explore the role of sub-surface structure on adhesion in general, Majumder et al. [46] embedded micro-channels in polydimethylsiloxane (PDMS) films (Fig. 7.8A) [47] and filled those channels either with visoleastic liquid (silicone oil) or left them air-filled. The performance of the adhesive was tested in a displacement controlled peel test in which a flexible microscope cover slip was first brought in contact with the adhesive film and then peeled off from its free end, as shown in Fig. 7.8B. Area under the curve of peeling force, F against displacement of the free end of the cover slip Δ gives the estimation of total adhesion energy required to be supplied to debond.

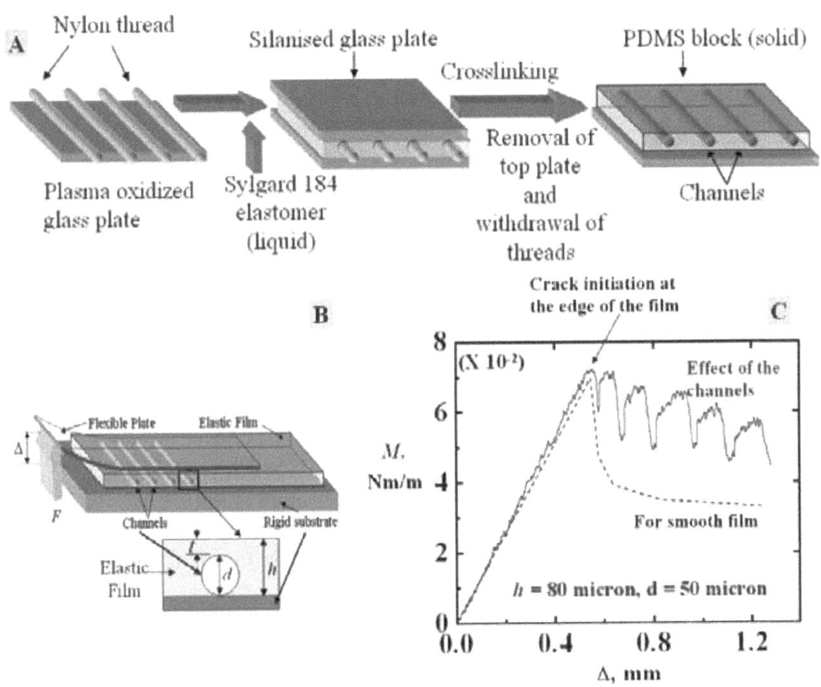

Fig. 7.8 (**A**). Template assisted 3D micropatterning technique. (Adapted in part with permission from Ref [46]. Copyright 2006 American Chemical Society). (**B**). Experimental scheme: Displacement controlled peeling of a flexible microscope cover slip off an elastic adhesive film with embedded micro-channels. (From ref [47]. Reprinted with permission from AAAS). (**C**). Presence of microchannels deter crack propagation and thus increases the torque required to peel a cover slip off an adhesive film

The peel test on an adhesive film with embedded micro-channels filled with air show that on such a film, the crack propagates with intermittent arrests and initiations at the location of channels. This stick-slip behavior is reflected in the plot of the peeling torque M against displacement Δ of the flexible plate in Fig. 7.8C. Here, $M = F \cdot a$, where F is the applied force (refer Fig. 7.8B) and a is the distance between the point of application of F and the crack front. The plot shows the existence of several peaks. While the first one corresponds to the initiation of crack at the edge of the film, the subsequent ones appear because of the embedded microchannels. The figure shows that each microchannel causes a considerable increase in peeling torque resulting enhanced adhesion. This increase in adhesion energy is similar to the crack arrest mechanism as mentioned in the Section 7.4.2. To initiate a crack, one needs to strain the material until it reaches a critical stress level. After that, when the crack propagates over the smooth surface, this critical stress level is maintained at the crack tip and for an elastic adhesive no further supply of strain energy is required. However, a discontinuity can disrupt this mechanism. In that situation stored strain energy cannot get transferred from one side of the discontinuity to the other. As a result, the crack gets arrested at the vicinity of the discontinuity and further supply of energy is required to re-initiate the crack. This suggests that the energy dissipation at a channel should be proportional to the void fraction αcaused by that channel. Indeed, the fractional increase in fracture energy derived from various experiments with different channel diameter, film thickness and rigidity of the adhering plate, was found to vary linearly with void fraction α.

The effect of the channels on adhesion is reported [46, 47] to increase further when the channels were filled with silicone oils of viscosity 5 – 50,000 cp and surface tension 22 mJ/m^2. The oil being a low surface tension liquid, creates a negative capillary pressure of magnitude $\Delta P = 4\gamma_{oil}/d$ inside the channel and also in close vicinity of its elastic wall. As a result, pressure at this location goes down. However, away from a channel, pressure in the film remains atmospheric. Due to this pressure difference, the thin skin layer on top of the channel gets pressed from both the sides and buckles up. By equating the elastic energy required to form a bump of height $\delta$with the surface energy gain from wetting of the channel wall by silicone oil, δ can be scaled as $\approx d\sqrt{2\pi\gamma_{PDMS}/\mu\,(h-d/2)}$, which matches well with the experimental trend. For thinner films, the deformation bulges appear as "spikes" with narrow peaks, which do not allow the plate to come in complete contact with the film. However, as the film thickness increases, the deformation flattens out resulting in a complete contact with the contactor.

Peeling test on these films gives typical $M - \Delta$ plots as presented in Fig. 7.9A. Curve 2 ($h = 300$ μm, $d = 50$ μm) in this plot shows when the channels are filled with oil of viscosity 380 cP the crack gets arrested at the vicinity of the channels and a large torque is required to re-initiate it. However, for a smooth film (curve 0) or when the same channels are filled with air, no such effect is visible (curve 1). This type of stick-slip propagation becomes more prominent for higher diameter channel (curve 3: $h = 750$ μm, $d = 710$ μm). The effect of oil can also be seen (Fig. 7.6B) in terms of fracture energy G which increases to about 1600 mJ/m^2

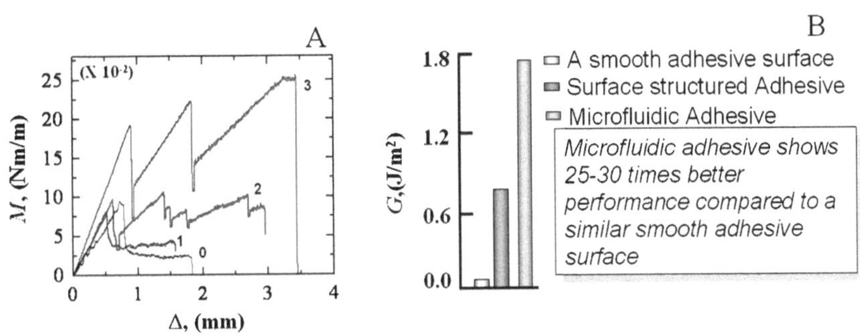

Fig. 7.9 (**A**). The peeling torque M vs. Δ. Curve 0 corresponds to a smooth film of thickness $h = 300\ \mu m$. Curve 1 and 2 represent film of $h = 300\ \mu m$ and $d = 50\ \mu m$ for channels filled with air and oil of viscosity $\eta = 380$ cP respectively. Curve 2 clearly shows higher peeling torque over curve 1. Curve 3 correspond to $\eta = 380$ cP, $h = 750\ \mu m$, and $d = 710\ \mu m$ which requires even higher peeling torque leading to high fracture energy. (**B**). Microfluidic adhesive shows higher adhesive strength compared to an otherwise similar smooth adhesive surface

when these channels are filled with an oil of intermediate viscosity. This remarkable enhancement in G by about a factor of 25, compared to otherwise similar but smooth adhesives, is achieved without incorporating any viscoelasticity in the adhesive but by simply manipulating the pressure inside the subsurface channels. As a result, the adhesive as a whole does not lose its adhering quality on repeated peeling. Interestingly, the same elastic layer can be used both as a strong adhesive and an easy release coating. To achieve that we embedded two layers of channel ($d = 50\ \mu m$) within the adhesive at two different vertical locations ($t_1 = 120\ \mu m$ and $t_2 = 300\ \mu m$). When the channels of the top layer are filled with oil while those at the bottom layer contain air at atmospheric pressure, the deformations at the surface of the film is too large to allow the plate to come in contact with the film. The adhesive then behaves like an easy release coating with effective adhesive energy to be zero. However, if the channels at the bottom layer are filled with oil instead of those at the top, the peel experiment yields the $M - \Delta$ plot with characteristic peaks at the location of the channels. This result shows that the same film can be used as a strong adhesive and a release coating without altering the intrinsic rheological or surface properties of the film.

This significant increase in adhesion energy can be explained in terms of critical stress that is to be maintained at the crack tip for crack propagation. As mentioned earlier, to initiate a crack, a critical tensile stress has to be reached. Once initiated, this stress level is maintained at the crack tip. However, when the crack meets a channel filled with oil, negative capillary pressure inside the channel frustrates the tensile stress. As a result effective stress at the crack tip goes down. To maintain the critical stress which is essential for crack propagation, further increase in applied external load is required which manifests itself in terms of enhanced adhesive strength.

Interestingly, a wide range of oil viscosities in the microcapillaries give similar adhesive strength, thus indicating that the dominant role of microcapillary oil is not in providing a dissipative element, but rather in generation of a stress field which counters the ease of crack initiation and propagation. Further, the capillary stress depends only on the channel diameter and the contact angle of the three phase contact line. Thus, such a microfluidic adhesive is expected to perform equally well in underwater adhesion because the contact angle of hydrophobic oil on hydrophobic capillary wall is also expected to be small when surrounded by water. Indeed, our recent tests [110] show the microfluidic adhesive to be equally effective and reversible both in dry and wet adhesion conditions.

7.7 Conclusions

A reusable yet strong adhesive can be very useful in microelectronics, robotics, biomedical and prosthetics, automobiles, sports and many other day-to-day applications [9–14]. The characteristics that are sought after in this futuristic adhesive are low pre-load, high debonding energy, resistance to particulate contamination and self-cleaning. Other than this, depending upon the application, the adhesive should be non-toxic, time independent and anisotropic too. Non-toxicity is required for any bio-medical application for obvious reason. For any long time-scale application like in automobiles, the adhesive should not creep/start to flow with time. Anisotropy is required where quick release is also an important issue. By anisotropy we mean that although in some direction the adhesive strength should be very high (for example in shear), in some other direction it should be easy to debond (for example under normal tension/peeling mode). This is an essential criterion for an adhesive to be used for wall-climbing which requires strong attachment yet easy and quick release. No present day man-made adhesive has all these qualities. However, adhesive pads found in insects and geckos satisfy all of these criteria with their amazing surface and sub-surface structures. An effort to mimic them is still in its infancy and only partial success has been achieved. Researchers have used top-down approach like photolithography and bottom-up approach like self-assembly of CNTs to fabricate intricately patterned adhesive surface and have shown that fibrillar adhesive gives higher adhesion than a smooth one. Self-cleaning property and directionality for synthetic adhesive has also been achieved to some extent. The key design parameters that have been identified are pattern dimension and spacing, its aspect ratio, fiber stiffness, its orientation or angle with the base, local geometry of the fiber tip and finally the design hierarchy. Many of these design criteria have been already adopted to fabricate gecko-like adhesive. The two most serious problems facing by all of these bio-inspired fibrillar adhesives are self-bunching and requirement of high pre-load. It seems that a complete understanding of the adhesion mechanism of a patterned/structured adhesive, optimization of different parameters and fabrication of large area with intricate structures are still a long way ahead.

Acknowledgments Support from Department of Science and Technology through an IRHPA grant is gratefully acknowledged.

References

1. Maderson PFA (1964) Keratinized epidermal derivatives as an aid to climbing in gekkonid lizards. Nature 203:780–781.
2. Walker G (1993) Adhesion to smooth surfaces by insects-a review. Int. J. Adhesion Adhesives 13:3–7.
3. http://en.wikipedia.org/wiki/Tanaji_Malusare.
4. Frantsevich L and Gorb S (2004) Structure and mechanism of the tarsal chain in the hornet, *Vespa crabro* (Hymenoptera, Vespidae): implications on the attachment mechanism. Arthr. Struc. Develop. 33:67–89.
5. Dai Z, Gorb SN, and Schwarz U (2002) Roughness-dependant friction force of the tarsal claw system in the beetle *Pachnoda marginata* (Coleoptera, Scrabaeidae). J. Exp. Biol. 205:2479–2488.
6. Goodwyn PP, Peressadko AP, Schwarz H et al. (2006) Material structure, stiffness and adhesion: why attachment pads of the grasshopper (*Tettigonia viridissima*) adhere more strongly than those of the locust (*Locusta migratoria*) (Insecta:Orthopetra). J. Comp. Physiol. A. 192:1233–1243.
7. Barnes W and Jon P (2007) Biomimetic solutions to sticky problems. Science 318:203–204.
8. Northen MT and Turner KL (2005) A batch fabricated biomimetic dry adhesive. Nanotechnology 16:1159–1166.
9. Northen MT and Turner KL (2005) Multi-scale compliant structure for use as a chip-scale dry adhesive. The 13th international conference on solid-state sensors, Actuators and Microsystems. Digest of Technical Papers. TRANSDUCERS '05 2:2044–2047.
10. Sitti M and Fearing RS (2003) Synthetic gecko foot-hair micro/nano structures for future wall-climbing robots. International conference on robotics and automation, 1164–1170.
11. Menon C, Murphy M, and Sitti M (2004) Gecko inspired surface climbing robots. Proceedings of the 2004 IEEE: International Conference on Robotics and Biomimetics, 431–436.
12. Cheung E, Karagozler ME, Park S et al. (2005) A new endoscopic microcapsule robot using beetle inspired microfibrillar adhesive. Proceedings of the 2005 IEEE/ASME International Conference on Advanced Intelligent Mechatronics, 551–557.
13. Daltorio KA, Gorb S, Peressadko A et al. (2005) A robot that climbs walls using micro-structured polymer feet. International conference on climbing and walking robots. International Conference on Climbing and Walking Robots, London.
14. Mahdavi A, Ferreira L, Sundback C et al. (2007) A biodegradable and biocompatible gecko-inspired tissue adhesive, Proc. Natl. Acad. Sci. 105:2305–2312.
15. Gay C (2002) Stickiness-some fundamentals of adhesion. Integr.Comp. Biol. 42:1123–1126.
16. Creton C and Papon E (2003) Materials science of adhesives: how to bond things together. MRS Bull. June:419–423.
17. Yarusso DJ (2002) Effect of rheology on PSA performance. In: Dillard DA and Pocius AV (eds.) The Mechanics of Adhesion, 1st edn. Elsevier, Amsterdam.
18. Taljsten B (2005) The importance of bonding-An historic overview and future possibilities. Proceedings of the International Symposium on Bond Behaviour of FRP in Structures 1–10.
19. Dahlquist CA (1969) Pressure Sensitive Adhesives. Vol. 2, Patrick RL (ed) Marcel Dekker, New York.
20. Creton C and Fabre P (2002) Tack. In: Dillard DA and Pocius AV (eds) The Mechanics of Adhesion, 1st edn. Elsevier, Amsterdam.
21. Chiche A and Creton C (2004) Cavitation in a soft adhesive, Proceedings of the 27th annual adhesion society meeting, Wilmington, 296–298.

22. Ghatak A, Chaudhury M, Shenoy V et al. (2000) Meniscus instability in confined thin elastic films. Phys. Rev. Lett. 85:4329–4332.
23. Chung JY, Kim K, Chaudhury MK et al. (2006) Confinement-induced instability and adhesive failure between dissimilar thin elastic films. Euro. Phys. J. E–Soft Matter 20:47–53.
24. Shenoy V and Sharma A (2001) Pattern formation in a thin solid film with interactions. Phys. Rev. Lett. 86:119–122.
25. Sarkar J, Shenoy V, and Sharma A (2004) Patterns, forces and metastable pathways in debonding of elastic films. Phys. Rev. Lett. 93:018302.
26. Gonuguntala M, Sharma A, Sarkar J et al. (2006) Contact Instability in adhesion and debonding of thin elastic films. Phys. Rev. Lett. 97:018303.
27. Creton C (2003) Pressure-sensitive adhesives: an introductory course. MRS Bull. June: 434–439.
28. Chan EP, Greiner C, Arzt E et al. (2007) Design model systems for enhanced adhesion. MRS Bull. 32:496–503.
29. Jagota A and Bennison SJ (2002) Mechanics of adhesion through a fibrillar microstructure. Integr. Comp. Biol. 42:1140–1145.
30. Creton C and Schach R (2009) Diffusion and adhesion. In: Benedek I and Feldstein MM (eds) Handbook of pressure-sensitive adhesive and products, Fundamentals of pressure sensitivity, 1st edn. Taylor and Francis, New York..
31. Creton C and Gorb S (2007) Sticky feet: From animals to materials. MRS Bull. 32:466–472.
32. Arzt E, Gorb S, and Spolenak R (2003) From micro to nano contacts in biological attachment devices. Proc. Natl. Acad. Sci. 100:10603–10606.
33. Ruibal R and Ernst V (1965) The structure of digital setae of lizards. J. Morphol. 117: 271–294.
34. Wainwright SA, Biggs WD, Currey JD et al. (1982) Mechanical design in organisms. Princeton University Press, Princeton, NJ.
35. Russel AP (1986) The morphological basis of weight-bearing in the scansors of the *Tokay gecko* (Reptilia, Sauria). Can. J. Zool. 64:948–955.
36. Schleich HH and Kastle W (1986) Ultrastrukturen an Gecko-Zehen (Reptilia, Sauria: Gekkonidae). Amphibia-Reptilia. 7:141–166.
37. Autumn K and Peattie AM (2002) Mechanisms of adhesion in geckos. Integr. Comp. Biol. 42:1081–1090.
38. Irschick DJ, Austin CC, Petren K et al. (1996) A comparative analysis of clinging ability among pad-bearing lizards. Biol. J. Linn. Soc. 59:21–35.
39. Autumn K and Hansen W (2006) Ultrahydrophobicity indicates a non-adhesive default state in gecko setae. J. Comp. Physiol. A. 192:1205–1212.
40. Autumn K, Hsieh ST, Dudek DM et al. (2006) Dynamics of geckos running vertically. J. Exp. Biol. 209:260–272.
41. Gillett JD and Wigglesworth VB (1932) The climbing organ of an insect, *Rhodnius prolixus* (Hemiptera, Reduviidae). Proc. R. Soc. B. 111:364–376.
42. Gorb S, Jiao Y, and Scherge M (2000) Ultrastructural architecture and mechanical properties of attachment pads in *Tettigonia viridissima* (Orthoptera, Tettigonia). J. Comp. Physiol. A. 186:821–834.
43. Jiao Y, Gorb S, and Scherge M (2000) Adhesion measured on the attachement pads of *Tettigonia Viridissima* (Orthoptera, Insecta). J. Exp. Biol. 203:1887–1895.
44. Gorb S and Scherge M (2000) Biological microtribology: anisotropy in frictional forces of orthopteran attachment pads reflects the ultrastructure of a highly deformable material. Proc. R. Soc. Lond. B. 267:1239–1244.
45. Gorb S, Beutel RG, Gorb EV et al. (2002) Structural design and biomechanics of friction-based releasable attachment devices in insects. Integr. Comp. Biol. 42:1127–1139.
46. Majumder A, Ghatak A, and Sharma A (2007) Microfluidic adhesion induced by subsurface microstructure. Science 318:258–261.

47. Verma MKS, Majumder A, and Ghatak A (2006) Embedded template assisted fabrication of complex microchannels in PDMS and design of a microfluidic adhesive. Langmuir 22:10291–10295.
48. Edward A and Ghatak A (2009) Bioinspired design of a hierarchically structured adhesive. Langmuir 25:611–617.
49. Hora SL (1923) The adhesive apparatus on the toe of certain geckos and tree frogs. J. Proc. Asiat. Soc. Bengal 9:137–145.
50. Mahendra BC (1941) Contributions to the bionomics, anatomy, reproduction and development of the Indian house gecko *Hemidactylus flaviviridis* Ru¨ppell. Part II. The problem of locomotion. Proc. Indian Acad. Sci. Sec. B. 13:288–306.
51. Hiller U (1975) Comparative studies on the functional morphology of two gekkonid lizards. J. Bombay Nat. Hist. Soc. 73:278–282.
52. Gennaro JGJ (1969) The gecko grip. Nat. Hist. 78:36–43.
53. Russell AP (1975) A contribution to the functional morphology of the foot of the tokay, *Gekko gecko* (Reptilia, Gekkonidae). J. Zool. Lond. 176:437–476.
54. Williams EE and Peterson JA (1982) Convergent and alternative designs in the digital adhesive pads of scincid lizards. Science 215:1509–1511.
55. Stork NE (1980) Experimental analysis of adhesion of *Chrysolina polita* (Chrysomelidae: Coleoptera) on a variety of surfaces. J. Exp. Biol. 88:91–108.
56. Autumn K, Liang YA, Hsieh ST et al. (2000) Adhesive force of a single gecko foot-hair. Nature 405:681–685.
57. Autumn K, Sitti M, Liang YA et al. (2002) Evidence for van der Waals adhesion in gecko setae. Proc. Natl. Acad. Sci. 99:12252–12256.
58. Federle W, Baumgartner W, and Holldobler B (2004) Biomechanics of ant adhesive pads: frictional forces are rate and temperature dependent. J. Exp. Biol. 206:67–74.
59. Huber G, Mantz H, Spolenak R et al. (2005) Evidence for capillarity contributions to gecko adhesion from single spatula nanomechanical measurements. Proc. Natl. Acad. Sci. 102:16293–16296.
60. Tian Y, Pesika N, Zenget H et al. (2006) Adhesion and friction in gecko toe attachment and detachment. Proc. Natl. Acad. Sci. 103:19320–19325.
61. Persson BNJ (2003) On the mechanism of adhesion in biological systems. J. Chem. Phys. 118:7614–7621.
62. Persson BNJ (2007) Biological adhesion for locomotion on rough surfaces: Basic principle and a theorist's view. MRS Bull. 32: 486–490.
63. Persson BNJ and Gorb S (2003) The effect of surface roughness on the adhesion of elastic plates with application to biological systems. J.Chem. Phys. 119:11437–11444.
64. Autumn K (2007) Gecko Adhesion: Structure, function and applications. MRS Bull. 32: 473–478.
65. Johnson KL, Kendall K, and Roberts AD (1971) Surface energy and the contact of elastic solids. Proc. R. Soc. Lond. A. 324:301–313.
66. Spolenak R, Gorb S, and Arzt E (2005) Ahesion design maps for bio-inspired attachment systems. Acta Biomaterialia. 1:5–13.
67. Varenberg M, Peressadko A, Gorb S et al. (2006) Effect of real contact geometry on adhesion. Appl. Phys. Lett. 89:121905–121907.
68. Campo AD, Greiner C, and Arzt E (2007) Contact shape controls adhesion of bio-inspired fibrillar surfaces. Langmuir. 23:10235–10243.
69. Spolenak R, Gorb S, Gao H et al. (2005) Effect of contact shape on the scaling of biological attachments. Proc. R. Soc. A. 461:305–319.
70. Spuskanyuk AV, McMeeking RM, Deshpande VS et al. (2008) The effect of shape on the adhesion of fibrillar surfaces. Acta Biomaterialia. 4:1669–1676.
71. Gao H and Yao H (2004) Shape insensitive optimal adhesion of nanoscale fibrillar structures. Proc. Natl. Acad. Sci. 21:7851–7856.

72. Hui CY, Glassmaker NJ, Tang T et al. (2004) Design of biomimetic fibrillar interfaces: 2. Mechanics of enhanced adhesion. J. R. Soc. Lond. Inter. 1:35–48.
73. Tang T, Hui CK, and Glassmaker NJ (2005) Can a fibrillar interface be stronger than a non-fibrillar one? J. R. Soc. Inter. 2:505–516.
74. Ghatak A, Mahadevan L, Chung JY et al. (2004) Peeling from a biomimetically patterned thin elastic film. Proc. R. Soc. Lond. A. 460:2725–2735.
75. Chung JY and Chaudhury MK (2005) Roles of discontinuities in bio-inspired adhesive pads. J. R. Soc. Inter. 2:55–61.
76. Glassmaker NJ, Jagota A, and Hui CY (2005) Adhesion enhancement in a biomimetic fibrillar interface. Acta Biomaterialia 1:367–375.
77. Gay C and Leibler L (1999) Theory of tackiness. Phys. Rev. Lett. 82:936–939.
78. Federle W (2006) Why are so many adhesive pads hairy? J. Exp. Biol. 209:2611–2621.
79 Lake GJ and Thomas AG (1967) The strength of highly elastic materials. Proc. R. Soc. Lond. A. 300:108–119.
80. Campolo D, Jones S, Fearing RS et al. (2003) Fabrication of Gecko foot-hair like nano structures and adhesion to random rough surfaces. IEEE Nano 2003, San Fransisco.
81. Gao H, Wang X, Yao H et al. (2005) Mechanics of hierarchical adhesion structures of geckos. Mech. mater. 37:275–285.
82. Majidi C, Groff R, and Fearing RS (2004) Clumping and packing of hair arrays manufactured by nanocasting. ASME International Mechancal Engineering Congress and Exposition. IMECE, California.
83. Sitti M and Fearing RS (2002) Nanomolding based fabrication of synthetic gecko foot hair micro/nanostructures. Proceedings of the IEEE Nanotechnology Conference. Washington 137–140.
84. Shah G and Sitti M (2004) Modeling and design of biomimetic adhesive inspired by gecko foot-hairs. IEEE International Conference on Robotics and Bio-Mimetics.
85. Geim AK, Dubonos SV, Grigorieva IV et al. (2003) Microfabricated adhesive mimicking gecko foot-hair. Nat. Mater. 2:461–463.
86. Hui CY, Jagota A, Lin YY et al. (2002) Constraints on micro-contact printing imposed by stamp deformation. Langmuir. 18:1394–1407.
87. Glassmaker NJ, Jagota A, Hui CY et al. (2004) Design of biomimetic fibrillar interfaces: 1 Making contact. J. R. Soc. Inter. 1:23–33.
88. Greiner C, Arzt E, and Campo A (2009) Hierarchical gecko-like adhesives. Adv. Mater. 21:479–482.
89. Kustandi TS, Samper VD, Yi DK et al. (2007) Fabrication of a gecko-like hierarchical fibril array using a bonded porous alumina template. J. Micromech. Microeng. 17: N75–N81.
90. Yao H and Gao H (2006) Bio-inspired mechanics of robust and releasable adhesion on rough surface. J. Mech. Phys. Solids. 54:1120–1146.
91. Bhushan B, Peressadko AG, and Kim TW (2006) Adhesion analysis of two-level hierarchical morphology in natural attachment systems for "smart adhesion". J. Adhesion Sci. Tech. 20:1475–1491.
92. Yao H and Gao H (2007) Mechanical principles of robust and releasable adhesion of gecko. J. Adhesion Sci. Tech. 21:1185–1212.
93. Porwal PK and Hui CY (2007) Strength statistics of adhesive contact between a fibrillar structure and a rough substrate. J. R. Soc. Inter. 5:441–448.
94. Crosby AJ, Hageman M, and Duncan A (2005) Controlling polymer adhesion with "Pancakes". Langmuir 21:11738–11743.
95. Lee H, Lee BP, and Messersmith PB (2007) A reversible wet/dry adhesive inspired by mussels and geckos. Nature 448:338–341.
96. Lamblet M, Verneuil E, and Vilmin T (2007) Adhesion enhancement through micropatterning at polydimethylsiloxane-acrylic adhesive interfaces. Langmuir 23:6966–6974.
97. Reddy S, Arzt E, and Campo A (2007) Bioinspired surfaces with switchable adhesion. Adv. Mater. 19:3833–3837.

98. Northen MT, Greiner C, Arzt E et al. (2008) A gecko inspired reversible adhesive. Adv. Mater. 20:1–5.
99. Gorb S, Varenberg M, Peressadko A et al. (2007) Biomimetic mushroom shaped fibrillar adhesive microstructure. J. R. Soc. Inter. 4:271–275.
100. Varenberg M and Gorb S (2008) A beetle-inspired soulution for underwater adhesion. J.R. Soc. Inter. 5:383–385.
101. Glassmaker NJ, Jagota A, Hui CY et al. (2007) Biologically inspired crack trapping for enhanced adhesion. Proc. Natl. Acad. Sci. 104:10786–10791.
102. Noderer WL, Shen L, Vajpayee S et al. (2007) Enhanced adhesion and compliance of film-terminated fibrillar surfaces. Proc. R. Soc. A 463:2631–2654.
103. Yurdumakan B, Raravikar NR, Ajayan PM et al. (2005) Synthetic gecko goot-hairs from multiwalled carbon nanotubes. Chem. Commun. 3799:3801.
104. Zhao Y, Tong T, Delzeit L et al. (2006) Interfacial energy and strength of multiwalled carbon nanotube based dry adhesive. J. Vac. Sci. Technol. 24:331–335.
105. Ge L, Sethi S, Ci L et al. (2007) Carbon nanotube based synthetic gecko tapes. Proc. Natl. Acad. Sci. 104:10792–10795.
106. Qu L, Dai L, Stone M et al. (2008) Carbon nanotube arrays with strong binding-on and easy normal lifting-off. Science 322:238–242.
107. Sethi S, Ge L, Ci L et al. (2008) Gecko-inspired carbon nanotube-based self-cleaning adhesives. Nano Lett. 8:822–825.
108. Dickrell PL, Sinnott SB, Hahn DW et al. (2005) Frictional anisotropy of oriented carbon nanotube surfaces. Tribol. Lett. 18:59–62.
109. Kinoshita H, Kuma I, Tagawa M et al. (2004) High friction of a vertically aligned carbon-nanotube film in microtribology. App. Phys. Lett. 85:2780–2781.
110. Majumder A, Sharma A, and Ghatak A (2009) A bio-inspired wet/dry microfluidic adhesive for aqueous environment. Langmuir. DOI: 10. 1021/la9021849.

Chapter 8
Molecular Simulation: Can it Help in the Development of Micro and Nano Devices?

Jayant K. Singh

Abstract Molecular modeling and simulations is gaining popularity as a mean to investigate equilibrium and non-equilibrium properties of fluids near solid and polymeric surfaces, and under confinement in nano- and meso-pores. In this chapter, we focus on advanced Monte Carlo and molecular dynamics techniques to study thermodynamics and transport phenomena of fluids near surfaces.

The state of the art in the field is demonstrated by reviewing selected results of our recent computer simulations. We present Monte Carlo studies of phase equilibria of geometrically restricted fluids, wetting and prewetting transitions of fluids on a substrate. Further, we demonstrate molecular dynamics techniques to investigate the wettability of fluids on surfaces and fluid flow in nano-pores.

Keywords Monte carlo · Molecular dynamics · Free-energy · Contact angle · Confined fluid

8.1 Introduction

Technological advances in recent years have made it possible to imprint solid surfaces with well-defined physical and chemical structures. Constructions of labs-on-a-chip can integrate chemical and biochemical analyzers on the micrometer and even on nanometer scale. Suitable compartments of micro and nano dimensions for the confinement of very small amounts of liquids and chemical reagents are necessary for appropriate miniaturization. These micro and nano-devices offer the capability to work with much smaller reagent volumes with shorter reaction times; and hence, it can dramatically affect the daily processing in life sciences. Such ability may lead to enormous parallel- processing. Structured surface created may

J.K. Singh (✉)
Department of Chemical Engineering, Indian Institute of Technology Kanpur, Kanpur-208016, India
e-mail: jayantks@iitk.ac.in

S. Chakraborty (ed.), *Microfluidics and Microfabrication*,
DOI 10.1007/978-1-4419-1543-6_8, © Springer Science+Business Media, LLC 2010

have position dependent free-energy. Thus these surfaces can modulate the shape of fluid layer and can control the microscopic flow of fluid on designated chemical channels. This facilitates the fabrication of chemical channels, which may have significant affect on micro, nano-fluidics and advanced separation technologies. Thus, understanding the fluid structure and dynamics near patterned and textured surfaces is essential for the appropriate design of micro and nano-devices. Further, significant drag reduction in fluid flow would have enormous economic and technological impact in micro and nano fluidics and biomedical analysis [1]. It is commonly believed that hydrophobic surface can result in the reduction of friction drag [1, 2]. A high contact angle is the main criterion, which material scientists use for the search of surfaces with low friction drag.

Variety of tools is available to investigate the nature of fluids near surfaces. Molecular simulation is one of the tools, whose application and impact have increased in recent years to study various properties. These tools are evolving with the development of new algorithms and computer hardware. We present a brief introduction to molecular simulation methodologies in subsequent sub-section followed by some recent activities on fluid behavior near solid surfaces and in nano-pores.

8.2 Molecular Modeling and Simulation

Molecular modeling and simulation, in general, refers to the following:

- Computational quantum chemistry: ab-initio methods [3, 4].
- Molecular structure-property correlations: QSAR/QSPRs [4, 5].
- Molecular Simulation: Monte Carlo and molecular dynamics [6, 7].
- Mesoscale modeling for material domain: kinetic Monte Carlo and dissipative particle dynamics [7, 8].

This chapter, however, will focus on molecular simulations, which can broadly be divided into two categories in terms of methods: Monte Carlo simulation and molecular dynamics. These methods invariably use the statistical mechanical formalism, which connects thermophysical properties such as pressure, energy and other macroscopic properties to the collection of microstates of systems of molecules, all having in common extensive properties, also called ensemble [6, 7]. A *microstate* of a system of molecules is a complete specification of all positions and momenta of all molecules. Most commonly, we encounter the total energy, the total volume, and/or the total number of molecules (of one or more species, if a mixture) as extensive properties.

Some examples of ensembles are:

- Micro-canonical ensemble
- Canonical ensemble
- Isothermal-Isobaric ensemble
- Grand-canonical ensemble

Molecular dynamics simulations consider a system of molecules and integrates the governing equation of motion through time i.e., successive configurations of the system are generated by integrating Newton's laws of motion.

The equation of motion can be written as a set of first order differential equations

$$\begin{aligned}\frac{dr_i}{dt} &= p_i \\ \frac{dp_i}{dt} &= F_i\end{aligned} \tag{8.1}$$

where, r_i, and p_i are coordinate and momentum of particle i, respectively. F_i is the force acting on particle i.

The usual approach to integrating a set of first-order differential equations is to advance the system variables through a discrete step in time, δt, by approximating the action of the derivative via finite differences. Methods vary in several ways and detailed algorithms are described in many textbooks [6, 7]. The property of interest, A, are collected as a time average of microscopic properties of independent configurations generated over a long finite time, t:

$$\langle A \rangle = \lim_{t \to \infty} \frac{1}{t} \sum_t A_i \Delta t_i \tag{8.2}$$

where A_i is the instantaneous property of the configuration i and Δt_i is an integral multiple of the time step, δt.

Monte Carlo simulations in molecular simulation, on the other hand, generates configurations of molecules using equilibrium probability distribution function for a statistical mechanical ensemble. For example, for canonical ensemble, probability of obtaining configuration of N molecules, $\{r_1, r_2, ..., r_N\} \equiv r^N$ i.e., probability density is defined as:

$$\pi\left(r^N\right) = \frac{\exp\left[-U\left(r^N\right)/k_{\mathrm{B}}T\right]}{\int \exp\left[-U\left(r^N\right)/k_{\mathrm{B}}T\right]dr^N} = \frac{\exp\left[-U\left(r^N\right)/k_{\mathrm{B}}T\right]}{Z} \tag{8.3}$$

where, U is the total potential energy of the configuration, Z is the configurational part of the partition function [9], k_{B} is the Boltzmann's constant and T is the temperature.

Most elementary procedure in the Monte Carlo simulation involves generating a trial configuration by making a perturbation to the configuration of molecules by selecting a molecule randomly and is moved by a small amount from its present position. Subsequently, the ratio of probabilities of the trial and original configurations is computed and a trial configuration would be accepted according to Metropolis algorithm [10]:

$$p_{\mathrm{acc}} = \min\left[1, \exp\left(-\beta \Delta U\right)\right], \tag{8.4}$$

where ΔU is the change in potential energy between the trial and original configuration, $\beta = 1/k_B T$.

In case $p_{acc} = 1$, we accept the trial otherwise, a random number is generated, r and, from a uniform distribution in the interval [0,1]. Trial configuration is accepted if $P_{acc} > r$. If the trial is accepted, the new configuration is taken as the next state in the Markov chain [6]; otherwise, the original configuration is taken as the next state in the chain. Averages are collected over the many configurations as shown below:

$$\langle A \rangle = \frac{\sum_{i=1}^{N\text{trial}} A\left(r^N\right) \exp\left[-\beta U\left(r^N\right)\right]}{\sum_{i=1}^{N\text{trial}} \exp\left[-\beta U\left(r^N\right)\right]} \tag{8.5}$$

Molecular simulation has become an invaluable tool to study the phase equilibria, surface tension, diffusivity and other dynamical properties. Phase equilibria calculation, in a crudest way, may be carried out with direct interfacial setup, where it requires to simulate the system with an explicit interface [11] with more than one phase in contact with other phase. However, such simulations are difficult to perform due to difficulty in maintaining the interface. The difficulty was alleviated by Panagiotopoulos, who introduced a method called Gibbs ensemble Monte Carlo [12]. The idea behind the method is based on the condition of equilibrium for two phases. Phase α and phase β are in equilibrium when the system has the following conditions:

$$\begin{aligned} T^\alpha &= T^\beta \\ P^\alpha &= P^\beta \\ \mu^\alpha &= \mu^\beta \end{aligned} \tag{8.6}$$

where, P is the pressure and μ is the chemical potential.

Panagiotopoulos realized that the equilibrium condition can be achieved without explicit interface if we satisfy a few conditions. The recipe, to obtain the equilibrium state at a constant temperature (i.e. $T^\alpha = T^\beta$) is as follows:

The method, formally known as Gibbs ensemble Monte Carlo (GEMC) technique, uses two basic simulation boxes, which contain certain number of molecules, without forming any interface. These boxes are located within two coexisting phases, which satisfy the conditions given in Eq. (8.6), and surrounded by the periodic images [6]. System temperature is specified a priori. The Monte Carlo technique uses three types of *moves* (*trials*) to satisfy the equilibrium condition. The usual particle displacement (to satisfy internal equilibrium) in each box is done using Metropolis algorithm [10]. To equalize pressure a combined volume-move is required where volume of one box changes by ΔV and volume of the other box changes by $-\Delta V$. To equalize chemical potential, there is a combined attempt of insert/delete move, in which a randomly chosen particle is extracted from one box and randomly placed in the other box.

The GEMC method has been extended to mixtures [13] and complex systems [14] with the help of advanced Monte Carlo techniques [15]. It has been used to study numerous systems [15, 16]. The method has significant advantages in computational speed over conventional free energy calculations of phase boundaries. The method relies on particle exchange between the coexistence phases for keeping the phases at the same chemical potential. There exist however shortcomings in this method. It is known that particle exchange has inherent problem in its simplest form at high density [7]. As a result, the GEMC method is not very efficient for studying phase equilibria involving very high dense to solid phase. Also, since there is no interface formed in this method, it is not useful for obtaining the interfacial properties. Nevertheless, it has changed the state of molecular simulation and is widely used for investigating the bulk behavior and properties of system involving multi-phases.

In 1992, inspired by GEMC, Kofke invented a method, which he named Gibbs-Duhem Integration (GDI) [17, 18] for direct evaluation of coexistence phases. GDI also does not need any establishment of interface. It does not rely on particle exchange, which is its key strength. Hence, it is useful for all kinds of systems. The method uses the concept of thermodynamic integration for evaluating the free energy and locating the transition point. The Clausius-Clapeyron differential equation for the coexistence line is the core of GDI [19]:

$$\left(\frac{dP}{dT}\right)_{\sigma} = \frac{\Delta H}{T\Delta V} \tag{8.7}$$

where, ΔH and ΔV are the differences in molar enthalpy and volume, respectively. The path along the coexistence line is represented by the subscript σ.

A predictor-corrector method may be used for performing the integration; however, more complicated algorithms have been used [7]. The right hand side of Eq. (8.7) involves "mechanical" quantities, which can be calculated in the course of the Monte Carlo or molecular dynamics simulation. Given a single point of coexistence curve (e.g. from GEMC), this method has the potential to calculate the complete phase diagram from a series of constant pressure simulation. The main limitation of GDI method is obtaining an initial point of coexistence in order to start the integration procedure. Error related to numerical integration is also of concern [20]. The method is quite broad. Any other field of variables can be used instead of temperature and pressure, and modifying the right side of Eq. (8.7) appropriately. GDI is a popular choice for system containing solid phase, where GEMC and insertion based methods fail. GDI also finds application in multicomponent systems [21].

Another popular method, which has find applications in the calculation of thermophysical properties, is grand canonical ensemble Monte Carlo (GCMC). In GCMC, volume, temperature, and chemical potential are fixed. In this method, number of particles fluctuates and the average value of it is determined by the fixed chemical potential and the temperature. By tuning the value of temperature and the chemical potential one can determine the phase diagram. With the advent of transition matrix Monte Carlo technique (TMMC) [22], GCMC simulations have become extremely powerful for phase equilibria calculations. Combination of the two methods is now referred as GC-TMMC.

GC-TMMC [23] simulations are conducted in a grand-canonical ensemble at constant chemical potential μ, volume V, and temperature T. Microstate probability density distribution for this ensemble is given by,

$$\pi_{\mathrm{s}} = \frac{1}{\Xi} \frac{V^{N_{\mathrm{s}}}}{\Lambda^{3N_{\mathrm{s}}} N_{\mathrm{s}}!} \exp\left[-\beta(U_{\mathrm{s}} - \mu N_{\mathrm{s}})\right] \tag{8.8}$$

where $\beta = 1/k_{\mathrm{B}}T$ is the inverse temperature, k_{B} is the Boltzmann's constant, Λ is the de Broglie wavelength , Ξ is the grand partition function, U_{s} is the interaction energies of particles, N_{s}, is the microstate, s. The macrostate probability, $\Pi(N)$, is calculated by summing all microstate probabilities at a constant number of molecules, N. The mathematical formula can be expressed as,

$$\Pi(N) = \sum_{N_{\mathrm{s}}=N} \pi_{\mathrm{s}} \tag{8.9}$$

Book keeping scheme of transition matrix is employed to obtain the macrostate probability. For example, for a move from a microstate s with N number molecules to a microstate t with M number molecules, the acceptance probability is defined as,

$$a\,(\mathrm{s} \to \mathrm{t}) = \min\left[1,\ \pi_{\mathrm{t}}/\pi_{\mathrm{s}}\right] \tag{8.10}$$

In this scheme, acceptance probability is recorded in a matrix C, for each MC move, as shown in Eqs. (8.11) and (8.12), regardless of whether the move is being accepted or not.

$$C(N \to M) = C(N \to M) + a(\mathrm{s} \to \mathrm{t}) \tag{8.11}$$

and

$$C(N \to N) = C(N \to N) + 1 - a(\mathrm{s} \to \mathrm{t}) \tag{8.12}$$

Macrostate transition probability then can be obtained from the matrix C at any time using the following expression,

$$P(N \to M) = \frac{C(N \to M)}{\sum_{O} C(N \to O)} \tag{8.13}$$

Detailed balance expression [6, 7] can be utilized to obtain the macrostate probabilities:

$$\Pi(N)P(N \to M) = \Pi(M)P(M \to N) \tag{8.14}$$

Configuration space is sampled in GCMC simulation by incorporating various MC moves. For pure simple fluid simulation, only deletion, addition, and displacement moves are employed considering one molecule at a time. The possible state change will be from one of the following choices: $N\rightarrow N$, $N\rightarrow N-1$, $N\rightarrow N+1$, hence the transition probability matrix is tri-diagonal. In such conditions, a sequential approach would be an efficient and simple way to obtain the macrostate probabilities,

$$\ln \Pi(N+1) = \ln \Pi(N) - \ln\left[\frac{P(N+1\rightarrow N)}{P(N\rightarrow N+1)}\right] \quad (8.15)$$

A typical probability distribution at vapor-liquid coexistence chemical potential is shown in Fig. 8.1.

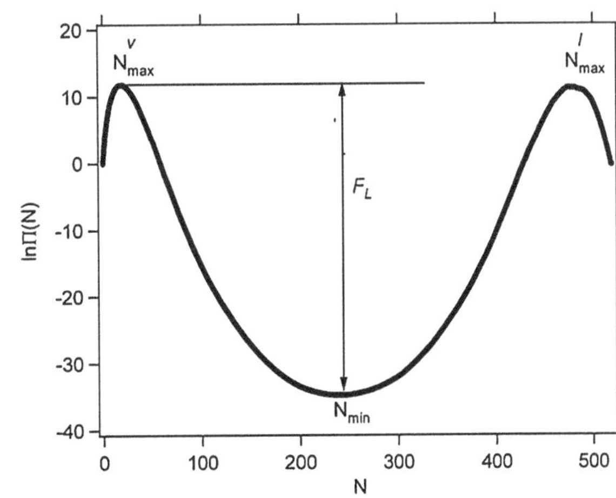

Fig. 8.1 Probability, Π(N) of observing number of particle, N, in a typical grand-canonical Monte Carlo simulation. The peaks represent vapor and liquid phase and trough represents the interface region. F_L is the finite size free energy of the interface

GCMC simulations are conducted at a specified value of the chemical potential, which is not necessarily close to the saturation value. In practice, coexistence chemical potential is usually not known a priori. To determine the phase-coexistence value of the chemical potential, the histogram reweighting method of Ferrenberg and Swendson [24] is generally used. This method enables one to shift the probability distribution obtained from a simulation at chemical potential μ_0 to a probability distribution corresponding to a chemical potential μ using the relation,

$$\ln \Pi(N;\mu) = \ln \Pi(N;\mu_o) + \beta(\mu - \mu_o)N. \quad (8.16)$$

The above relation is usually used to find the chemical potential that produces a probability distribution $\Pi_c(N)$ where the areas under the vapor and liquid regions are equal. Saturated densities are related to the first moments of the vapor and liquid peaks of the coexistence probability distribution. A typical phase equilibria calculated from GC-TMMC is displayed in Fig. 8.2, for square-well fluids, which is represented by the following potential [11]:

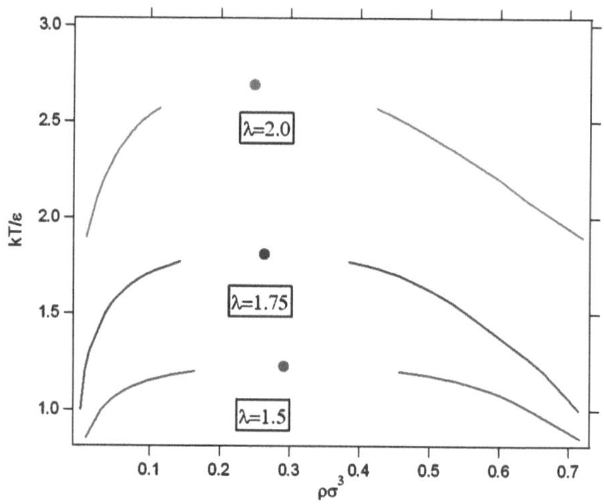

Fig. 8.2 Vapor–liquid phase diagram of variable square-well fluids (adapted from [11])

$$u(r_{ij}) = \begin{cases} \infty, & 0 < r_{ij} < \sigma, \\ -\varepsilon, & \sigma \le r_{ij} < \lambda\sigma, \\ 0, & \lambda\sigma \le r_{ij}, \end{cases} \tag{8.17}$$

where r_{ij}, is the distance between two particles, σ is the diameter of the hard-core repulsive interaction, $\lambda\sigma$ is the potential well extent, and ε is the depth of the isotropic well.

To calculate the saturation pressure, P, following expression is used [25],

$$\beta PV = \ln\left(\sum_N \Pi_c(N)/\Pi_c(0)\right) - \ln(2). \tag{8.18}$$

A typical behavior of saturation pressure against temperature for a bulk vapor–liquid is represented in Fig. 8.3.

The interfacial free energy F_L for a finite-size system with a simulation box of edge length, L, is determined from the maximum likelihood in the liquid ($\Pi^l_{\max}$) and vapor ($\Pi^v_{\max}$) regions and the minimum likelihood in the interface ($\Pi_{\min}$) region,

$$\beta F_L = \frac{1}{2}\left(\ln \Pi^l_{\max} + \ln \Pi^v_{\max}\right) - \ln \Pi_{\min} \tag{8.19}$$

Infinite size surface tension is obtained by evaluating series of finite size surface tension and extrapolating it to infinite system size as shown in Fig. 8.4.

Surface tension as calculated from finite size scaling has been applied to various systems including simple fluids [11, 23], associating fluid [27], metal [28], n-alkane [29] and colloids [26]. Typical vapor–liquid surface tension behavior for pure fluid as a function of temperature is shown in Fig. 8.5.

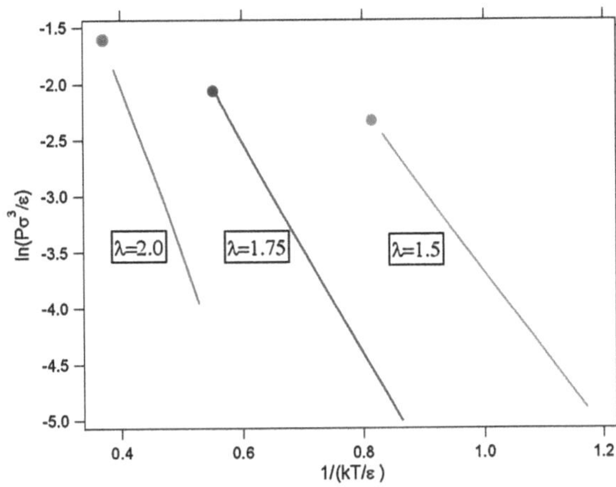

Fig. 8.3 Pressure vs. inverse temperature of variable square-well fluids (adapted from [11])

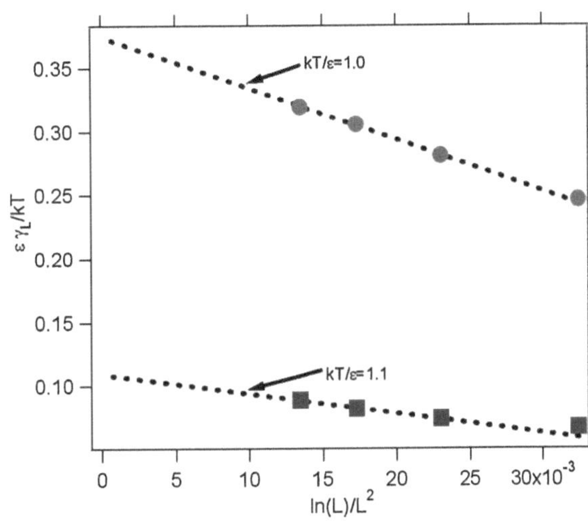

Fig. 8.4 Typical system size dependence of surface tension of hard-core Yukawa fluid with interaction range 1.8, for two different temperatures. The *dashed lines* show the linear extrapolation to infinite system size. L is the edge length of the cubic simulation box (adapted from [26])

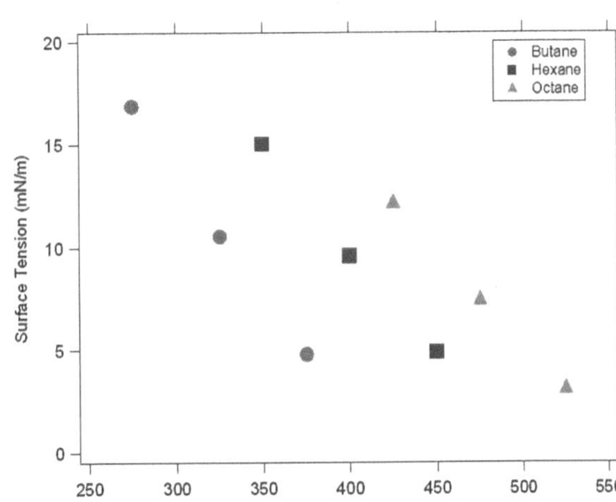

Fig. 8.5 Surface tension vs. temperature of n-alkane. *Symbols* represent predictions from the exponential-6 model [23] (adapted from [29])

8.3 Wetting Transition of Fluid Near Surfaces

Wetting behavior of fluid–solid interfaces is of practical interest to technologically important areas such as sensors and coating. Wetting of patterned surfaces by liquids plays a key role in the field of nano-fluidics and biophysics. Increase in demand of new nano-based technologies requires understanding the wetting behavior on functional surfaces. Particular, the nature of functionalization and how the wetting behavior is affected is of considerable importance to the development of new materials. Competition between surface-fluid and fluid-fluid interactions can cause various phase transitions such as prewetting, layering, and capillary condensation [30]. Coexistence phases of vapor and liquid in contact with an attractive solid surface can induce two different kinds of phenomena. For weakly adsorbed molecules, if interactions of the fluid molecules are stronger than the surface–fluid interaction, partial wetting occurs in the adsorbed film, which consists of a layer of vapor bubbles and liquid drops (see Fig. 8.6a) [31]. For strong attractive surface, the liquid molecules spread across the surface with a thick film, leading to complete *wetting* of the surface (see Fig. 8.6b); on the contrary a drying state is seen for extremely strong repulsive surface (see Fig. 8.6c).

Fig. 8.6 Wetting of a liquid drop on a flat substrate: **a** partial wetting state; **b** complete wetting state; **c** complete drying state

Liquid drop shape (see Fig. 8.6a) is directly related to the contact angle, which it makes with the surface. Partial wetting is most popularly represented in terms of Young's formalism:

$$\gamma_{\mathrm{sv}} - \gamma_{\mathrm{sl}} = \gamma_{\mathrm{lv}} \cos\theta, \tag{8.20}$$

where, γ_{sv}, γ_{sl} and γ_{lv} are the interfacial tensions between solid and vapor phases, solid and liquid phases and liquid and vapor phases, respectively.

The above equation is oversimplified and strictly valid for atomically smooth and chemically homogeneous substrate. However, for rough surface, two regimes are identified depending on the level of surface roughness. In case the surface is made of small protrusions, it cannot be filled by the liquid and hence, is filled by the air (see Fig. 8.7a). Such case is referred to as Cassie–Baxter regime [32] and the contact angle equation is modified accordingly:

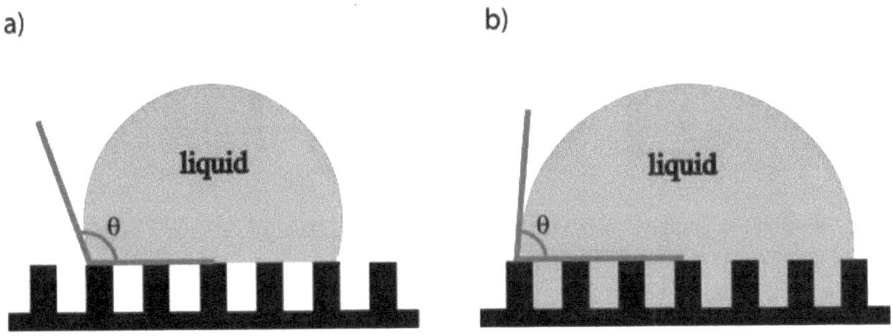

Fig. 8.7 Wetting of a liquid drop on a textured surface: **a** Cassie–Baxter regime; **b** Wenzel regime

$$\cos\left(\theta^{*}\right) = f - 1 + f\left[\cos\left(\theta\right)\right] \tag{8.21}$$

where, θ^* and θ are measured and true contact angles, respectively; f is the fraction of the surface that is in contact with the liquid.

Wenzel regime [33] (see Fig. 8.7b) identifies the state where the liquid wets the surface; however, the measured contact angle is different from the true contact angle, as given by:

$$\cos\left(\theta^{*}\right) = R\cos\left(\theta\right) \tag{8.22}$$

where, R is the ratio between the actual surface area of the rough/textured surface and the projected area.

Superhydrobhocity is primarily due to roughness on the surface. For smooth hydrophobic surface, contact angle greater than 120° is rare [34]. However, it is well known that contact angle of water on a lotus leaf is as high as 160° due to hierarchal structure of lotus leaf [35]. The rapid development of micro and nano processing techniques has made it possible to fabricate superhydrophobic surfaces with various microstructural patterns [36–38]. In recent years, molecular simulation, on the other hand, has evolved and used heavily to understand effect of various micro and nano patterns on the wettability of water and other fluids [39–41].

Contact angle calculation from molecular simulation can be done using the route taken in experiments. In such simulation, a drop consisting of liquid particles near the surface, few Angstrom above the top of the surface is placed. Surface molecules, usually, are kept frozen during the simulation. Figure 8.8 presents a typical case of water drop on a graphite surface at room temperature, which was performed using DLPOLY [42]. Simulation was conducted using molecular dynamics with 2000 water molecules. Equilibration run was done for 1 ns and contact angle was calculated for 600 ps during the production run using two approaches, which are described below.

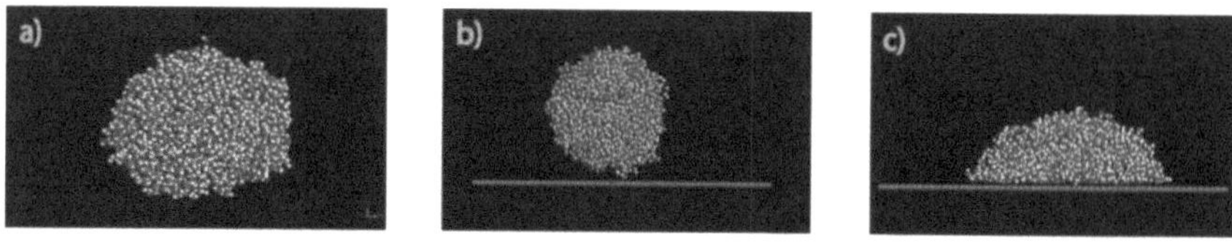

Fig. 8.8 Water on a graphite surface **a**: Free drop; **b** initial stage of drop on a graphite sheet; **c** drop on a graphite sheet after 200 ps

8.3.1 Fitting Method

In this method, using molecular dynamics trajectories or Monte Carlo configurations, fluid isochore profiles are obtained from the simulation by introducing a cylindrical binning, which uses the top most solid surface layer as zero reference level and the surface normal through the center of mass of the droplet as reference axis. For example, for the water drop on a graphite case presented in this chapter, the bins have a height of 0.5 Å and are of equal volume, i.e., the radial bin boundaries are located at $r_i = \sqrt{\frac{i\delta A}{\pi}}$ for $i = 1, ..., N_{\mathrm{bin}}$ with a base area per bin of $\delta A = 95$ Å^2. To extract the contact angle from such a profile, a two-step procedure is adopted [43] First, the location of the equimolar dividing surface is determined within every single horizontal layer of the binned drop. Second, a circular best fit through these points is extrapolated to the substrate surface where the microscopic contact angle θ is measured. Note that the points of the equimolar surface below a height of 8 Å from the substrate surface are not taken into account for the fit, to avoid the influence from density fluctuations at the liquid–solid interface. Figure 8.9 presents a typical drop surface used to obtain the contact angle for water on a graphite surface at 300 K.

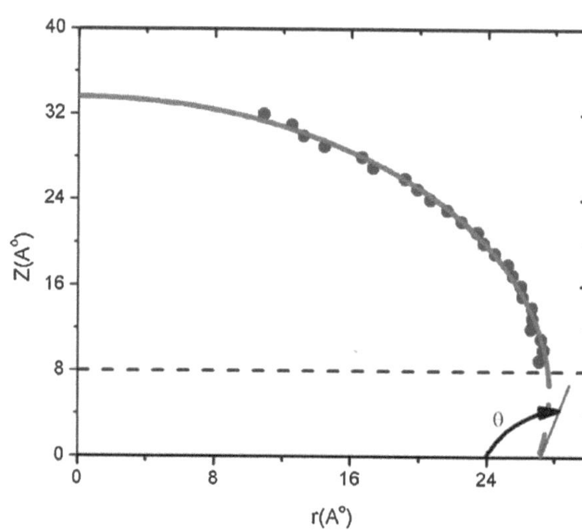

Fig. 8.9 Contact angle measured by fitting a circle to the point of equimolar dividing plane, represented by circles, with Z > 8 Å to exclude region near the substrate

8.3.2 Center of Mass Method

In this method, a microscopic contact angle is calculated by comparing the average height of the centre of mass, of the liquid drop to that of an ideal sessile drop in the shape of a sphere intersecting the surface plane. The position of the sphere relative to the plane is determined by the centre-of-mass position, $z_{c.m.}$, and the condition that the volume of the sphere in the half space above the surface plane contains the correct number of liquid molecules (assuming a uniform density in the idealized drop equal to that of bulk liquid).

The equation employed is

$$\langle z_{c.m} \rangle = (2)^{-4/3} R_0 \left(\frac{1 - \cos\theta}{2 + \cos\theta} \right)^{1/3} \left(\frac{3 + \cos\theta}{2 + \cos\theta} \right) \tag{8.23}$$

where θ is the contact angle, $<z_{c.m.}>$, the average height of centre of mass, is measured relative to the planar surface and $R_0 = (3N/4\pi\rho_o)^{1/3}$ is the radius of a free spherical drop of N molecules at uniform bulk density ρ_o.

The contact angle for water-graphite system at 300 K, studied in this work, based on the fitting approach is 102°; on the other hand, contact angle calculation using center of mass (COM) method is 110°. COM method is particularly more approximate compared to the rigorous fitting the surface of the drop. Various studies of contact angle have been performed mainly using molecular dynamics. Examples include a Lennard-Jones droplet on a structureless substrate [44], water on polar and non-polar surfaces [45], and hexadecane and water on a self-assembled monolayer [46]. One drawback of the method is that the values for the contact angle are often sensitive to the size of the droplet [44]. In fact, macroscopic (infinite size drop) contact angle, θ_∞, is related to microscopic (finite size drop) contact angle, θ, through the following modified Young-Dupre's equation [47–49] :

$$\cos\theta_\infty = \cos\theta + \frac{\tau}{r_B \gamma_{lv}}, \tag{8.24}$$

where r_B is the base radius of the liquid drop, τ is the line tension.

Macroscopic contact angle can be obtained by evaluating series of microscopic contact angle with different drop size and extrapolating it to infinite system size according to Eq. (8.24), as shown by Hirvi et al. [50].

The other class of approach is due to the density of state method. In this approach, GC-TMMC, can be used to obtain the contact angle as recently shown by Grzelek and Errington [51]. In the partial wetting regime, GCMC simulation, on a substrate with an area of A, at the bulk saturated chemical potential would lead to a macrostate probability distribution as shown in Fig. 8.10. A peak emerges at vapor-like surface densities, which corresponds to a stable phase at bulk saturation condition.

The barrier between the vapor–solid peak with respect to the plateau (vapor–liquid and liquid-solid region) can be plugged in Eq. (8.20) to obtain the contact angle. This methodology can easily be used to obtain, for a variety of substrates,

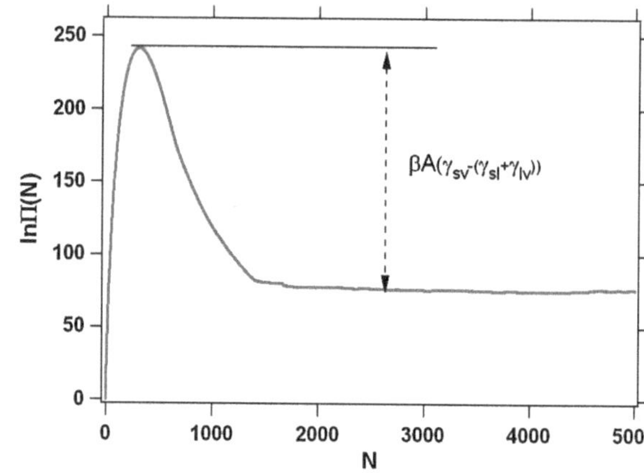

Fig. 8.10 Probability distribution of square-well fluid with well width, $\lambda_{\text{ff}} = 1.5$, on a weak square-well substrate with well width, $\lambda_{\text{wf}} = 1.0$ and $\varepsilon_{\text{wf}} = \varepsilon_{\text{ff}}$ at the bulk saturation chemical potential value and reduced temperature $k_{\text{B}}T/\varepsilon_{\text{ff}} = 1.0$

the drying and wetting temperatures, for which contact angles are 180° and 0°, respectively. For example, square-well fluid of well extent $\lambda_{\text{ff}} = 1.5$ on a substrate of different wall-fluid interaction range displays distinct wetting transitions as a function of substrate-fluid interaction strength, as shown in Fig. 8.11.

Wetting transition is closely associated with a temperature called wetting temperature, T_{w}, at which adsorption state transforms from partial wetting ($\theta > 0$) to complete wetting ($\theta = 0$). Below the wetting temperature the thickness of the film adsorbed on a surface remains finite at all pressures, which are below the bulk saturation pressure. Above wetting temperature, prewetting transition might be observed between two surface states differed by thickness. Prewetting transition stems from the saturation curve at the wetting temperature (see Fig. 8.12) and terminates at the prewetting critical point, T_{cpw}, where thin and thick films (surface phase states) become indistinguishable.

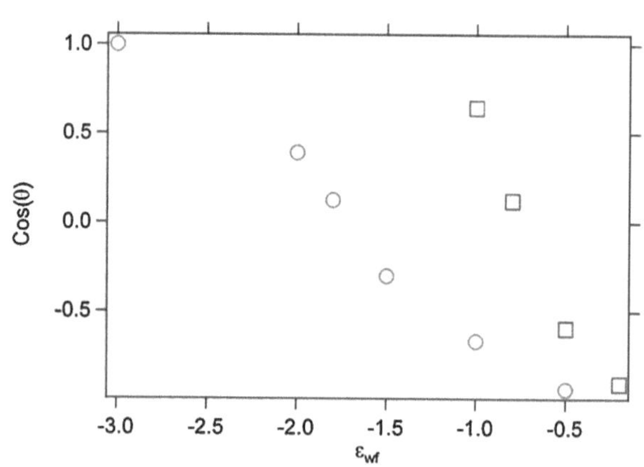

Fig. 8.11 Contact angle vs. wall–fluid interaction strength for square–well fluid on square–well substrates of two wall–fluid interaction range. Symbol circle represents the data for substrate-fluid interaction range, $\lambda_{\text{wf}} = 1.0$; symbol square represents the data for $\lambda_{\text{wf}} = 1.8$

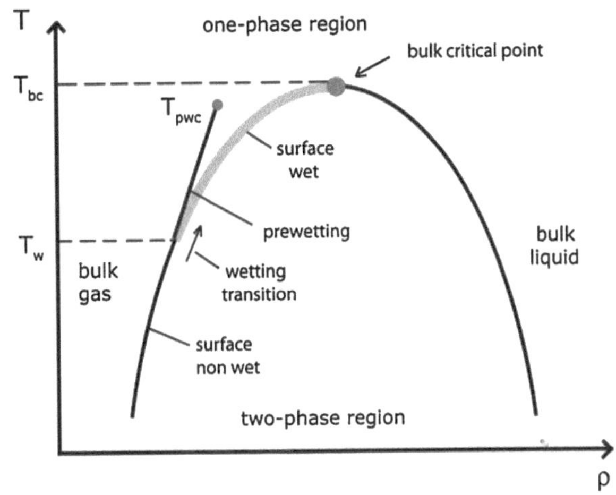

Fig. 8.12 Bulk phase diagram with prewetting and wetting curves

In 1977, Cahn [31] predicted the existence of wetting transition through a two-phase mixture of fluids near a third phase, surface. Independently, Ebnar and Saam [52] also predicted wetting and prewetting transition of argon film adsorbed onto a weakly attractive solid carbon-dioxide surface. The authors used density functional theory (DFT) for the prediction of wetting transition. Since then, DFT has been widely used to study wetting phenomena. Experimental evidences, which came much later, support the prediction of prewetting transitions. Examples can be found for helium adsorption on Cs [53, 54] and Rb [55], liquid hydrogen on various substrates [56] and acetone on graphite [57].

In particular, argon on solid carbon-dioxide has been studied by numerous investigators by various means [58–60]. There are other model systems, though, for which prewetting transition has been observed [61–64]. The first Monte Carlo (MC) simulation on prewetting transition was done by Finn and Monson on model argon on solid carbon-dioxide surface [65]. Subsequently, many different systems have been tried to investigate the prewetting and wetting transitions using Monte Carlo techniques [61, 63, 66]; specially, extensive work has been done for simple gases on alkali metal surfaces by Curtarolo and co-workers [67–70]. Omata and Yonezawa [71], in particular, studied the effect of fluid-substrate interaction on the prewetting transition. While, majority of the work has been done on planar surfaces, Bohlen and Schoen [72] however, studied the prewetting transitions on nonplanar surfaces.

In recent years with the development of advanced methodology such as GC-TMMC [25] it is feasible to investigate more precisely the first order phase transition [11, 27, 73] including prewetting transition as shown by Errington and co-workers for model argon on solid carbon surface [74, 75]. Compared to Monte Carlo techniques, molecular dynamic (MD) is less utilized to predict the prewetting transition. Nonetheless, in conjunction with Monte Carlo techniques, MD is quite useful to predict boundary tension as shown recently by Singh and co-workers [76]. A typical thin-thick film at coexistence (at a pressure less than bulk saturation pressure) is shown in Fig. 8.13.

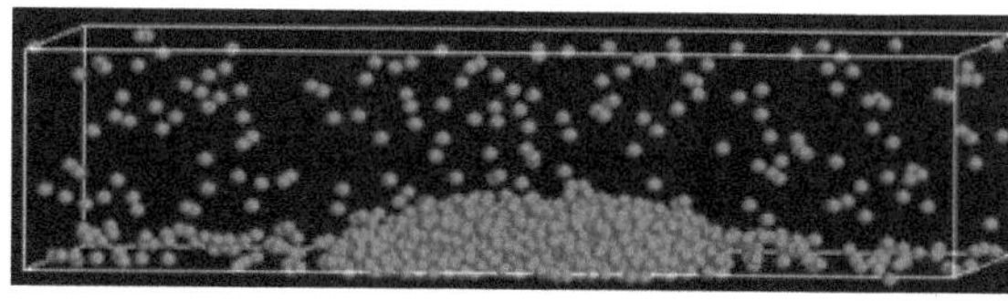

Fig. 8.13 Typical phase separation of thin-thick film for SW fluid on a SW substrate

Nevertheless, investigation of prewetting transition is difficult as the region of occurrence of such transition is limited as shown by Errington and co-workers [77] due to which, such transition is extremly difficult to observe in experiments. For example, using GCMC simulations Zhao [63] observed prewetting transitions for water on a graphite surface, which is yet to be seen in experiments.

8.4 Fluid in Nanopores

8.4.1 Phase Equilibria Under Confinement

Fluid confined in nanopores is a common occurrence in sensors and devices. In such cases, understanding the phase equilibria and having knowledge of structural and transport properties of confined fluids are important in the development of new technologies for manufacturing and in the modification of current methods. Phase behaviors of fluid in porous material are dramatically different from those of the bulk fluid because of the competition of fluid-fluid and fluid-wall interaction energies. Moreover, geometry of the adsorbate can make the adsorbent to behave as a two dimensional fluid in nanotubes or one dimensional fluid as for gases stuck in the corners of rectangular pores [78, 79]. These geometrical constraints and the presence of external forces are the primary source for different phase transition such as layering, prewetting and capillary condensation [30].

A typical vapor–liquid equilibria under confinement is shown in Fig. 8.14 for *n*-alkane [80]. It is well known that capillary critical temperature is suppressed under confinement [81, 82]. Capillary critical temperature in confined geometries is reported experimentally for SF_6 in porous glass and several gases in MCM-41 by Thommes et al. [83] and Morishage et al. [84], respectively.

Mean field theory suggests that the shift in the critical temperature of fluids, under confinement, has a linear dependence on the inverse pore width for slit pores [85]. Vishnyakov et al. [86] performed Monte-Carlo simulations on carbon slit pore and obtained similar results as seen experimentally [83, 84]; however, the simulations were limited to five molecular diameter in pore size. Recent work of Vörtler [87] on a square-well fluid in a hard slit-pore suggests, on the other hand, non-linear dependence of shift in critical temperature as a more generic behavior in nanopores

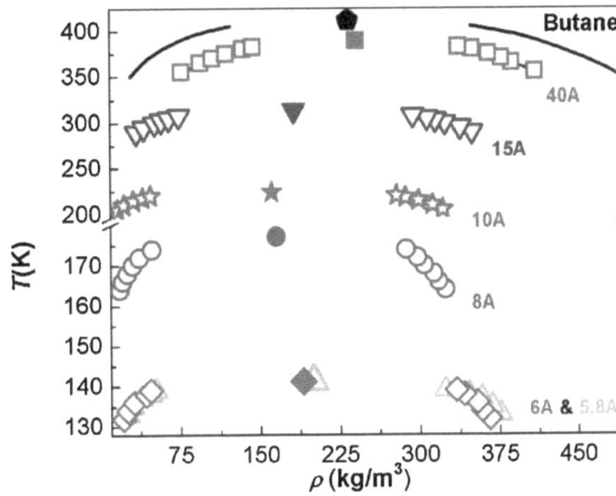

Fig. 8.14 Phase equilibria of *n*-butane under slit pore of graphite with varying slit width. *Filled symbols* represent the critical point. *Solid line* represent the data of bulk *n*-alkane (adapted from [80])

for variable pore sizes. As focused on variable pore-size in the aforementioned works, there are other studies pertaining to the effect of wall-fluid interaction on the critical temperature. Zhang and Wang [88] studied the shift in the critical temperature of square-well fluids in a cylindrical pore for various wall-fluid interaction strengths and found a non-monotonic behavior. For square-well fluids in slit-pore Singh et al. [89] investigated, in detail, the shift in critical temperature for a range of slit widths of different substrate strength, $\varepsilon_{\mathrm{wf}}$ (see Fig. 8.15), and found that the shift in critical temperature goes through various regimes as a function of inverse of the slit width, ranging from quasi-3D to 2D.

Similar to the critical temperature, critical pressure and density behave interestingly under confinement [80].

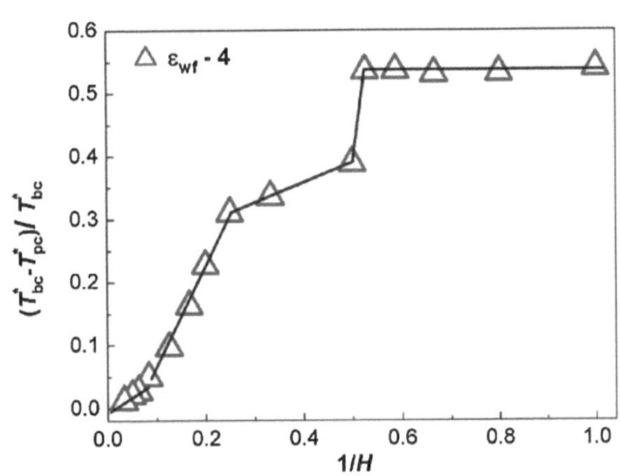

Fig. 8.15 Shift in critical temperature as a function of inverse slit width, $1/H$. T_{bc} and T_{pc} represent bulk and pore critical temperatures, respectively

8.4.2 Flow Properties of Fluids in Nano-Channels

Understanding the flow of fluids in nano channels is important for the development of nanofluidic devices. Despite its significance for so many processes such as in nanolithography [90], nanolubrication [91], and the fluid transport in living organism [92], the flow of fluid under confinement of nano scale are not well understood. The intriguing aspects of the flow in confined region, for example, water rise against gravity in plants has led to controversial debate for more than 100 years [93, 94]. Molecular simulation can help to shed some ambiguity and give insight to the flow behavior of fluid under nano-scale. Considering the advantage of molecular dynamics for the investigation of flow behavior, various groups are actively utilizing the tool to study the flow properties of fluid in nano channels. The flow behavior is found to be puzzling when the molecular diameter is in the same order of the pore size. For example, Hummer and co-workers [95] observed a pulse and burst transition of water in CNT of 13.4 Å long and of diameter 8.1 Å, after the tube is solvated in a water reservoir.

The behavior of fluid rise in a nanotube is fundamentally explained by Washburn formalism [96]. The rise of fluid in t time at a zero external pressure is given by

$$H^2(t) = \left(\frac{\gamma_{lv} R \cos\theta}{2\eta}\right) t + H_0^2 \tag{8.25}$$

where, H is the length of fluid flow in the pore/tube in time t, R is the radius of the tube, θ is the contact angle, η is the shear viscosity and H_0 is a constant.

The above formalism holds after some transient period when inertial effect is vanished. Fascinating behavior is seen during the transient period by Quirke and co-workers [97, 98] based on molecular dynamics for simple fluids, where $H(t)$ is found to linearly rise with t. Similar behavior was observed by Martic et al. [99] in their MD simulations of polymer melt. Supple and Quirke [97, 100] further questioned about the validity of Eq. (8.25) and their result implied linear law of H in t on the nanoscale. More rigorous simulations were conducted by Dimitrov et al. [101] and found the relation in Eq. (8.25) to hold for simple fluid. However, for polymeric fluids slip-length, δ, is not negligible and in such cases the following modification of Eq. (8.25) is necessary [101]:

$$H^2(t) = \left(\frac{\gamma_{lv} (R+\delta)^2 \cos\theta}{2R\eta}\right) t + H_0^2 \tag{8.26}$$

Apparently, there are several questions, which are yet to be addressed related to fluid flow near surfaces and through nanopores. For example, there is hardly any clarity on suitable surfaces for surface electrophoresis applications [102]. Certainly in such cases molecular modeling and simulation would be immensely useful. It is envisaged that screening various possible surfaces via means of molecular simulation would expedite the process, with relatively less cost, in the development of appropriate devices. Further, one can also use these techniques to understand how to control the flow properties of the fluid in the nanopore by a suitable modification of the surface forces acting from the pore walls on the liquid particles.

8.5 Conclusions

Molecular simulations methods have experienced remarkable advances over the last decades. The examples presented in this chapter serve to illustrate that Monte Carlo and Molecular Dynamics can now be used to address a wide variety of problems in nano-fluidics including phase separation, wetting, flow and electrophoresis.

Despite these advances, important challenges however, persist; hence, advanced molecular simulation methodologies must still be devised to simulate complex molecules particularly related to flowing systems. Additional advances are needed in the development of coarse grain models and methods for coarse graining, which would expand the range of problems amenable to direct simulations, and considerably increase our understanding of the structure and dynamics of complex fluids.

Acknowledgments This work is supported by the Department of Science and Technology and Department of Atomic Energy of India.

References

1. Ou J, Perot B, and Rothstein JP (2004) Laminar drag reduction in microchannels using ultrahydrophobic surfaces. Phys. Fluids. 16:4635.
2. Min TG and Kim J (2004) Effects of hydrophobic surface on skin-friction drag. Phys. Fluids. 16:L55.
3. Cramer JC (2002) Essentials of Computational Chemistry, John Wiley & Sons, NewYork.
4. Leach AR (2001) Molecular Modelling: Principles and Applications, Prentice Hall, USA.
5. Cohen NC (1996) Guidebook on Molecular Modeling in Drug Design Academic Press, NewYork.
6. Allen MP and Tildesley DJ (1987) Computer Simulation of Liquids. Clarendon Press, Oxford.
7. Frenkel D and Smit B (2002) Understanding Molecular Simulation: From Algorithms to Applications, Academic Press, NewYork.
8. Chatterjee A and Vlachos DG (2007) An overview of spatial microscopic and accelerated kinetic Monte Carlo methods. J. Computer-Aided Mater. Des. 14:253.
9. McQuarrie DA (1976) Statistical Mechanics. Harper & Row, New York.
10. Metropolis N, Rosenbluth AW, Rosenbluth MN, Teller AH, and Teller E (1953) Equation of state calculations by fast computing machines. J. Chem. Phys. 21:1087–1092.
11. Singh JK, Kofke DA, and Errington JR (2003) Surface tension and vapor–liquid phase coexistence of the square-well fluid. J. Chem. Phys. 119:3405.
12. Panagiotopoulos AZ (1987) Direct determination of phase coexistence properties of fluids by Monte Carlo simulation in a new ensemble. Mol. Phys. 61:813.
13. Panagiotopoulos AZ, Quirke N, Stapleton M, and Tildesley DJ (1988) Phase equilibria by simulation in the Gibbs ensemble: Alternative derivation, generalization and application to mixture and membrane equilibria. Mol. Phys. 63:527–545.
14. Smit B, Karaborni S, and Siepmann JI (1995) Computer simulation of vapor–liquid phase equilibria of *n*-alkanes. J. Chem. Phys. 102:2126.

15. Siepmann JI, Karaborni S, and Smit B (1993). Simulating the critical behavior of complex fluids. Nature. 365:330–332.
16. Escobedo F and de Pablo J (1996) Expanded grand-canonical and Gibbs ensemble Monte Carlo simulations of polymers. J. Chem. Phys. 105:4391.
17. Kofke DA (1993) Gibbs-Duhem integration: A new method for direct evaluation of phase coexistence by molecular simulation. Molec. Phys. 78: 1331–1336.
18. Kofke DA (1993) Direct evaluation of phase coexistence by molecular simulation via integration along the saturation line. J. Chem. Phys. 98: 4149–4162.
19. Denbigh K (1971) Principles of Chemical Equilibrium. Cambridge University, Cambridge.
20. Frenkel D and Smit B (1996) Understanding Molecular Simulation: From Algorithms to Applications, Academic Press, Cambridge.
21. Mehta M and Kofke DA (1994) Coexistence diagrams of mixtures by molecular simulation. Chem. Eng. Sci. 49:2633–2645.
22. Fitzgerald M, Picard RR, and Silver RN (1999) Canonical transition probabilities for adaptive Metropolis simulation. Europhys. Lett. 46:282–287.
23. Errington JR (2003) Evaluating Surface Tension using grand-canonical transition-matrix Monte Carlo simulation and finite-size scaling. Phys. Rev. E. 67:012102.
24. Ferrenberg AM and Swendsen RH (1988) New Monte Carlo technique for studying phase transitions. Phys. Rev. Lett. 61:2635–2638.
25. Errington JR (2003) Direct calculation of liquid–vapor phase equilibria from transition matrix Monte Carlo simulation. J. Chem. Phys. 118:9915.
26. Singh JK (2009) Surface tension and vapour-liquid phase coexistence of variablerange hard-core attractive Yukawa fluids. Mol. Sim. 35:880.
27. Singh JK and Kofke DA (2004) Effect of Molecular Association on Interfacial Properties: A Monte Carlo Study. J. Chem. Phys. 121:9574.
28. Singh JK, Adhikari J, and Kwak SK (2006) Vapor–liquid phase coexistence curves for Morse fluids. Fluid. Phase. Equil. 248:1–6.
29. Singh JK and Errington JR (2006) Calculation of phase coexistence properties and surface tensions of *n*-alkanes with grand-canonical transition Monte Carlo simulation and finite-size scaling. J. Phys. Chem. B 116:1369.
30. Gelb LD, Gubbins KE,Radhakrishnan R, and Bartkowiak MS (1999) Phase separation in confined systems. Rep. Prog. Phys. 62:1573–1659.
31. Cahn JW (1977) J. Chem. Phys. 66:3667.
32. Cassie ABD and Baxter S (1944) Wettability of porous surfaces. Trans. Faraday Soc. 40:546–551.
33. Wenzel RN (1936) Resistance of solid surfaces to wetting by water. Ind. Eng. Chem. Res. 28:988–994.
34. Lafuma A and Quere D (2003) Nat. Mat. 2:457–460.
35. Barthlott W and Neinhuis C (1997) Planta 2002:1–8.
36. Yoshimitsu Z, Nakajima A, Watanabe T, and Hashimoto K (2002). Langmuir 18:5818.
37. Lau KKS, Bico J, Teo KBK, Chhowalla M, Amaratunga GAJ, Milne WI, McKinley GH, and Gleason KK (2003) Nano Lett. 3:1701.
38. Shirtcliffe NJ, McHale G, Newton MI, and Perry CC (2005) Langmuir. 21:937.
39. Halverson JD, Maldarelli C, Couzis A, and Koplik J (2008) A molecular dynamics study of the motion of a nanodroplet of pure liquid on a wetting gradient. J. Chem. Phys. 129:164708.
40. Shen Y, Couzis A, Koplik J, Maldarelli C, and Tomassone MS (2005) Molecular dynamics study of the influence of surfactant structure on surfactant-facilitated spreading of droplets on solid surfaces. Langmuir 21:12160.
41. Werder T, Walther JH, Jaffe RL, Halicioglu T, and Koumoutsakos P (2003) On the water–carbon interaction for use in molecular dynamics simulations of graphite and carbon nanotubes. J. Phys. Chem. B. 107:1345.
42. Smith W, Forester TR, and Todorov IT (2009) The DLPOLY 2 user manual.

43. de Ruijter MJ, Blake TD, and De Coninck J (1999) Langmuir 15:7836–7847.
44. Ingebrigtsen T and Toxvaerd S (2007) Contact angles of Lennard-Jones liquids and droplets on planer surfaces. J. phys. Chem. C 111:8518–8523.
45. Hautman J and Klein ML (1991) Microscopic wetting phenomena. Phys. Rev. Lett. 67: 1763–1766.
46. Srivastava P, Chapman WG, and Laibinis PE (2005) Odd–even variations in the wettability of *n*-alkanethiolate monolayers on gold by water and hexadecane: a molecular dynamics simulation study. Langmuir. 21:12171–12178.
47. Rowlinson JS and Widom B (1982) Molecular Theory of Capillarity. Oxford, Oxford.
48. Wang JY, Betelu S, and Law BM (1999) Line tension effects near first-order wetting transitions. Phys. Rev. Lett. 83:3677–3680.
49. Wang JY, Betelu S, and Law BM (2001) Line tension approaching a first-order wetting transition: Experimental results from contact angle measurements. Phys. Rev. E 63:031601-1–031601-11.
50. Hirvi JT and Pakkanen TA (2006) Molecular dynamics simulations of water droplets on polymer surfaces. J. Chem. Phys. 144712:144712-1–144712-11.
51. Grzelak EM and Errington JR (2008) Computation of interfacial properties via grand canonical transition matrix Monte Carlo simulation. J. Chem. Phys. 128:014710.
52. Ebner C and Saam WF (1977) Physical Review Letters 38:1486.
53. Rutledge JE and Taborek P (1992) Prewetting phase diagram of 4He on Cesium. Phys. Rev. Lett. 69:937.
54. Hallock RB (1995) Review of some of the experimental evidence for the novel wetting of helium on alkali metals. J. Low. Temp. Phys. 101:31.
55. Phillips JA, Ross D, Taborek P, and Rutledge JE (1998) Superfluid onset and prewetting of 4He on rubidium. Phys. Rev. B. 58:3361.
56. Cheng GME, Lee HC, Chan MHW, Cole MW, Carraro C, Saam WF, and Toigo F (1993) Wetting transitions of liquid hydrogen films Phys. Rev. Lett. 70:1854–1857
57. Kruchten F and Knorr K (2003) Multilayer Adsorption and Wetting of Acetone on Graphite. Phys. Rev. Lett. 91:085502.
58. Sokolowski S and Fischer J (1990) Wetting transitions at the argon-solid-Co2 interface: Molecular-dynamics studies. J. Phys. Rev. A. 41:6866.
59. Evans R and Tarazona P (1983) Wetting and thick-thin film transitions in a model of argon at a solid CO2 substrate. Phys. Rev. A. 28:1864.
60. Shi W, Zhao X, and Johnson J (2002) Phase transitions of adsorbed fluids computed from multiple-histogram reweighting. Mol. Phys. 100:2139.
61. Malo BM, Huerta A, Pizio O, and Sokolowski S (2000) Phase behavior of associating two- and four-bonding sites lennard-jones fluid in contact with solid surfaces. J. Phys. Chem. B. 104:7756–7763.
62. Gatica SM, Johnson JK, Zhao XC, and Cole MW (2004) Wetting transition of water on graphite and other surfaces. J. Phys. Chem. B. 108:11704.
63. Zhao X (2007) Wetting transition of water on graphite: Monte Carlo simulations. Phys. Rev. B. 76:041402.
64. Sacquin S and Schoen M (2003) Fluid phase transitions at chemically heterogeous, nonplanar solid substrates: Surface versus confinement effects. J. Chem. Phys. 118:1453.
65. Finn JE and Monson PA (1989) Prewetting at a fluid-solid interface via Monte Carlo Simulation. Phys. Rev. A. 39:6402.
66. Finn JE and Monson PA (1993) Further studies of prewetting transitions via Monte Carlo simulations. J. Chem. Phys. 99:6897.
67. Bojan MJ, Stan G, Curtarolo S, Steele WA, and Cole MW (1999) Wetting transitions of Ne. Physical Rev. E. 59:864.
68. Curtarolo S, Stan G, Cole MW, Bojan MJ, and Steele WA (1999) Computer simulations of the wetting properties of neon on heterogeneous surfaces. Physical Rev. E. 59:4402.

69. Curtarolo S, Stan G, Bojan MJ, Cole MW, and Steele WA (2000) Threshold criterion for wetting at the triple point. Physical Rev. E 61:1670.
70. Curtarolo S, Cole MW, and Diehl RD (2004) Wetting transition behaviour of Xe on Cs and Cs/graphite. Phys. Rev. B. 70:115403.
71. Omata K and Yonezawa F (1998) Prewetting and density fluctuations in the prewetting supercritical phase. J. Phys.: Condens. Matter. 10:9431.
72. Bohlen H and Schoen M (2004) Aspects of prewtting at nonplanar surfaces. J. Chem. Phys. 120:6691.
73. Kwak SK, Singh JK, and Adhikari J (2007) Molecular simulation study of vapor-liquid equilibrium of morse fluids. Chemical Product and Process Model. 2, part, no. 8
74. Errington JR (2004) Prewetting transitions for a model argon on solid carbon dioxide system. Langmuir. 20:3798.
75. Errington JR and Wilbert DW (2005) Prewetting boundary tensions from Monte Carlo simulation. Phys. Rev. Lett. 95:226107.
76. Singh JK Sarma G, and Kwak SK (2008) Thin-thick surface phase coexistence and boundary tension of the square-well fluid on a weak attractive surface. J. Chem. Phys. 128: 044708.
77. Sellers MS and Errington JR (2008) Influence of Substrate Strength on Wetting Behavior. J. Phys. Chem. C. 112:12905.
78. Bojan MJ and WA. Steele (1998) Computer simulation in pores with rectangular cross-sections. Carbon. 36:1417.
79. Davies GM and Seaton NA (1998) The effect of the choice of pore model on the characterization of the internal structure of microporous carbons using pore size distributions. Carbon 36:1473.
80. Singh SK, Sinha A, Deo G, and Singh JK (2009) Vapor-liquid phase coexistence, critical properties and surface tension of confined alkanes. J. Phys. Chem. C. 113:7170.
81. Fisher ME and Nakanishi H (1981) Scaling theory for the criticality of fluids between plates. J. Chem. Phys. 75:5857.
82. Nakanishi H and Fisher ME (1983) Critical point shifts in films. J. Chem. Phys. 78:3279.
83. Thommes M and Findenegg GH (1994) Pore condensation and critical-point shift of a fluid in controlled-pore glass. Langmuir 10:4270.
84. Morishige K and Shikimi M (1998) Capillary critical point of argon, nitrogen, oxygen, ethylene, and carbon dioxide in MCM-41. J. Chem. Phys. 108:7821.
85. Evans R (1990) Fluids adsorbed in narrow pores: phase equilibria and structure. J. Phys.: Condens. Matter. 2:8989.
86. Vishnyakov A, Piotrovskaya EM, Brodskaya EN, Votyakov EV, and Tovbin YK (2001) Critical propertiees of Lennard-Jones fluids in narrow slit-shaped pores. Langmuir 17: 4451.
87. Vortler HL (2008) Simulation of fluid phase equilibria in square-well fluids: From three to two dimensions Collect. Czech. Chem. Commun. 73:518.
88. Zhang X and Wang W (2006) Square-well fluids in confined space with discretely attractive wall-fluid potentials: Critical point shift. Phys. Rev. E 74:062601.
89. Singh JK and Kwak SK (2007) Surface tension and vapor liquid phase coexistence of confined square well fluid. J. Chem. Phys. 126:024702.
90. Piner RD, Zhu J, Xu F, Hong S, and Mirkin CA (1999) Dip-pen nano lithography. Science. 283:661.
91. Klein J and Kumacheva E (1995) Confinement-induced phase transitions in simple liquids. Science 269:816.
92. Alawa M, Dube M, and Rost M (2004) Imbibition in disordered media. Adv. Phys. 53:83.
93. Zimmermann U, Schneider H, Wegner LH, Wagner HJ, Szimtenings M, Haase A, and Bentrup FW (2002) What are the driving forces for water lifting in xylem conduit. Physiol. Plant. 114:372.
94. Zimmermann U, Schneider H, Wegner LH, and Haase A (2004) Water ascent in tall trees: Does evolution of land plants rely on a highly metastable state? New Phytol. 162:575.

95. Gummer G, Rasaiah JC, and Noworyta JP (2001) Water conduction through the hydrophobic channel of a carbon nanotube. Nature. 414:188.
96. Warshburn EW (1921) The dynamics of capillary flow. phys. Rev XVII:273.
97. Supple S and Quirke N (2003) Rapid imbibition of fluids in carbon nanotubes. Phys. Rev. Lett. 90:214501.
98. Sokhan VP, Nicholson D, and Quirke N (2004) Transport properties of nitrogen in single walled carbon nanotubes. J. Chem. Phys. 120:3855.
99. Martic G, Gentner F, Seveno D, Coulon D, De Coninck J, and Blake TD (2002) A molecular dynamics simulation of capillary imbibition. Langmuir. 18:7971.
100. Supple S and Quirke N (2004) Molecular dynamics of transient oil flows in nanopores I: Imbibition speeds for single wall carbon nanotubes. J. Chem. Phys. 121:8571.
101. Dimitrov DI, Milchev A, and Binder K (2007) Capillary rise in nanopores: Molecular dynamics evidence for the Lucas-Washburn equation. Phys. Rev. Lett. 99:054501.
102. Pernodet N, Samuilov V, Shin K, Sokolov J, Rafailovich MH, Gersappe D, and Chu B (2000) DNA Electrophoresis on a Flat Surface. Phys. Rev. Lett. 85:5651.

Chapter 9
Fabrication of Spring Steel and PDMS Grippers for the Micromanipulation of Biological Cells

G.K. Ananthasuresh, Nandan Maheswari, A. Narayana Reddy, and Deepak Sahu

Abstract A biological cell is considered and studied as a mechanical entity today with important implications in diagnostics and therapeutics. This article describes the opportunities and challenges in using mechanical grippers in manipulating and characterizing isolated biological cells and embryos. Direct mechanical manipulation of cells has some advantages over a wide variety of techniques used in the study of cell-mechanics. Hence, we begin with a brief overview of currently used techniques and then describe the gripper-based cell-manipulation. Five aspects concerning the miniature grippers for cells are addressed here: design, materials, fabrication, force-sensing, and actuation. It is shown that systematic topology and shape optimization techniques help in designing the stiffness of the grippers to match that of the cells. Materials considered here are metals and polymers rather than just silicon. This choice of material warrants new manufacturing techniques at the meso (100 μm to cm) scale. While economic fabrication without the expensive overhead of silicon-based fabrication is the motivating factor for developing meso-scale manufacturing techniques for metals and polymers, we argue here that there are actually some other advantages in terms of functionality and performance. They arise in the context of actuating the gripper in cell's native aqueous environment and sensing the forces exerted on the cell. We describe here a minimally intrusive force-sensing technique that does not need any specialized sensor other than the gripper itself. All of the above are made possible by compliant mechanisms, which are simply elastically deformable structures. They can be designed systematically to specifications; made with any flexible material; manufactured at any size; actuated in many ways and easily; and used as sensors. The last aspect, the sensing, needs computation as it uses the visually measured deformations to deduce the forces. Two inverse problems in elasticity are mentioned in this regard. While the first inverse problem helps in estimating forces, the second one helps in estimating the inhomogeneous mechanical properties inside the cell. It is argued that this

G.K. Ananthasuresh (✉)
Mechanical Engineering, Indian Institute of Science, Bangalore, India
e-mail: suresh@mecheng.iisc.ernet.in

S. Chakraborty (ed.), *Microfluidics and Microfabrication*,
DOI 10.1007/978-1-4419-1543-6_9, © Springer Science+Business Media, LLC 2010

requires dexterous manipulation of cells—a capability not usually shared by other cell-manipulation techniques.

Keywords Microgrippers · micromanipulation · vision-based force sensor · topology optimization

9.1 Introduction

A biological cell, the basic unit of all living creatures, is a complex machine. It is self-contained with provisions for input and output between itself and its environment. It can sense a variety of external signals, which at some level can be thought of as mechanical stimuli. It has means to nourish, repair, and defend itself. A cell can propel and move if its particular type needs it. It can replicate by cell-division. It has internal mechanisms for information-exchange. It maintains its shape and can change it as necessary. With so many machine-like characteristics and more, it is only natural that a cell is studied today as a mechanical entity. Biochemists also have begun to appreciate the mechanical processes that take place in a cell and treat it like a "clockwork device than a reaction vessel" [1].

There is a large body of literature and numerous studies on the mechanical aspects of biological cells. It is mostly experimental while the theoretical work is also underway. This article does not aim to review all that work. While it does provide a brief overview, it also takes a viewpoint that is somewhat different from the majority of the studies undertaken on cells. The first distinct viewpoint is the use of mechanical grippers. Miniature grippers are not new but they do not seem to be in vogue in today's studies on cell-mechanics. We make a case in support of the grippers here. The second viewpoint is towards developing manufacturing techniques that are different from those derived from the mainstream micro and nano technologies. In fact, the techniques described in this article have a leaning towards macro manufacturing techniques rather than micro and nano fabrication. This view opens up opportunities for materials other than silicon. It also leads to some advantages that arguably are not offered by current techniques used in mechanical studies on cells.

9.2 Why Mechanical Manipulation of Cells?

When a system is known to be composed of individual units, it is natural to study those units to get a better understanding of the system. A cell is the basic unit and hence the interest in studying it is natural. Conventionally, investigations on cells depended on observations through microscopes to study the size, shape, morphology, and bio-chemical expressions. Much has been learnt about cells in this manner. But what is not completely understood yet is what the cells actually sense: is it stress, strain, strain energy, strain rate or something else [2]? The line of research that addresses these issues enriches not only our understanding of cells and cell-processes but also the mechanics itself. Therefore, there is a need to subject cells to mechanical testing. This requires mechanical manipulation at the single cell level as well as the organelle.

Currently, many techniques have been developed for grasping and manipulating single cells. Of these, aspiration using pipettes is perhaps the most widely used [3]. It involves making a fine hollow needle (the pipette) and applying a negative pressure to partially suck the cell into the hollow tube; or hold it in place if the cell is much bigger than the exit diameter of the pipette and the cell membrane is sufficiently stiff. Although it is a simple and effective technique to hold the cell in place, the implication of the large mechanical forces applied on the cell is not fully understood. The other popular techniques use magnetic or dielectric particles that get attached to the cells and then magnetic, electric, or optical fields are varied to hold, move, and manipulate cells [4–7]. These techniques are intrusive. Atomic force microscope (AFM) or other such probes are also used to test cells mechanically [8] but with these techniques, it is the cell-membrane that is studied and not the entire cell. Using flow channels of appropriate shape to squeeze the cells is another technique that has fewer disadvantages than the others. This type of technique is promising from the mechanical viewpoint but this is mostly passive in the sense that a particular flow channel will be able to test a certain characteristic of a cell's mechanical response such as a stretch or a squeeze.

None of the techniques described above have the versatile manipulation capability that mechanical grasping fingers offer. Using micromechanical grippers with fingers has already been attempted [9]. In light of growing interest in mechanical response of cells, there are reasons that favour microgrippers. First, the microfabrication technology and microactuation are sufficiently well-developed. Sensing forces and displacements found in the cellular processes is possible today. Second, with grippers, it is possible not only just grasp the cell but also manipulate it to make it undergo gross motions such as rigid-body translation and rotation but also required deformations—elastic or plastic. Third, all this can be done with forces and displacements measured in real time. A fourth advantage is that miniature grippers help characterize the bulk response (such as stiffness, strength, relaxation time, etc.) as well as the inhomogeneous variation of the intrinsic properties of the cell-material. The latter requires manipulation of the cell in ways that the other techniques might not be able to do. This is explained at the end of the paper. The rest of the paper is organized as follows.

We begin by introducing miniature compliant grippers in the next section. This is followed by a discussion of materials that can be used and then the design and fabrication methods. The micromanipulation setup and the method of actuating the grippers are explained after that. Subsequently, testing on egg cell using spring steel and polydimethylsiloxane (PDMS) miniature grippers is described. The method of sensing force with the gripper itself without any additional force-sensor is described and how an extension of this helps in characterizing the mechanical properties of the inside of the cell is commented upon at the end.

9.3 Miniature Compliant Grippers

Making miniature grippers with joints and assembly, although possible, is not economically viable. A miniature gripper that has joints also suffers from friction and wear both of which are more pronounced at the small sizes due to the scaling effect. Hence, grippers that are devoid of joints are preferred. Such a solution is offered by *compliant mechanisms*. Compliant mechanisms are single-piece structures that serve the functions of a jointed mechanism using elastic deformation [10, 11]. A number of advantages arise due to the absence of joints. It is easy to realize them in practice at any size because no assembly is required. Many types of actuation, often embedded into the gripper mechanism, are possible. One more distinct advantage with compliant mechanisms is that it is possible to measure forces with its visually captured deformation data [12, 13].

Figure 9.1a shows a compliant gripper. It has two handles at the bottom, which when drawn together open or close the gripper jaws to grasp an object. The jaws are along the center line of the symmetric gripper. The jaws move along the center

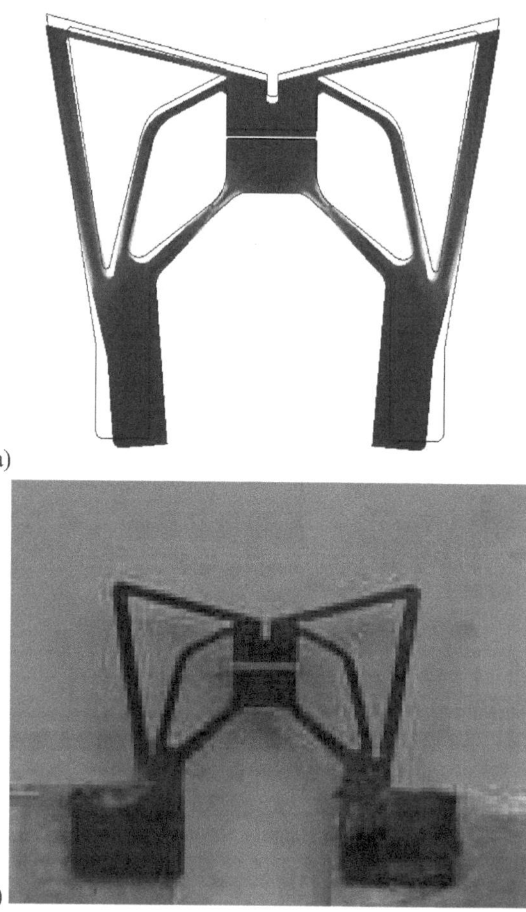

(a)

(b)

Fig. 9.1 (**a**) A compliant gripper, (**b**) a miniature spring steel prototype line vertically. The outline in Fig. 9.1a shows the undeformed gripper and the filled contour shows the deformed gripper with relative levels of stress in it. It is important to note that the two jaws of the gripper move in a purely vertical translation without any arc-like motion that is common in most grippers. Thus, this gripper has a parallel jaw motion. A practical implementation of this design is shown in Fig. 9.1b. It is made of spring steel using wire-cut electro discharge machining (EDM). It has a size of 11 mm × 8 mm × 0.5 mm. The handle portion of the gripper is slightly modified to create large square plates to enable easy actuation by moving them. This modification, however, does not change the behavior of the gripper.

Many other designs are possible as one can imagine that there are numerous shapes possible to get gripper-like motion. An important point is to match the stiffness of the gripper to that of the biological cell being grasped. If not, there could be damage during grasping and manipulation if the stiffness of the gripper is more or there would be ineffective grasping if the stiffness of the gripper is insufficiently low. There are also other requirements that a micro gripper should meet. All the desired features, not just of the micro gripper but of cell-manipulation in general, are listed below.

(i) Applying forces and displacements in the required range and with sufficient resolution.
(ii) Measuring forces and displacements with sufficient accuracy.
(iii) Controlling the motion as needed.
(iv) Manipulating the objects dexterously.
(v) Imaging the manipulating environment including the gripper and objects.
(vi) Working in aqueous environment.
(vii) Estimating the material properties of the objects.
(viii) Automating the entire procedure.
(ix) Providing a haptic interface.

Some of the above are not absolutely essential but it will be useful to have them. Haptic interface and automation are two examples. Haptic interface helps the human user to have a tactile feel in delicate manipulation tasks where direct force feedback is beneficial. Automation is necessary when there are routine but tedious tasks such as intraplasmic cell injection into many cells at once.

9.4 Materials

Biological cells typically have micrometer size. Egg cells are bigger and their sizes vary from species to species but they too are usually less than a millimeter. Therefore, macro-sized grippers are not suitable for cell-manipulation. Micron-size grippers are possible if we use microfabrication techniques to make them. Since silicon still remains the choice material for microfabrication, silicon grippers are commonly used in micro-manipulation [13]. Silicon is also compatible for this task because it is bio-inert. But there is a difficulty with silicon grippers when it comes to actuation in aqueous environment. Most actuating techniques in silicon-based micro-devices use electric current or charge and hence are not suitable operating in water. Some actuation techniques such as electro-thermal actuation also tend to heat the actuating environment. A bigger problem is that silicon-based grippers tend to be expensive because the market for micro grippers is not large enough. A third difficulty is that very thick planar grippers need sophisticated microfabrication equipment such as deep reactive ion etching in order to have sufficient out-of-plane stiffness. Hence, other materials need to be considered.

Metals and polymers are competing alternatives to silicon. The suitability of polymers for microfluidic tasks is well known [14, 15] because of polymers such as polydimethylsiloxane (PDMS), SU-8, polymethylmethacrylate (PMMA), polyimide, etc. But not all polymers are suitable for making micro grippers. PMMA and SU8 are examples of unsuitable materials because they are too brittle and weak for acting like compliant grippers. PDMS and polyimide are sufficiently flexible materials. Between the two, PDMS is more preferred because it can be easily cast into almost any shape. In this work, we use PDMS.

Metals are not to be set aside when it comes to micro grippers. Etching metals is not yet convenient in realizing micro-scale devices because anisotropic deep chemical etching of them is not possible. Laser machining would be too expensive and slow. However, there is no need to make the micro grippers in the micron size. Just as we can pick up very small objects such as a mustard seed with our fingers, meso-scale (say, hundreds of microns to even centimeter size) grippers can handle micron-sized objects. Micromachining of metals in the meso-scale is possible by adapting macro machining techniques. These include milling and wire-cut EDM. We need ductile metals that have high yield strength and have large repeatable elastic range of motion. For this, titanium, beryllium-copper, and spring steel are suitable. Spring steel, being inexpensive and easily accessible, is used in this work for making miniature grippers.

9.5 Design

Our goal is to design a miniature gripper that can grasp and dexterously manipulate biological cells with minimal intrusiveness. Force feedback and haptic interface are also desired. For this, we chose to use the concept of joint-free compliant mechanisms. As noted in an earlier section, compliant mechanisms are capable of giving the desired motion with a single input force if suitable geometric form is designed in the monolithic elastic body that forms the compliant gripper mechanism. Thus, the goal of design is also to minimize the number of actuation points so that controlling the gripper becomes easy. The stiffness of the gripper is to be such that it matches the requirements from the cells being manipulated. Finally, we need to keep the constraints imposed by the material and manufacturing technique. These are explained in this section.

9.5.1 Designing the Stiffness

Both geometric form and material properties decide the stiffness of a planar compliant gripper. Its geometric form is decided by many factors: the topology (how many holes are there?), the shape of its segments, the sizes of the segments, the out-of-plane thickness, etc. Of all these, topology is the crucial aspect in deciding the functionality of the gripper. That is, the output point moves in a desired direction when a force or displacement is applied at the input point in a specified direction. Once this basic functionality is ensured, we can move onto shape and size design. After that material and put-of-plane thickness can also be changed as needed and as possible. Clearly, there is enough freedom for designing the stiffness of the gripper. Table 9.1 shows the details of seven grippers whose photographs are shown in Fig. 9.2. These are of vastly different sizes and their stiffness values also change vastly by five orders of magnitude while their size varies only two orders of magnitude. Three different materials are used here. Manufacturing techniques also differ.

Table 9.1 Data concerning grippers of various geometric form, material, and size

S. No.	Gripper material	Size (mm)	Minimum feature size (mm)	Gap between the jaws (mm)	Stiffness of the gripper (N/m)
1	Polypropylene	$100 \times 100 \times 3$	0.5	25.0	9.76
2	Spring steel 1	$50 \times 50 \times 1$	0.1	12.5	23437.5
3	PDMS 1	$50 \times 50 \times 3$	1.1	11.1	1.5
4	PDMS 2	$25 \times 25 \times 3$	0.77	2.0	33.33
5	PDMS 3	$12 \times 12 \times 2$	0.5	0.95	80.49
6	PDMS 4	$6 \times 6 \times 1$	0.5	0.5	406.57
7	Spring steel 2	$11 \times 8 \times 0.5$	0.1	0.5	3367.0

Overall, what this data confirms is the assertion made above about the freedom available for the designer. When the permitted minimum feature size is small enough, 1 N/m stiffness is also realizable with PDMS grippers of small size. Manufacturing is crucially linked here because that decides the minimum feature size and hence the stiffness. The first six grippers in Table 9.1 and Fig. 9.2 were designed using topology optimization, a technique widely used in structural design [16, 17]. Only the last one was designed intuitively. In what follows, we briefly describe how topology optimization was done by incorporating manufacturing considerations.

9.5.2 Topology Optimization with Manufacturing Constraints

Topology optimization is a systematic method to obtain a compliant mechanism for any specific objective with regard to stiffness, flexibility, strength, etc. [17]. The objective of the mechanism in this case is a gripper. We used frame finite elements to model the mechanism and the ground structure of the design as shown in Fig. 9.3.

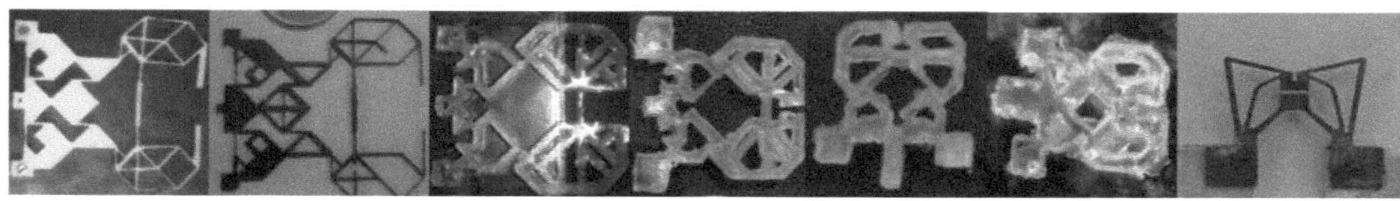

Fig. 9.2 Prototypes of several grippers of various sizes whose data is shown in Table 9.1. All photographs were re-sized to the same size in this picture. The order follows that shown in Table 9.1: Polypropylene, Spring steel 1, PDMS 1, PDMS 2, PDMS 3, PDMS 4, and Spring steel 2

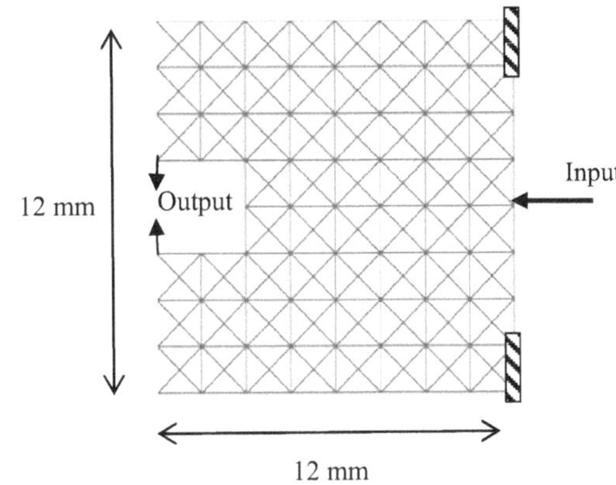

Fig. 9.3 Specifications and ground structure for topology optimization of a miniature gripper

The ground structure consists of many frame elements. All of them have the same out-of-plane thickness. The widths are to be adjusted by the optimization algorithm. Thus, these are the design variables. If the width of any frame element goes to the lower bound (a value close to zero), that element is assumed to be not present in the final topology determined by the optimization algorithm. An element with a value larger than the lower bound will stay in the topology. Thus, ground-structure based topology optimization determines the design of the gripper.

The width of the frame elements that stay cannot assume any value. They should not go below a certain value. If they do, it might not be possible to manufacture them. The constraint posed by the minimum feature of the manufacturing technique was implicitly taken into account in the design process. Here, we have used wire-cut EDM as the manufacturing technique. Since the wire diameter of our machine is 250 μm and the minimum beam width we could reliably and repeatably cut is 100 μm, we set that as the minimum width of the beam.

The statement of the topology optimized problem is as follows:

$$\begin{aligned}&\underset{\mathbf{w}}{\text{Maximize}}\ \frac{MSE}{SE} = \frac{\mathbf{V}^T\mathbf{KU}}{0.5\mathbf{U}^T\mathbf{KU}}\\&\text{Subject to}\\&\qquad \sum_{i=1}^{NELEM} tw_i l_i - V^* \leq 0\\&\qquad \mathbf{KU} = \mathbf{F}\\&\qquad \mathbf{KV} = \mathbf{F}_d\end{aligned}$$

where *MSE* is the mutual strain energy that is numerically equal to the output displacement, *SE* is the strain energy that is a measure of stiffness, **K** is the stiffness matrix of the finite element model, **U** and **V** are displacements for the actual load **F** applied at the input degree of freedom and unit dummy load $\mathbf{F}_d$ applied at the output degree of freedom, w_{i} contains the widths of the rectangular cross-section of the frame elements, l_{i} contains the length of the frame elements, t is the thickness, and finally V^* is the allowed volume of material to be used by the mechanism. We have

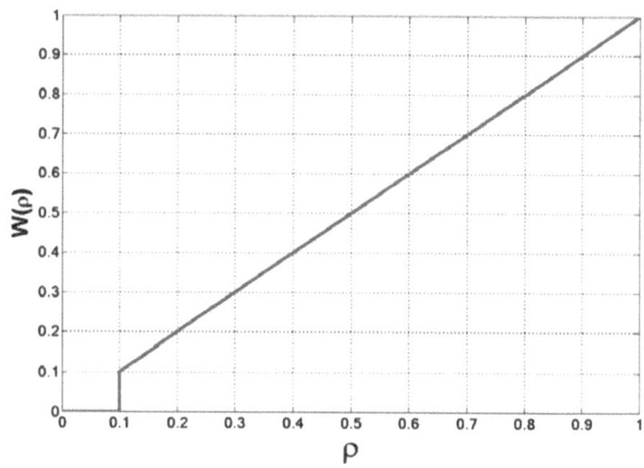

Fig. 9.4 A plot of the width of a frame element vs. the design variable. There is no non-smoothness at the corner because the Heaviside function is smoothly approximated

used design variables as Heaviside function convoluted with a ramp function, which takes care of manufacturing constraints [18]. The smoothened Heaviside function is shown in Eq. (9.1) and the plot of the function is shown in Fig. 9.4.

$$w(\rho) = \frac{\rho}{1 + e^{-a(\rho - c)}} \tag{9.1}$$

where ρ is a design variable and w is width of a frame element. The reason for defining another variable (ρ) needs further explanation. As noted above, if the width is the variable in optimization, is must be allowed to assume a zero value in order to eliminate some frame elements in the ground structure and thereby decide a topology. But this means that some elements that stay will be above the zero value but below the manufacturable lower limit. If we set the manufacturable limit as the lower bound for all the width variables, then no element will disappear and there would be no topology coming out of optimization. Hence, we define $w(\rho)$ as a function of ρ as shown in Eq. (9.1) and Fig. (9.4). Here, it should be noticed that until ρ equals 100 μm, width of frame element will remain zero and soon after it goes above 100 μm, the width will become 100 μm and will then increase linearly with ρ. Since we used a smooth approximation for the Heaviside function, we ensure that the optimization problem can be solved using continuous optimization methods. Parameter C is Eq. (9.1) is 100 μm and a is the tuning parameter that determines the sharpness of the transition from 0 to 100 μm.

A design obtained using topology optimization without manufacturing constraint (and hence without using Eq. (9.1)) is shown in Fig. 9.5. Only the symmetric top-half of the gripper is shown in the picture. It has many elements that "remained" in the final topology but have widths smaller than 100 μm. These are shown as dashed lines in the figure. Hence, this is not a useful topology. In fact, as can be seen in the inset, a few such elements are indeed crucial for the functionality of the gripper. Hence, the same problem was solved again by incorporating the manufacturable

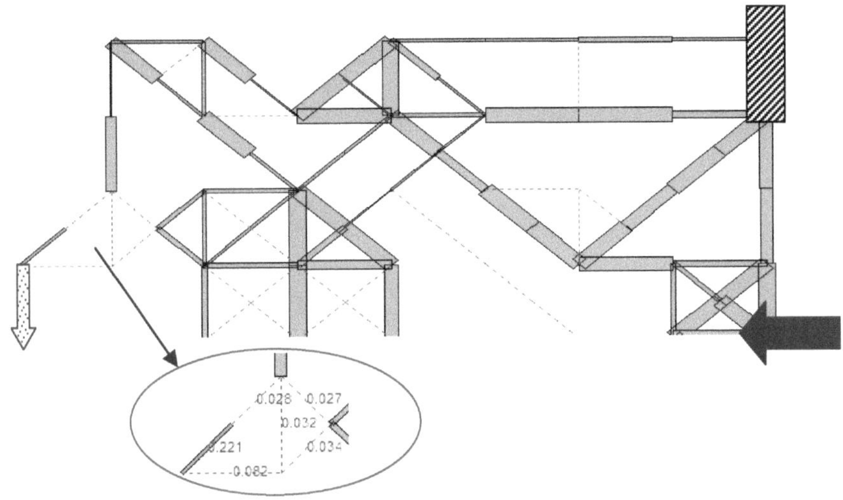

Fig. 9.5 A result of optimized topology for the gripper without using the manufacturing constraint. The dashed lines stayed in the optimization but their widths are below the manufacturable limit of 100 μm

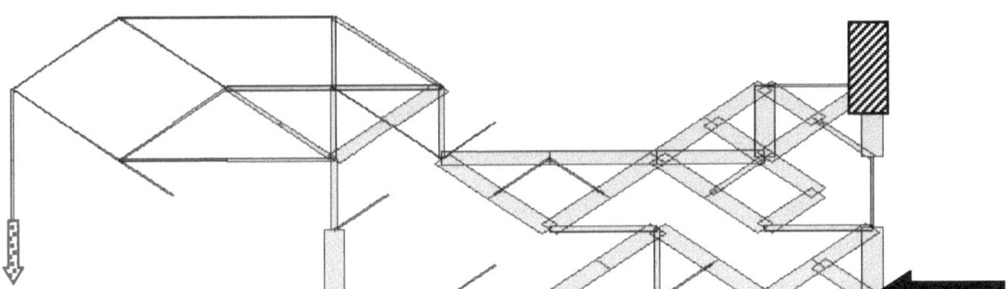

Fig. 9.6 A result of optimized topology for the gripper after including the manufacturing consideration. Now, all the remaining elements have a width of more than or equal to 100 μm

lower limit as per Eq. (9.1). The result is show in Fig. 9.6. All the frame elements in this have widths greater than 100 μm. There is, of course, one element that is not connected to any other element. Such dangling elements are numerical artifacts that the optimization algorithm is unable to remove at the end. They do not affect the performance significantly.

Figure 9.7 shows the complete mechanism whose top half is shown in Fig. 9.6. It shows the undeformed and deformed configurations of the beam finite element analysis. A result of the plane-stress finite element analysis done using COMSOL MultiPhysics software and is shown in Fig. 9.8.

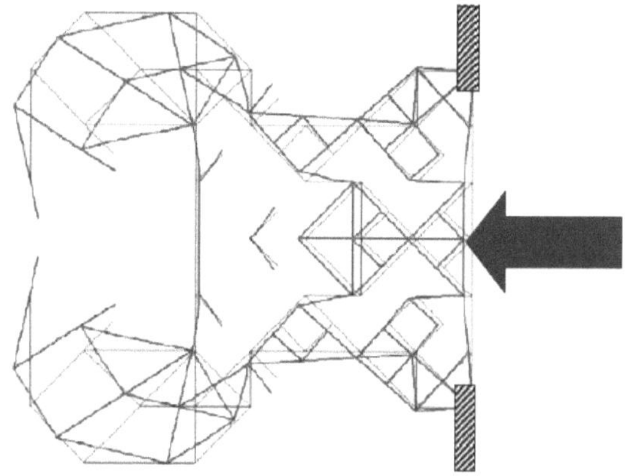

Fig. 9.7 The full result of the gripper design. Undeformed (*blue and thin*) and deformed (*red and thick*) configurations

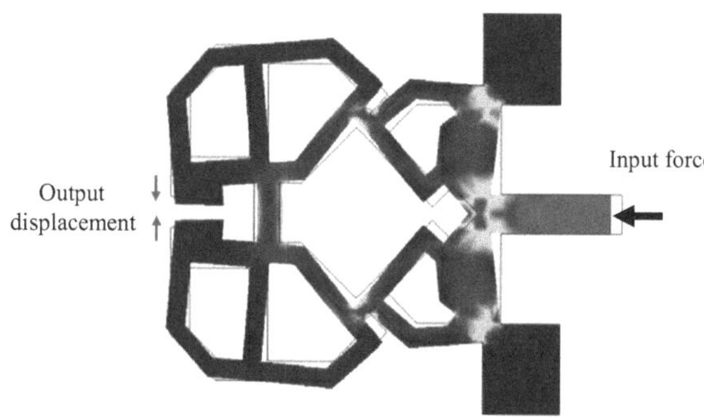

Fig. 9.8 The von Mises stress plot in the deformed configuration of the topology-optimized gripper. This is the manufacturable form of the gripper

9.6 Fabrication

Spring steel is a good material for fabricating the gripper because it has uniform material properties and there are standard manufacturing techniques available for machining it. We used wire-cut EDM to fabricate the grippers of overall dimensions of 11 mm × 11 mm × 0.5 mm as shown in Fig. 9.1b. By using wire-cut EDM, it is possible to manufacture the grippers within a dimensional tolerance of ± 20 μm. It is also possible to use micro-milling where nanometer roughness levels can be achieved [19]. But such precision is not required for the grippers.

PDMS is another preferred material for grippers because of its optical clarity, biocompatibility and low stiffness as compared to spring steel. For manufacturing PDMS grippers, a spring steel mould for vacuum-casting PDMS was made using wire-cut EDM (see Fig. 9.9a, b). Here, we took a sheet of metal, either spring steel or aluminum, and cut the mould shape of the gripper in it. This results in several loose pieces for a reasonably intricate topologies of the grippers as shown in Fig. 9.9a. The pieces wee then arranged into a mould as shown in Fig. 9.9b. The PDMS gel and 10% binder were mixed and poured in the mould after degassing in a vacuum

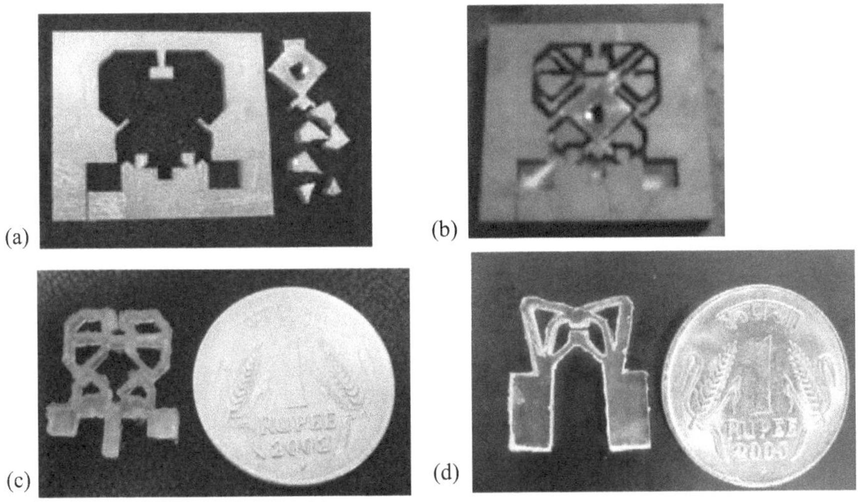

Fig. 9.9 (**a**) Wire-cut EDM used for making the mould piece in spring steel, (**b**) an arranged mould using the pieces from (**a**), (**c**) the vacuum-cast PDMS gripper, and (**d**) another vacuum-cast PDMS gripper

chamber (MK Technology GmbH D-53501 Grafschaft) for 30 min. PDMS was then cured in an oven (Memmert W 8540 Schwabach) at 100°C for 6–7 h. After curing, the PDMS gripper was ejected out from the mould. The resultant prototype grippers are shown in Fig. 9.9c. The thickness of the PDMS gripper was set at 2 mm to minimize sagging due to its own weight. Figure 9.9d shows a similarly fabricated PDMS gripper of another kind. The whole process is not only expensive but is also quite fast.

The stiffness of the PDMS gripper shown in Fig. 9.9d is 80.5 N/m and that of the one in Fig. 9.9c is 0.024 N/m. As compared with the stiffnesses of spring steel grippers (see Fig. 9.2), the PDMS grippers have orders of magnitude lower stiffness. By changing the overall size, thickness, and beam width, it is possible to achieve 1 N/m stiffness with PDMS. This is roughly the stiffness of the cell membranes. Thus, it is possible to tailor the stiffness as per the stiffness of the objects grasped or manipulated.

9.7 Actuation

Actuation, as noted before, must work in aqueous medium when we handle biological cells. By avoiding any transduction-based microactuation near the gripper, we used a simple but effective technique. Here, we attach a commercial XYZ stage (SUTTER INTRUMENT COMPANY MP285) to a point on the compliant gripper if it has a single point actuation. The fixed portion(s) of the gripper are attached to a strip of metal and it is attached to another XYZ stage. Two long but sufficiently rigid rods are used for attachment between the gripper and the XYZ stages. This setup

(a) (b)

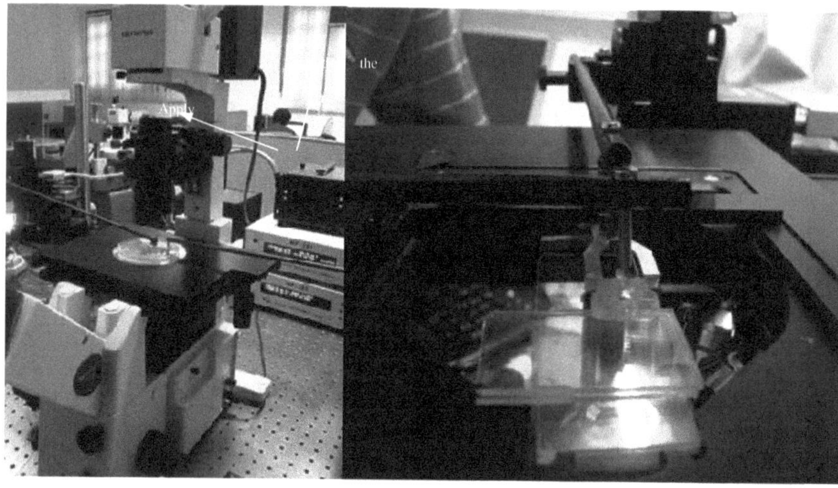

Fig. 9.10. (**a**) Experimental setup for micromanipulation, (**b**) a close-up view of the gripper

(a) (b)

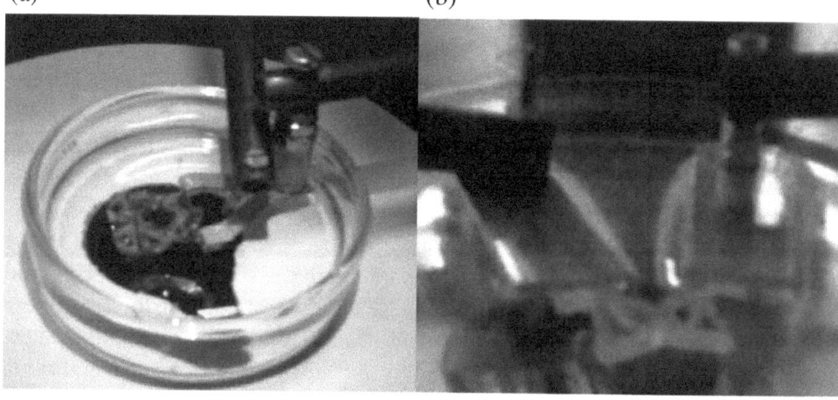

Fig. 9.11 (**a**) Single-point actuation for the gripper, (**b**) two-point actuation for another gripper

can be seen in Fig. 9.10a in which using OLYMUS IX71 inverted microscope can also be seen. Figure 9.10b shows a close-up view of the setup and the gripper that needs a single-point actuation. In Fig. 9.11a, we can see this gripper placed inside a Petri dish. Figure 9.11b shows another gripper, also placed in Petri dish. This needs a two-point actuation. Hence, two of its parts are attached to long rods, which are in turn attached to the XYZ stages. By moving the stages, we can actuate the gripper as needed. Computer-based control is also possible in this manner. A very large force is also available and the permissible displacements are also large. No on-chip microactuation can give all these advantages so easily.

9.8 Force-Sensing

As discussed in the introduction, we use vision-based force-sensing for estimating the forces of manipulation. The formulation of Cauchy's problem is presented in this section. The gripper being an elastic body, it can be schematically represented as shown in Fig. 9.12.

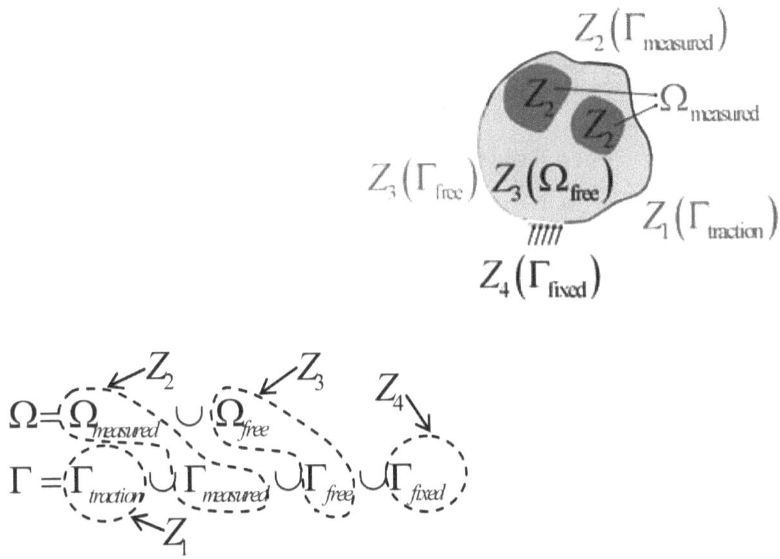

Fig. 9.12 Schematically shown elastic body with partitions of the domain and its boundary

Domain Ω and boundary Γ of the elastic body are partitioned into four regions: the first region is the one where forces act (Γ_{traction}); the second region is where displacements are measured ($\Omega_{\text{measured}} \cup \Gamma_{\text{measured}}$); the third region is where we do not know displacements but we know forces ($\Omega_{\text{free}} \cup \Gamma_{\text{free}}$); and the fourth region is where the displacements are fixed. All these partitions are shown in Fig. 9.12.

The mathematical statement of the Cauchy's problem is as follows.

$$\nabla \cdot \tilde{\mathbf{T}} = 0 \ \text{ in } \ \Omega \tag{9.2a}$$

$$\tilde{\mathbf{T}} = \lambda \text{tr} \left(\tilde{\mathbf{E}} \right) \tilde{\mathbf{I}} + 2\mu \tilde{\mathbf{E}} \tag{9.2b}$$

$$\tilde{\mathbf{E}} = \frac{1}{2} \left(\nabla \mathbf{U} + \nabla \mathbf{U}^{\mathrm{T}} + \nabla \mathbf{U}^{\mathrm{T}} \nabla \mathbf{U} \right) \tag{9.2c}$$

$$\mathbf{U} = 0 \ \text{ on } \ \Gamma_{\text{fixed}} \tag{9.2d}$$

$$\mathbf{t} = 0 \ \text{ on } \ \Gamma_{\text{measured}} \cup \Gamma_{\text{free}} \tag{9.2e}$$

$$U = U_{\text{measured}} \ \text{ on } \ \Omega_{\text{measured}} \cup \Gamma_{\text{measured}} \tag{9.2f}$$

where $\tilde{\mathbf{T}}$ is the second Piola-Kirchoff stress tensor, $\tilde{\mathbf{I}}$ the identity tensor, $\tilde{\mathbf{E}}$ the Green strain tensor, λ and μ the Láme constants of the material, $\mathbf{U}$ the displacement vector, $\mathbf{t}$ the traction, and $\mathbf{U}_{\text{measured}}$ the measured displacement vector.

Static equilibrium is achieved when internal forces balance the external forces. It is mathematically shown in Eq. (9.3).

$$\mathbf{F}_{\text{int}} = \mathbf{F}_{\text{ext}} \tag{9.3}$$

The internal force ($\mathbf{F}_{\text{int}}$) is a nonlinear function of displacements because we are considering geometric nonlinearity in the problem. By using the first order Taylor series for $\mathbf{F}_{\text{int}}$, we get:

$$\mathbf{F}_{\text{int}} + \frac{\partial \mathbf{F}_{\text{int}}}{\partial \mathbf{u}} \Delta\mathbf{u} = \mathbf{F}_{\text{int}} + \mathbf{K}_t \Delta\mathbf{u} = \mathbf{F}_{\text{ext}} \tag{9.4}$$

where $\mathbf{K}_t$ is the tangent stiffness matrix, which depends on $\mathbf{u}$ in geometrically nonlinear problems.

$$\begin{aligned} &\mathbf{K}_t \Delta\mathbf{u} = \mathbf{F}_{\text{ext}} - \mathbf{F}_{\text{int}} \Rightarrow \\ &\begin{bmatrix} \mathbf{K}_{11} & \mathbf{K}_{12} & \mathbf{K}_{13} \\ \mathbf{K}_{21} & \mathbf{K}_{22} & \mathbf{K}_{23} \\ \mathbf{K}_{31} & \mathbf{K}_{32} & \mathbf{K}_{33} \end{bmatrix} \begin{Bmatrix} \Delta\mathbf{u}_1 \\ \Delta\mathbf{u}_2 \\ \Delta\mathbf{u}_3 \end{Bmatrix} = \begin{Bmatrix} \mathbf{F}_1 \\ \mathbf{0} \\ \mathbf{0} \end{Bmatrix}_{\text{ext}} - \begin{Bmatrix} \mathbf{F}_1 \\ \mathbf{F}_2 \\ \mathbf{F}_3 \end{Bmatrix}_{\text{int}} = \begin{Bmatrix} \Delta\mathbf{F}_{1c} \\ \Delta\mathbf{F}_{2c} \\ \Delta\mathbf{F}_{3c} \end{Bmatrix} \end{aligned} \tag{9.5}$$

By rearranging matrix equations we get

$$\begin{aligned} &\left[\mathbf{K}_{21} - \mathbf{K}_{23}\mathbf{K}_{33}{}^{-1}\mathbf{K}_{31}\right]\{\Delta\mathbf{u}_1\} \\ &= \Delta\mathbf{F}_{2c} - \mathbf{K}_{23}\mathbf{K}_{33}{}^{-1}\Delta\mathbf{F}_{3c} = -\mathbf{F}_{2\text{int}} + \mathbf{K}_{23}\mathbf{K}_{33}{}^{-1}\mathbf{F}_{3\text{int}} \end{aligned} \tag{9.6}$$

We use pseudo-inverse of $\left[\mathbf{K}_{21} - \mathbf{K}_{23}\mathbf{K}_{33}{}^{-1}\mathbf{K}_{31}\right]$ to solve Eq. (9.6). However, this does not guarantee zero forces at the measured locations. For further details refer to [20, 21]. Here, we use Newton's method as an alternate to constrained optimization as presented below.

$$\mathbf{F}_2(\mathbf{u}_2 + \Delta\mathbf{u}_2) = \mathbf{F}_2(\mathbf{u}_2) + \frac{\partial \mathbf{F}_2}{\partial \mathbf{u}_2}\Delta\mathbf{u}_2 \tag{9.7}$$

By using the second and third rows of Eq. (9.5), we obtain

$$\begin{aligned} \Delta\mathbf{F}_{2c} &= \left[\mathbf{K}_{22} - \mathbf{K}_{23}\mathbf{K}_{33}{}^{-1}\mathbf{K}_{32}\right]\Delta\mathbf{u}_2 \\ &+ \left[\mathbf{K}_{21} - \mathbf{K}_{23}\mathbf{K}_{33}{}^{-1}\mathbf{K}_{31}\right]\Delta\mathbf{u}_1 \\ &+ \mathbf{K}_{23}\mathbf{K}_{33}{}^{-1}\Delta\mathbf{F}_{3c} \end{aligned} \tag{9.8}$$

The gradient of the spurious forces can be expressed as

$$\frac{\partial \mathbf{F}_2}{\partial \mathbf{u}_2} = \left[\mathbf{K}_{22} - \mathbf{K}_{23}\mathbf{K}_{33}{}^{-1}\mathbf{K}_{32}\right] \tag{9.9}$$

We use these gradients in the Newton's method to drive the spurious forces to zero. This method converges to a local minimum; hence, we do not need to specify bounds on variable $\mathbf{u}_2$. From Eqs. (9.7) and (9.9), we can get corrections to measured displacements as follows.

$$\Delta \mathbf{u}_2 = \left[\frac{\partial \mathbf{F}_2}{\partial \mathbf{u}_2} \right]^{-1} \{\mathbf{F}_2(\mathbf{u}_2)\} \tag{9.10}$$

Figure 9.13a, b show the gripper holding an object, a ball of yeast cells that measures about 1 mm in diameter. Ink dots were placed on the gripper to serve as measurement points for the above force-sensing algorithm. By image processing, we can compute the displacements of these points. Figure 9.14 shows the beam finite element model of the gripper with its marked. This model is used to estimate the forces acting on the grasped object as well as the on the gripper itself. Figure 9.15 shows another gripper set up in the same manner for force-measurement.

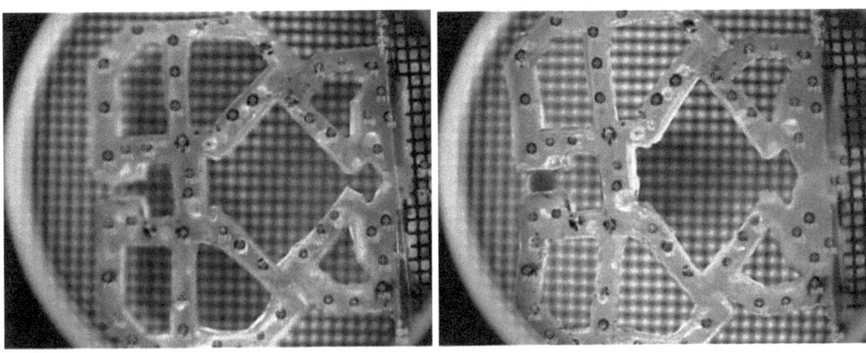

Fig. 9.13 Undeformed and deformed configurations of PDMS gripper

9.9 Experimentation

Figure 9.16 shows the gripper with two actuation points positioned just above a glass slide. The objective of the inverted microscope is right under the glass slide. We tested the grippers on spherical-shaped zebra fish egg-cells (~0.7 mm in diameter), ellipsoidal-shaped drosophila (fruit fly) embryos (~0.2 mm wide and ~0.5 mm long), yeast ball (less than 1 mm in diameter), and hibiscus pollen (0.1 mm in diameter). All but the yeast ball are shown in Fig. 9.17. The drosophila embryo is studied in biology as a model organism because its genome resembles the human genome [22]. In case of zebra fish, which is another popular model organism, the embryo develops organs that are similar to human central nervous system and pancreas [23]. Furthermore, both embryos have good optical clarity.

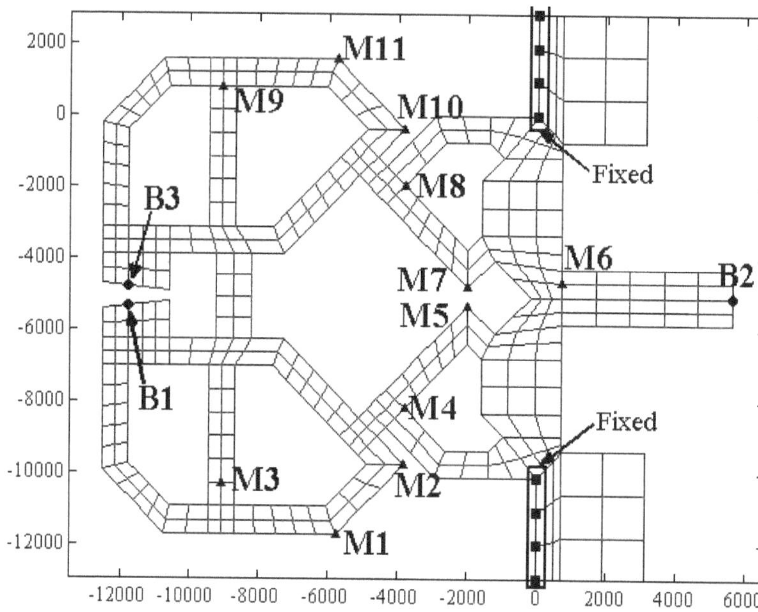

Fig. 9.14 Undeformed and deformed configurations of PDMS gripper

Fig. 9.15 Undeformed and deformed configurations of PDMS gripper

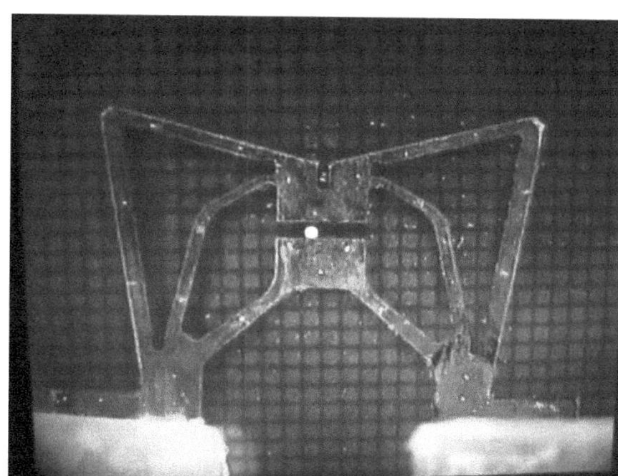

We were able to grasp, roll, squeeze, position, move, and pick-and-place the aforementioned biological objects along three mutually perpendicular axes in aqueous medium. We have achieved a stroke of 0.3 mm using the spring steel gripper and hence we have used it for manipulating drosophila embryo and pollen which are of the comparable size. The spring steel gripper is actuated at two points by attaching

Fig. 9.16 A two-point actuated spring steel gripper placed above a glass slide. The objective of the inverted microscope is visible underneath the glass slide. The cells are sometimes placed on the glass slide

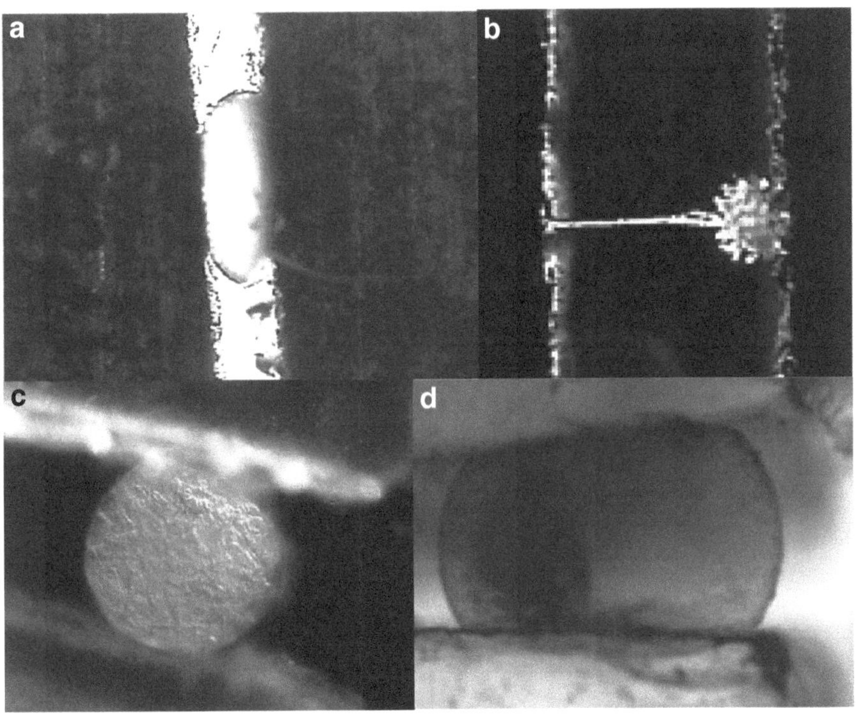

Fig. 9.17 (**a**) Drosophila embryo grasped between the gripper jaws, (**b**) Hibiscus pollen squeezed and then drawn between the gripper jaws, (**c**) Undeformed Zebra fish egg cell between the two jaws of the PDMS gripper, (**d**) Deformed Zebra fish egg cell by the PDMS gripper

the ends of the gripper to two different xyz stages using an aluminum strip as shown in Fig. 9.9c. The biological objects are grasped by first moving the gripper from the top to the plane of the object and later gripper is moved in the plane so as to get the object within the jaws. In the case of the drosophila embryo (see Fig. 9.17a), we were able to draw it into the space between the jaws at one end and let it exit at the other by modulating the motion of the jaws. In the case of pollen, we were able to separate single pollen from the cluster and crush it by repeated action. The contents of the pollen grain that oozed out and stretched between the two jaws can be seen in Fig. 9.17b.

We used PDMS gripper for zebra fish egg cells as the stroke of the gripper is around 1 mm. The PDMS gripper used here requires one point actuation as discussed before. The gripper is fixed on an aluminum plate and attached to one of the XYZ stage. The actuation is done using another XYZ stages as described in the previous section. The biological objects are grasped in a similar manner as mentioned above. Figures 9.17c, d show the zebra fish egg cells grasped and then squeezed.

9.10 Discussion

The experiments described above are preliminary in that they need to be calibrated and validated using independent force measurements. We have conducted some macro-scale experiments where it is easier and validated the force-sensing technique. Thus, the results presented here should be deemed as proof-of-concept.

An important point to note is that this technique offers an easy way to characterize the bulk stiffness of a biological object. More importantly, it also offers a method to characterize the inhomogeneous properties of the grasped and manipulated object. This requires the solution of another inverse problem. This technique, not described here, needs several sets of force-displacement data on the boundary of the object. By virtue of the manipulative capability of our technique, it is possible to generate such data. This forms the ongoing work in our research group. Further reduction in size of the gripper to be able to handle cells that are only a few microns in size as well as providing haptic interface are also in progress.

9.11 Closure

In this article, we have given an overview of various aspects of our recent work related to mechanical manipulation of biological cells using miniature grippers. We argued that gripper-based actuation has certain advantages over other widely used techniques such as pipette aspiration, particle-adhesion and tunable fields, lasers, etc. The advantages of gripper-based actuation are: dexterous manipulation, stiffness tuning, in-built force-sensing, and easy and inexpensive fabrication. Here, we used compliant micro gripper concept. We made a case for making the grippers using spring steel and PDMS. The manufacturing techniques we used are adapted from the macro scale. They enabled us to make meso-scale (100 μm to a cm) grippers, which have movable jaws that can hold and manipulate biological cells. The most significant advantage is that actuation in aqueous medium is possible and there is sufficiently large range of forces and displacements. We also described a computational technique for estimating the forces using visually measured displacement data. Preliminary experiments conducted are described. Analysis of the data and the measured stiffness properties of the biological objects is in progress.

Acknowledgments This article is based on the work done by many students in the author's research group. The authors thanks all of them, and in particular, G. Ramu, A. Ravi Kumar, B.M. Vinod Kumar, V. Mallikarjuna Rao, Sudarshan Hegde, and M. Manjunath. The financial support from the Swarnajayanthi Fellowship of the Department of Science and Technology (DST), Government of India, as well as the DST Centre for Mathematical Biology (grant No. SR/S4/MS: 419/07) in the Indian Institute of Science, Bangalore, is also gratefully acknowledged.

References

1. Bustamante C, Chemla YR, Forde NR, and Izhaky D (2004) Mechanical Processes in Biochemistray. Annu. Rev. Biochem. 73:705–748.
2. Humphrey JD (2003) Continuum Biomechanics of Soft Biological Tissues. Proc. Math. Phys. and Eng. Sci. 459(2029):3–46.
3. Wang W, Liu X, Gelinas D, Ciruna B, and Sun Y (2007) PLoS ONE. 9:e862.
4. Crick F and Hughes A (1950) Physical Properties of Cytoplasm: A study by Means of the Magnetic Particle Method. Exp. Cell Res. 1:37.
5. Butler JP and Kelly SM (1998) A Model for cytoplasmic rheology consistent with magnetic twisting cytometry. Biorheol. 35:193.
6. Dai J and Sheetz MP (1995) Mechanical properties of neuronal growth cone membranes studied by thehter formation with laser optical tweezers. Bio. Phys. J. 68:988–996.
7. Guck J, Ananthakrishnan R, Mahmood H, Moon TJ, Cunningham CC, and Kas J (2001) The optical stretcher: A novel laser tool to micromanipulate cells. Biophys. J. 81(2):767–784.
8. Ethier CR and Simmons CA (2007) Introductory Biomechanics: from Cells and Organisms, Cambridge University Press, Cambridge.
9. Caille O, Thoumine O, Tardy Y, and Meister JJ (2002) Contribution of the nucleus to the mechanical properties of endothelial cells. J. Biomech. 35(2):177–187.
10. Ananthasuresh GK and Kota S (1995) Designing compliant mechanisms. Mech. Engrg. 117(11):93–96.
11. Howell LL (2001) Compliant Mechanisms, John Wiley and Sons, New York.
12. Wang X, Ostrowski J, and Ananthasuresh GK (2001) Vision-based sensing of forces in elastic objects. Sens. Actu. A Phys. 94(3):142–156.
13. Yu S and Nelson BJ (2001) Autonomous injection of biological cells using visual servoing. Exp. Rob. VII, LNCIS. 271:169–178.
14. Gad-el-Hak M (ed.) (2006) MEMS: Applications, CRC Press, Boca Raton
15. Ginachandani Y, Tabata O, and Zappe, H (2008) Comprehensive Microsystems, Vol. 1, Elsevier, Amsterdam
16. Bendsoe MP and Sigmund O (2003) Topology Optimization: Theory, Methods, and Applications, Springer, New York.
17. Ananthasuresh GK (2004) Optimal Synthesis Methods for MEMS, Kluwer Academic Publishers, New York.
18. Sahu DK (2008) Topology optimization and prototyping of miniature compliant grippers for micromanipulation. Master of Engineering Report, Mechanical Engineering, Indian Institute of Science, Bangalore
19. Liu X, DeVor RE, and Kapoor SG (2004) The mechanics of Machining at the microscale: Assessment of the current state of the science. J. Manuf. Sci. Eng. 126(4):666–679
20. Kamiyama K, Vlack K, Mizota T, Kajimoto H, Kawakami N, and Tachi S (2005)"Vision-based sensor for real time measuring of surface traction", IEEE Comput. Graph. Appl., 25: 68–75.

21. Reddy AN and Ananthasuresh GK (2008) On computing the forces from the noisy displacement data of an elastic body. Int. J. Num. Meth. Eng. 76:1645–1677.
22. Bernstein RW, Zhang X, Zappe S, Fish M, Scott M, and Solgaard O (2004) Characterization of fluidic microassembly for immobilization and positioning of DROSOPHILA embryos in 2-D arrays. Sens. Actu. 114:191–196.
23. Kim DH, Yun S, and Kim B (2004) Mechanical Force Response of Single Living Cells Using a Microrobotic System. IEEE International Conference on Robotics and Automation, Hilton New Orleans Riverside, New Orleans, LA, USA, Vol. 5:5013–5018.

Subject Index

S. Chakraborty (ed.), *Microfluidics and Microfabrication*,
DOI 10.1007/978-1-4419-1543-6, © Springer Science+Business Media, LLC 2010

FSC
www.fsc.org
MIX
Papier | Fördert
gute Waldnutzung
FSC® C083411

Kolibri 360 GmbH
Ferdinand-Jühlke-Straße 7
99095 Erfurt, Deutschland
produktsicherheit@kolibri360.de